Agricultural Yield and Production: Theory, Techniques and Technology

Agricultural Yield and Production: Theory, Techniques and Technology

Edited by Matthew Ryder

MURPHY & MOORE
www.murphy-moorepublishing.com

Published by Murphy & Moore Publishing,
1 Rockefeller Plaza,
New York City, NY 10020, USA
www.murphy-moorepublishing.com

Agricultural Yield and Production: Theory, Techniques and Technology
Edited by Matthew Ryder

International Standard Book Number: 978-1-63987-033-2 (Hardback)

Cataloging-in-Publication Data

Agricultural yield and production : theory, techniques and technology / edited by Matthew Ryder.
p. cm.
Includes bibliographical references and index.
ISBN 978-1-63987-033-2
1. Agricultural productivity. 2. Crop yields. 3. Soil productivity.
4. Agricultural productivity--Technological innovations. I. Ryder, Matthew.
S494.5.P75 A37 2021
338.16--dc23

TABLE OF CONTENTS

PREFACE

Agriculture is referred to as the science of cultivation of crops. A crop can be both a plant and animal product which is primarily harvested for earning profit. Crop yield is defined as the measure of production of yield of crop per unit area and generation of seeds of the plant itself. It is also known as agricultural output. The primary categories of crop are horticulture, floriculture and industrial crops. Examples of horticultural crops include tomatoes, oranges, cucumbers, etc. Floriculture crops include houseplants, cut flowers and bedding plants. Industrial crops include those crops which are grown for clothing and biofuel such as cotton, elephant grass, etc. The growth and production of forest and agricultural crops for profit is known as agricultural production. The techniques used in measuring crop yield are seed ratio and area ratio. Seed ratio is defined as the number of grain harvested for each grain seeded whereas area ratio is the number of grain harvested per unit land. This book discusses the fundamentals as well as modern approaches of agricultural and crop yield. It also unravels the recent studies in this field that unfold the innovative aspects of crop yield which will be crucial for the progress of this field in the future. The readers would gain knowledge that would broaden their perspective about agricultural yield and production.

Various studies have approached the subject by analyzing it with a single perspective, but the present book provides diverse methodologies and techniques to address this field. This book contains theories and applications needed for understanding the subject from different perspectives. The aim is to keep the readers informed about the progresses in the field; therefore, the contributions were carefully examined to compile novel researches by specialists from across the globe.

Indeed, the job of the editor is the most crucial and challenging in compiling all chapters into a single book. In the end, I would extend my sincere thanks to the chapter authors for their profound work. I am also thankful for the support provided by my family and colleagues during the compilation of this book.

Editor

ECONOMIC COMPARISON OF DIRECT SEEDED AND TRANSPLANTED RICE: EVIDENCES FROM ADAPTIVE RESEARCH AREA OF PUNJAB PAKISTAN

Muhammad Younas[1*], Momna Abdul Rehman[2], Abdi Hussain[3], Liaqat Ali[4] and Masood Qadir Waqar[5]

[1]*Senior Subject Matter Specialist (FM&FE), Adaptive Research Farm, Sheikhupura,*
[2]*Assisant Research Officer (FM&FE), Adaptive Research Farm, Sheikhupura,*
[3]*Senior Scientific Officer, Social Sciences Research Institute, NARC, Islamabad,*
[4]*Assistant Research Officer, Adaptive Research Farm, Sheikhupura,*
[5]*Director of Agriculture, Adaptive Research Punjab, Lahore*

ABSTRACT

The study was conducted in three of the four districts of Adaptive Research Zone, Sheikhupura; which contains Sheikhupura, Lahore and Kasur districts for economic comparison of direct seeded (DSR) and transplanted super basmati rice during Kharif-2014. Average yield, cost of production, net economic benefits and benefit cost ratio for DSR and transplanted rice was calculated. Extent of area under DSR technology was also determined. Farmers growing both DSR and transplanted rice were selected using purposive sampling technique and sample size was 36. Data was collected by face to face interview by using a well-structured questionnaire having personal, farm related and cost variables. The results revealed that DSR technology was adopted on 22.8% of rice area on surveyed farms. Average paddy yield, total cost of production, net economic benefits per hectare and BCR for direct seeded rice were 3.09, Rs.112047, Rs.15014, 1.11 and for transplanted the figures were 3.19, Rs. 134882, Rs.-8433, 0.95 respectively. Thus, in the year 2014, farmers gained profit from practicing DSR technology and born loss from transplanted rice crop. It was found that farmers were growing DSR through seven different sowing methods and using different seed rates. Moreover, weeds infestation is a major problem in DSR technology that can be overcome through a combination of cultural as well chemical control methods. It was concluded that DSR is a promising technology subject to weeds management. Better coordination between research, extension and farming community can bring fruitful results with respect to adoption of DSR technology.

Keywords: DSR, Transplanted rice, Net Benefits, BCR

INTRODUCTION

Rice is an important food and cash crop across the globe. In Pakistan's agrarian economy, rice plays multifarious roles as it is the 2nd staple food after wheat and provides source of employment and foreign exchange. During the fiscal year 2013-14, Pakistan earned US$ 1.667 billion through exporting rice and it has 0.7 % share in national GDP and 3.1 % in value added services. (GOP, 2013-14). Pakistan is one of the largest rice producing countries of the world and produces about 6 million tons of rice every year.

In Pakistan rice is grown on 2.8 million hectares of arable land, with an annual production of about 6.8 million tonnes. In Punjab, rice is grown on 1.7 million ha of land with total production of 3.5 million tons that accounts 51% of total national production of rice in Pakistan (GOP, 2014). Gujranwala, Sheikhupura, Sialkot, Narrowal, Hafizabad, Mandi Bahaudin Din, Okara, and Jhang districts of Punjab account for more than 70 percent of Basmati rice production in the country (Abedullah et al,. 2007).

There has been a significant growth in rice production in the country since independence. Area under rice increased from 856,000 hectare in 1947 to 2.8 million hectares in 2014 whereas production increased from 0.7 to 6.8 million tons during the same period. Increase in total production came from increase in area along with high yielding varieties; improved agronomic practices supported with fertilizers and plant protection measures (Memon, 2013).

In Pakistan mostly rice is cultivated through traditional transplanting method. This method not only requires a lot of water, but also is laborious, cumbersome, time consuming and involves a lot of expenditure on raising nursery, its uprooting and transplanting etc. During the

*Corresponding author: e-mail: malik_younass@yahoo.com

sowing season, scarcity of farm labour coupled with higher labour cost results in relatively lower plant population, delayed transplanting and use of aged seedlings. All these issues directly lead to low paddy yield. Although transplanting is the common method of rice production but impending water scarcity, lack of farm labour along with the increased labour cost has compelled all the stake holders to think about the other alternatives to conventional ways of rice planting. Shifting from transplanting to direct seeding is a promising available alternative. Although rice cultivation by transplanting is generally considered superior to that by direct sowing, yet the latter is also reported as a successful method in some parts of the world (Adair et al., 1992). It saves labour (Murugaboopathi et al., 1991) and is more economical than transplanting (Awan et al., 2007).

In many countries of the world, direct seeding of rice is extensively used with profitable results (Naklang et al 1996, Sharma 1996, Anagadi 1993). It significantly gave higher yields than transplanted rice in some of the research trials when appropriate cultural practices were adopted (Awan et al., 2007). Considering the above scenario, a study was conducted to compare the economics of direct seeded rice and of traditionally transplanted rice. It also aimed to determine the extent of DSR in the above mentioned areas, so that some valuable recommendations be given to the farming community and extension workers for its successful adoption.

MATERIALS AND METHODS

The study was carried out in Adaptive Research Zone Sheikhupura which consists of four districts namely Lahore, Kasur, Nankana Sahib and Sheikhupura. Lahore, Sheikhupura and Kasur districts fall in rice zone of the Punjab having different cropping rotations such as rice - wheat, potato – wheat, sugarcane–wheat, Kharif fodder/ maize–wheat, mash–wheat, sunflower - Kharif fodder – Wheat (Khan, 2015). Before conducting the interviews of the respondent farmers, a list of such farmers who were growing super basmati through both the technologies i.e. Direct Seed Rice (DSR) as well as transplanted, was compiled with the help of Agriculture Extension Department at district level and Engro EXIMP working on DSR in the area. It was found that DSR technology is not being practiced in Nankana Sahib, so Nankana Sahib was excluded from the study region.

A well-structured questionnaire was developed and was pretested in the field before launching the actual data collection. A total of 36 respondent farmers practicing DSR as well as transplanting method were interviewed for collection of data. About 67% of the respondents were interviewed from Sheikhupura district (as promotion of DSR technology was more prominent by a private company, ENGRO) followed by Lahore district (19%) and Kasur district (14%).

Economic cost of production of super basmati rice was calculated by asking some direct questions from the respondent farmers and cost of production publication of Crop Reporting Service (CRS), Punjab for the year 2014-15 was also consulted (GOP, 2015). Data on some variables such as cost of bund making, cost of spreading fertilizer, cleaning of irrigation channel, cost of capital investment, management charges etc. were taken from this publication. The data was analyzed using statistical software package SPSS (17.0).

The procedure adopted by Ahmad and Chaudhry (1987) and Chaudhry et al., (1992) was used for cost estimation. Variable cost incurred per acre on land preparation, seed and seeding, plant protection measures, fertilizers, micronutrients, farm yard manure, harvesting and threshing labor used for different operations, management component, mark up on investment @ 9.5% for six months was summed up. Total economic cost of production per acre was estimated by adding the land rent to the variable cost.

By following Ahmad and Chaudhry (1987) and Chaudhry et al., 1995, the gross income/ gross value product (GVP) per acre was estimated by multiplying total paddy production with its unit price and then dividing the resultant by total acreage of super basmati. Value of rice straw if any was added to this income in order to reach to the gross revenue of the crop The procedure adopted by (Chaudhry et al., 1987) was used for calculation of net income according to which net income per acre is the money available to management after meeting all the crop production expenses and is calculated by deducting total cost of production including land rent from the gross income earned from one acre of land.

RESULTS AND DISCUSSION

Personal as well as farm level attributes of the farming families affect a lot in decision making with respect to adoption of new technologies and enhancing farm's productivity. Since the same person practicing both the technologies was interviewed, so mean values of the personal attribute are given accordingly. Average age of the respondent farmers involved in decision making at the sample farmers was 49 years with 9 years of schooling and family size of 8 individuals per family. Average length of farming experience of the sample farmers was 26 years. Out of 36 respondents interviewed for the purpose, 50% were large farmers having 25 acre or above operational holding followed by medium holding farmers (36.1%) having 12.5 to less than 25 acre of operational holding and small holding farmers (13.9%) having less than 12.5 acre of operational land.

Extent of DSR Technology

Out of total cropped area of 1896 acres of the respondent farmers, rice was being cultivated on 1465 acres equivalent to 77.2% of the total cropped area, whereas DSR technology was adopted on 334 acres out of 1465 acres equivalent to 22.8% of rice area on overall basis. Maximum extent of DSR technology across districts was calculated in district Kasur (52.2%) followed by Sheikhupura (23.5%) and Lahore (8.0%). Thus, extent of rice cultivation in Lahore and Kasur districts is less as compared to Sheikhupura.

Respondent farmers were using a varied seed rate (15-50 kg ha^{-1}) with a mean of 10.35 kg ha^{-1} and were seeding rice through seven different methods *i.e.* broadcasted soaked seed in "*a"(moisture)* soil condition (39%) followed by broadcasting of dry seed in dry soil and applying irrigation afterwards (27%), broadcasting of dry seed in wattar condition (11%) and broadcasting soaked seed in dry soil followed with immediate irrigation (8%), broadcasting sprouted seed in wattar condition (6%), drilling of dry seed in wattar condition (6%) and drilling of soaked seed in wattar condition (3%). Moreover, the respondent farmers planted directly seed rice on nine different time periods, with an interval of one week starting from 3rd week of May and ending at 3rd week of July. Whereas they transplanted super basmati on 10 different time periods, with an interval of one week starting from 2nd week of June to 3rd week of August. This shows that every farmer is practicing DSR in his own way. If proper route of technology dissemination had been adopted, different practices of DSR technology would not have been adopted by the farming community. It revealed the fact that the technology had gone to the end users without fulfilling the due course of evaluation and standardization by research and extension departments.

Yield Comparison

Paddy yield comparison of super basmati DSR with transplanted technologies across districts given in the Figure 1 showed that on an overall basis paddy yield with transplanted technology (3.19 tonnes per hectare) was 3.19% higher than under DSR (3.09 tonnes per hectare) which was not a significant difference. Yield comparison across districts indicated that paddy yield under DSR (3.25 tonnes per hectare) was slightly higher in district Sheikhupura than under transplanted methodology (3.23 tonnes per hectare), whereas opposite was true in case of Lahore (DSR 2.70 tonnes per hectare as compared to transplanted 3.01 tonnes per hectare) and Kasur districts (DSR 2.93 tonnes per hectare as compared to 3.23 tonnes per hectare under transplanting method. Super basmati paddy yield was 11.7% and 10.5% higher in case of transplanting technology as compared to DSR in Lahore and Kasur districts, respectively. Non-significant difference in yield gap between the technologies showed a greater scope for adoption of DSR technology, if technical issues of weeds infestation and germination of red rice locally known as "choba" are tackled. During the course of study, some of the farmers who used double pre-sowing irrigation concept as a cultural method of weeds management reported less weeds infestation. These results are in line with previous studies that direct seeding technology of rice is almost at par in yield with the conventional planted crop. It has great potential for adoption as a substitute for transplanted rice if the weeds are controlled properly (Awan et al., 2007; Ali et al., 2007).

Comparison of Economic Cost of Production

cost of production of DSR and transplanted rice are given in the Table 1, total economic cost of production per hectare of super basmati transplanted and directly seeded are Rs. 134882 and Rs. 112047, respectively i.e. economic cost of production of transplanted rice was 20.4% higher than DSR. Thus, comparison of economic cost of production variables in both technologies revealed that direct seeding enabled farmers to save 20% of the total cost of a transplanted crop. This result was similar to work of Nai-Kin and Romli, (2002), according to which direct seeding enabled farmers in Muda area of Malaysia to save 29% of the total cost of transplanted crop. The difference in cost saving percentage is due to the fact that Nai-Kin and Romli considered only cash expenditures, while in the present study under consideration economic costs are compared rather than financial costs. The data given in Table 1 show that DSR technology is cost saving in land preparation (15.08%), seed and sowing (66.93%) and irrigation (44.74%), whereas there is increase in cost of plant protection measures (weed and disease infestations on DSR (29.80%) is more as compared to transplanted rice crop), fertilizers (20.62%), FYM (38.75%) and micronutrients (14.04%). The data shows that major cost saving is in seed and seeding (66.93%) followed by irrigation (44.74%) and land preparation (15.08%) with an overall saving of 16.93%. The results show that cost saving in the DSR is more than that of increase in yield under transplanted that leads towards better economics of DSR technology.

Comparison of Net Economic Returns

When net returns were compared under DSR and transplanting techniques, t they were negative to the tune of Rs. 8433 per hectare in transplanted method whereas net returns were positive in case of DSR technology amounting to Rs. 15014 per hectare (Figure 2). The year 2014 was an abnormal year with respect to prices of rice paddy as almost 40% lower prices prevailed in the market when compared with last year (2013) prices. Even with low market prices, farmers practicing DSR technology gained profit on their investment in growing super basmati whereas those who adopted transplanted technique born losses. This result was in line with findings of Ali et al. (2012) as they reported highest net benefits of Rs.37235 per hectare under DSR (broadcasting of soaked seed in well prepared seed bed after rauni) as compared to conventionally transplanted rice of Rs.3299 per acre. Moreover, BCR of the above mentioned DSR and transplanted methods were 1.36 and 1.07 respectively.

Comparison of Benefit Cost Ratio

BCR for super basmati under DSR technology was 1.11 as compared to 0.95 in case of transplanting technology (Figure 3) meaning that growing of super basmati rice through DSR technology was better than transplanting method. This result was comparable with work of Awan et al. (2007) as they reported that BCR of DSR was better (1.29) than conventionally transplanted method (1.15). Thus comparison of economic analysis of Super Basmati under DSR and transplanted methodology revealed that DSR has more potential in reducing cost of production by reducing water and labor inputs compared with traditional open flooded and transplanted rice. However DSR technology faces several potential yield reducing problems such as heavy weeds infestation, lack of effective herbicide and poor crop management practices etc. that should be solved through proper management practices (Singh et al., 2002 and Pandey et al., 2002).

Conclusion and Recommendations

Based on study findings, it is concluded that DSR technology is better as compared to conventionally transplanted method of rice production if practiced properly as net economic returns and Benefit Cost Ratio were better under DSR technology than transplanting technique. In spite of better economics, weeds management is still a major obstacle in the way of adoption of DSR technology that leads to lower yield in DSR method as compared to transplanting technique as reported by 76% of the DSR respondents that effective weedicides are not available in the market which could control newly invading weeds such as Dhidhen grass (Echinochloa Crusgalli), Narru (Paspalum Distichum), Lumb grass / kallar grass (Lipotchloa Chinensis) etc.

The respondent farmers were practicing DSR technology on 22.8% of rice area on overall basis. Maximum extent of DSR technology across districts was calculated in district Kasur (52.2%) followed by Sheikhupura (23.5%) and Lahore (8.0%).

Weeds can be controlled by cultural as well as chemical control methods. If DSR technology is practiced on puddled soil, then the issue of weeds as well as that of germination of wild rice (chobba) that deteriorates the rice quality can be solved, although there would not be cost saving in land preparation but increase in yield because of less weeds infestation would lead to better economics of DSR technology.

Moreover, enhanced role of extension department in disseminating the standardized DSR technology especially with respect to land preparation, seed rate, time of sowing and plant protection measures can play a vital role for adoption of the technology on wider scale as farmers are eager to adopt new technologies for rice production in the context of high cost of production especially that of land preparation, irrigation and labor cost incurred on conventionally transplanted rice. The study results show that cost incurred on above mentioned three variables in case of conventional method of rice transplting is 2.34 times (Rs.46042/hectare) that of in case of DSR technology (Rs.19610).

Table – 1: Economic comparison of cost of production of DSR and transplanted rice crops

(Rs. per hectare)

Sr. No.	Detail of cost components	DSR	TPR	Savingsin DSR (%)
1	Land preparation	9476	11159	15.08
2	Nursery and transplanting/seed and sowing	3457	10455	66.93
3	Irrigation	13499	24428	44.74
4	Fertilizer	15666	12988	-20.62
5	Farm yard manure	2955	2130	-38.75
6	Micronutrients	2288	2006	-14.04
7	Plant protection measures (weeds, insect pests and disease control)	8772	6758	-29.80
8	**Total input cost**	**55148**	**69927**	21.14
9	Mark up on investment for six months @ 9.5% excluding water rates	2607	3311	21.27
10	Harvesting & threshing	8794	8881	0.97
11	Land rent for six months per acre	49998	50685	1.36
12	Management charges for six months @ Rs.14000 per 100 acres	2076	2076	0.00
13	**Total economic cost of production**	112047	134882	**16.93**

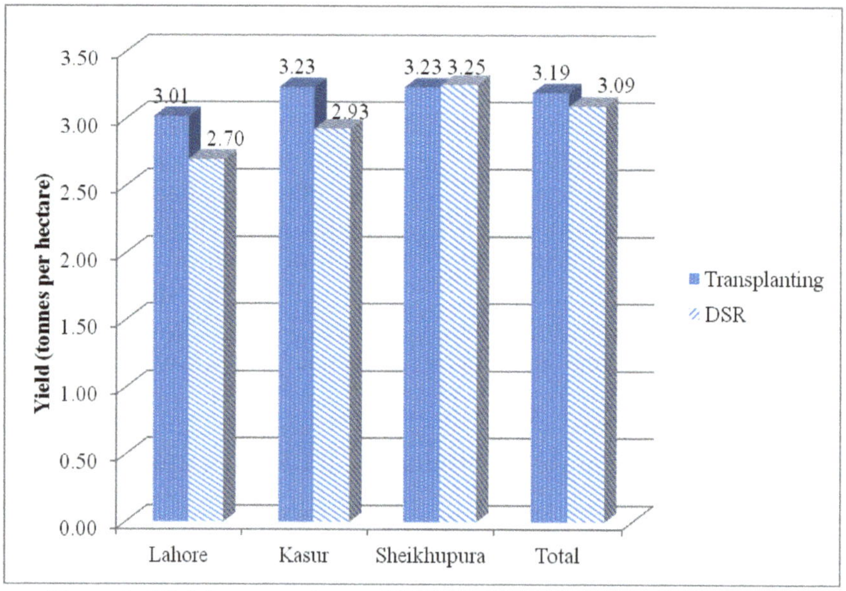

Fig. – 1: Yield comparison of DSR and transplanted rice crops

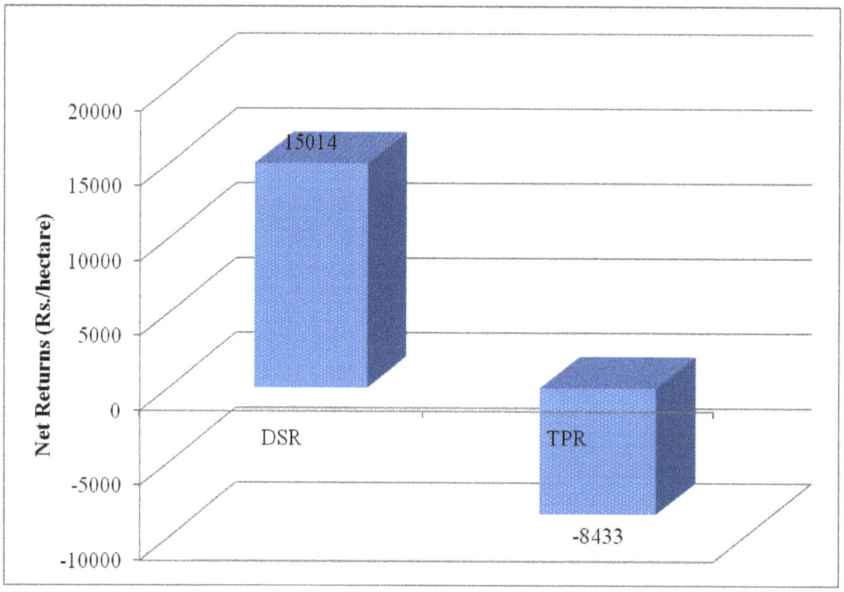

Fig. – 2: Comparison of net economic returns of DSR and transplanted rice crops

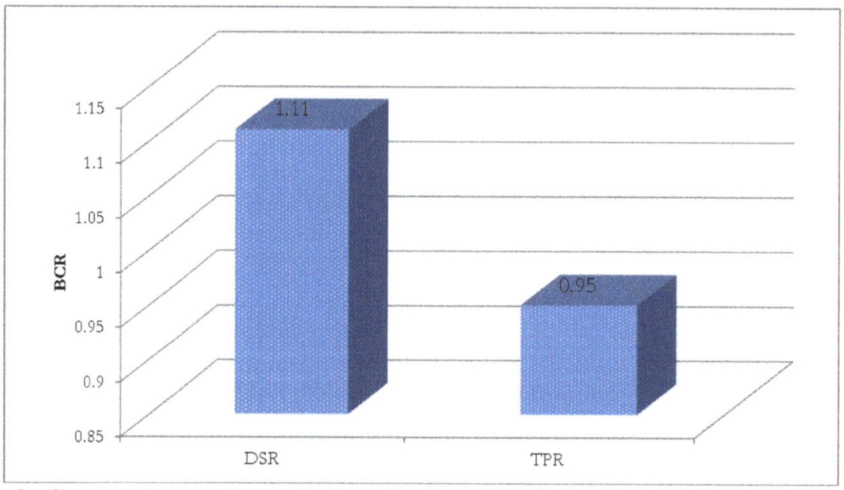

Fig. – 3: Comparison of Benefit Cost Ratios of DSR and transplanted rice crops

REFERENCE

Abedullah, Kouser S and Mushtaq K, 2007. Analysis of technical efficiency of rice production in Punjab (Pakistan). Pakistan Economic and Social Review 42(2):231-244.

Ahmad B and AM Chaudhry, 1987. Profitability of Pakistan's Agriculture. The Pakistan Development Review 26(4):459-469.

Ali RI, Awan TH, Manzoor Z, Ashraf MM, Safdar ME and Ahmad M, 2007. Screening of rice varieties suitable for direct seeding in Punjab. *J. Anim. Pl. Sci.* 17(1-2): (24-26)

Awan TH, Ali I, Safdar ME, Ashraf MM and Yaqub M, 2007. Economic Effect of different plant establishment techniques on Rice, Oryza Sativa production. *J. Agric. Res.,* 45(1):73-80

Ali RI, Iqbal N, Saleem MU and Akhtar M, 2012. Effect of different planting methods on economic yield and grain quality of rice. *Int. J. Agric. Appl. Sci.* 4(1):28-34.

Adair CR, Beachell HM, Jodon NE, Davis LL and Jones JW, 1942. Comparative Yields of Transplanted and Direct Sown Rice. J. Amer. Soc. Agron. 34(2): 129-137.

Anagadi VV, Umapathy PN, Raddor GD, Nadaf SK and Chittapur BM, 1993. Evaluation of Rice Planting Method for Rainfed Low Lands of Karnatake. Inter.Rice Res. Notes. 18(1): 44-45.

Chaudhry AM, Ahmad B and Chaudhry MA, 1992. Cost of Producing Major Crops in Pakistan 1991–92. Research Report, Department of farm management, University of Agriculture, Faisalabad.

Chaudry MA, Ahmad B and Sharif M, 1995. Hand Book for Economic Analysis of Experimental Data. Agri. Social Sciences Res. Centre, Faculty of Agri. Eco. and Rural Sociology, Uni. of Agri., Faisalabad, p 23-28.

Government of Pakistan, 2014. Economic Survey of Pakistan 2013-14. Finance Division, Islamabad.

Government of Punjab, 2015. Cost of production publication for Super Basmati 2014. Crop Reporting Service of Punjab, Lahore.

Khan MA, 2015. Wheat production technology in Punjab. Wheat producing zones. http://www.pakissan.com/english/allabout/crop/wheat/wheat.production.tech.punjab.shtml

Memon NA, 2013. Rice: Important cash crop of Pakistan. Pakistan Food Journal (Sep-Oct 2013).

Murugaboopathi C, Tomita M, Yamaji E and Koide S, 1991. Prospect of large size paddy field using direct seeding supported by surface irrigation system. Transaction of the ASAE. 34(5): 2040-2046.

Nai-Kin H. and Romli Z, 2002. Impact of direct seeding on rice cultivation: lessons from the Muda area of Malaysia, pp.87-98. In Proceedings of the International Workshop on Direct Seeding in Asian Rice Systems: Strategic Research Issues and Opportunities, 25-28 January 2000, Bangkok, Thailand. Los Baños (Philippines): International Rice Research Institute.

Naklange K, Fukai S and Nathabut K, 1996. Growth of rice cultivars by direct seeding and transplanting under upland and low land conditions. Field Crop Res. 48 (2, 3): 115-123.

Pandey S, Velasco LE, and Suphanchaimat N, 2002. Economics of direct seeding in northeast Thailand, pp. 139-160. In Proceedings of the International Workshop on Direct Seeding in Asian Rice Systems: Strategic Research Issues and Opportunities, 25-28 January 2000, Bangkok, Thailand. Los Baños (Philippines): International Rice Research Institute.

Singh AK, Choudhury BU, and Bouman BAM, 2002. Effects of rice establishment methods on crop performance, water use, and mineral nitrogen, pp. 237-246. In Proceedings of the International Workshop on Water-wise Rice Production, 8-11 April 2002, Los Baños, Philippines. Los Baños (Philippines): International Rice Research Institute.

Sharma, A. R. 1995. Direct seeding and transplanting for rice production under flood prone low land conditions. Field Crop Res. 44(2-3): 129-137

2

IMPACT OF SOWING TIME AND SEEDING DENSITY ON GRAIN YIELD OF WHEAT VARIETY GOMAL-08

Umer Farooq[1], Ejaz Ahmed Khan[1], Abdul Aziz Khakwani[1], Saeed Ahmed[2], Nazeer Ahmed*[3] and Gohar Zaman[4]

[1]*Department of Agronomy, Faculty of Agriculture, Gomal University, Dera Ismail Khan-Pakistan*
[2]*State University of Londrina Centre of Agriculture Sciences Londrina, Parana (PR) Brazil*
[3]*Key Laboratory of Applied Entomology, Northwest A&F University. Yangling, China*
[4]*Department of Environmental Science COMSATS University, Vehari campus Multan, Pakistan*

ABSTRACT

The research was performed, at Agronomic Research Area, Faculty of Agriculture, Gomal University Dera Ismail Khan, to assess the impact of sowing time and seeding density on grain yield of wheat variety Gomal-08. Treatments included three dates of sowing as D_1 (Oct-25), D_2 (Nov-20), D_3 (Dec-15) and three seeding rates (kg ha^{-1}) of S_1-100, S_2-120, S_3-150. Almost all yield parameters showed significant variability among treatments means. Maximum grain spike^{-1}(g), number of fertile tillers m^{-}, spike length (cm), economic yield (t ha^{-1}) and thousand grain weight (g) were recorded in D_1(Oct-25) and D_2 (Nov-20) with S_1 and S_2 as compared to late planted wheat with higher seed rate applied.

Keywords: Wheat, Plant Density, Sowing Date, Production

INTRODUCTION

Wheat (*Triticum aestivum* L.), a member of Gramineae family, is the most important staple food of the world including Pakistan. It was originated from Asia and then was spread to other parts of the globe i.e. Africa, America and Europe. It is consumed by major portion of Pakistan's population thus it occupies first position among the cereal crops growing in the country. Our country is the 7th largest wheat producer in the globe (FAO, 2008). But its average yield is less as compared to other countries due to several factors including less choice of varieties, improper seed rate and sowing time. Average yield in Khyber Pakhtunkhwa is 1286 kg ha^{-1} during five years 2001-05 (Anonymous, 2008) quite less than, 3210 kg ha^{-1}, the average economic yield of the world (Brown, 2011).

Factors like improper grain rate, late sowing, methods of sowing, soil nutrients deficiencies with inadequate population of tillers and lack of irrigation water at critical period of plant growth initiate the decrease of wheat grain yield. Selection of variety and grain rate are believed the most important yield reducing factors. Amongst these, seed rate and sowing time are very important which determine the appropriate crop establishment through pairing struggle and finally influence yield of wheat crop (Nakano and Morita, 2009; Kabesh et al., 2009). Early planting of wheat gives the highest yield than delay in planting due to long planting duration (Tanveer et al., 2003) and better growth development due to rapid seedling and uniform emergence (Kirby, 1993) with more combination of leaf area and fertile tillers (Regan et al., 1992; Kristo et al., 2007). By increasing seed rate, decreased root-shoot ratio and increased leaf growth unit^{-1} area, more production is channelled into increasing total dry matter and shoot growth.

The impact of seeding density on plant size and crops effectively has received concentration (Harper, 1977). Different researchers suggested different seeding rate from 70 to 100 kg ha^{-1} (Chila, 1993; Shah, 1994; Khan, 1993; Singh and Singh 1984). Delayed sowing is manipulated normally with high seed rate because late sowing decreases the growth of plants and number of tillers (Gooding and Davies 1917), (Satorr, 1999). The influence of late sowing of wheat crop at the gap of 10 days (Byerlee et al., 1984) showed that late sowing of wheat crop increased the hazards of hot weather and reduced time of tillering specially at grain filling stage which decreases the grain yield (Soomro et al., 2009). The sowing of crop from (early November to early December) produced higher grain yield as compared to late sowing. Late wheat crop sown from mid to end of December produced reduced yield from 27-59% (Iqbal et al., 2001) as it depended on

**Corresponding author:* e-mail: abuzardikhan@gmail.com

obtainable food energy to produce wheat Grain yield (Stoskopit, 1981). Such mechanism sketch on food energy and photosynthesis to enhance the grain yield of crop must be increased. If crop production practices considered how they will affect the photosynthesis then grain yield must be increased effectively. The present research was conducted to optimize the proper seed rate and best sowing time for wheat variety Gomal-8 under the agro-climatic condition of D.I. Khan.

MATERIALS AND METHODS:

The research was conducted to see the impact of seeding density and sowing time on grain yield of wheat variety Gomal-08 at Research Area of Faculty of Agriculture Gomal University D. I. Khan. The experiment was laid out with split plot arrangement in RCB-design having 4-replications. Three sowing date D_1 (Oct-25), D_2 (Nov-20), D_3 (Dec-15) were kept in main plots with three seeding densities 100 (kg ha^{-1}), 120 (kg ha^{-1}), 150 (kg ha^{-1}) in sub plots. The net plot size was kept 9m^2 with row-row distance of 0.3m. Land was well prepared by ploughing 3-4 time in order to make suitable seedbed for better germination. NPK fertilizers were applied to soil in form of Urea, DAP and MOP with recommended doses i.e., 100:120:150. At the time of sowing, all the phosphate and potassium was applied with nitrogen in split dozes i.e., half at first irrigation and remaining at second irrigation.
Following observations were recorded during the experiment.

Plant height (cm)
Data were recorded by selecting 5 plants randomly from each plot. Their height was measured from the soil surface to the tip of spike with the help of a meter rod and average height was calculated.

Number of tiller (m^{-2})
Number of tillers (m^{-2}) was counted randomly at three places from each plot.

Spike length (cm), grains spike^{-1}
Twenty spikes were selected from each plot and their length was measured to calculate grains spike^{-1}.

1000-grain weight (g), Grain yield (t ha^{-1})
Thousand grain samples were taken randomly from each plot and weighed using digital electronic balance and converted to grain yield (t ha^{-1}).

Biological yield (t ha^{-1})
The crop was harvested and bundles left in respective plots for sun-drying. After that weighed and converted into t ha^{1}.

Harvest index (%)
By using the following formula the harvest index of each plot was calculated;

$$H.I\ (\%) = \frac{\text{Economic yield}}{\text{Biological Yield}} \times 100$$

RESULTS AND DISCUSSION

Plant height (cm)
Maximum plant height (109.20) cm was on D_1 (October-25) with S_1 (100 kg ha^{-1}) followed by (106.55cm) (October-25) S_2 (120 kg ha^{-1}) seed rate and plant height of 105.19 cm was measured on October-25 sowing with the higher seed rate (S_3) (150 kg ha^{-1}). The lowest plant height (96.45cm) was observed on D_3 (December-15) with S_3 (150 kg ha^{-1}) seed rate followed by similar values (96.78) and (97.16) cm in seeding densities S_1-100 (kg ha^{-1}) and S_2-120 (kg ha^{-1}) on same date (Table 1). Minimum plant height in late sowing may be due to short growing period as observed previously that wheat planted on 25-October produce maximum plant height (Baloch et al., 2010).

Table 1: Effect of sowing time and seeding density on plant height (cm) of spring wheat

Seed Rate	Date of Sowing			Mean
	Oct-25	Nov-20	Dec-15	
S_1 (100 kg ha^{-1})	109.20 a	102.00 c	96.78 d	102.16 a
S_2 (120 kg ha^{-1})	106.55 b	102.31 c	97.16 d	102.01 a
S_3 (150 kg ha^{-1})	105.19 b	101.85 c	96.45 d	101.16 b
Mean	106.98 a	102.05 b	96.80 c	

Various characters (S) are likely to be followed by the level of statistical significance of 5% LSD..

$LSD_{0.05}$ for

Date of sowing = 1.607

Seed rate = 0.78

Interaction = 1.36

Number of tiller (m^{-2})

More fertile tillers (407.7) (m-2) were obtained on D1 (October-25) planted with seeding density S3-150 (kg ha-1) followed by statistically insignificant with 404.5 tiller m-2 with seeding density S_2-120 (kg ha^{-1}) on same planting time. The lower number of fertile tillers 278.75 (m^{-2}) were noted on D_3 (December-15) where S_1-100 (kg ha^{-1}) seed was used (Table 2). Maximum fertile tillers (m^{-2}) in early sowing might due to favourable temperature for germination while in delay sowing the numbers of tillers (m^{-2}) decreased due to low temperature as previously observed by Razzaq et al. (1986).

Table 2: Effect of sowing time and seeding density on number of tillers (m^{-2}) of spring wheat

Seed Rate	Date of Sowing			Mean
	25-Oct	20-Nov	15-Dec	
S_1 (100 kg ha^{-1})	399.75 b	371.25 d	278.75 g	349.92 c
S_2 (120 kg ha^{-1})	404.50 ab	379.00 c	320.50 f	368.00 b
S_3 (150 kg ha^{-1})	407.75 a	375.25 cd	341.25 e	374.75 a
Mean	404.00 a	375.17 b	313.50 c	

Various characters (S) are likely to be followed by the level of statistical significance of 5% LSD.

Date of sowing = 4.303

Seed rate = 4.201

Interaction = 7.276

Spike length (cm)

The highest spike length (8.92 cm) was recorded on D_1 (October-25) followed by (8.80 cm) D_2 (November-20) for the same seed rate S_1 (100 kg ha^{1}) while minimum spike length (7.02 cm) was obtained on D_3 (December-15). It was followed by statistically similar value (7.05 cm) for the seed rate S_2 (120 kg ha^{-1}) and (7.1 cm) for the S_3 (150 kg ha^{-1}) on the same date (Table 3). In early planting spike length increases due to longer time available for spike to develop. Our results are in agreement with Ahmad et al. (2000) who reported that spike length, grain $spike^{-1}$ and harvest index decreases when seed rate increases.

Table 3: Effect of seeding density and sowing time on Spike length (cm) of spring wheat

Seed Rate	Date of Sowing			Mean
	25-Oct	20-Nov	15-Dec	
S_1 (100 kg ha^{-1})	8.92 a	8.80 a	7.14 e	8.28 a
S_2 (120 kg ha^{-1})	8.38 b	8.09 c	7.05 e	7.84 b
S_3 (150 kg ha^{-1})	8.51 b	7.67 d	7.02 e	7.73 b
Mean	8.60 a	8.19 b	7.07 c	

Various characters (S) are likely to be followed by the level of statistical significance of 5% LSD.

$LSD_{0.05}$ for

Date of sowing = 0.120

Seed rate = 0.113

Interaction = 0.195

Grains per spike (g)

Grains $spike^{-1}$ depicted that sowing date D_1 (October-25) and D_2 (November-20) gave the highest (41.25) number grains at all the seeding rate of S_1-100 kg ha^{-1}, S_2-120 kg ha^{-1} and S_3-150 kg ha^{-1} and all were statistically insignificant (Table 4). In last date of sowing D_3 (December-15) minimum grains $spike^{-1}$ (31.75) were recorded with highest seeding density S_3 (150 kg ha^{-1}), followed (34.50) with

S_2 (120 kg ha^{-1}) and 35.25 in S_1 100 (kg ha^{-1}). The higher grains spike^{-1} in early planting might be due lengthy growing period in which more photosynthesis energy converted into grains as compared to shorter period available for grain development. Shafiq (2004) revealed that early planted wheat enhanced grains spike^{-1} over then delayed planting.

Table 4: Effect of sowing time and seeding density on number of grain (g) spike^{-1} of spring wheat

Seed Rate	Date of Sowing			Mean
	25-Oct	20-Nov	15-Dec	
S_1 (100 kg ha^{-1})	40.50 a	40.75 a	35.25 b	38.83 a
S_2 (120 kg ha^{-1})	41.25 a	40.75 a	34.50 b	38.83 a
S_3 (150 kg ha^{-1})	40.25 a	41.25 a	31.75 c	37.75 b
Mean	40.66 a	40.91 a	33.83 b	

Various characters (S) are likely to be followed by the level of statistical significance of 5% LSD.
LSD $_{0.05}$ for
Date of sowing = 1.552
Seed rate = 0.774
Interaction = 1.341

Thousand-grain weight (g)
Maximum (47.22 g) 1000-grain weight was obtained in D_2 (November-20) sowing date with lowest seed rate S_1 (100 kg ha^{-1}) followed by (46.74 g) with S_2 (120 kg ha^{-1}) (Table 5). The lowest (41.06 g) 1000-grain weight was noted in delayed sowing date D_3 (December-15) with highest seed rate used S_3 (150 kg ha^{-1}) followed by (43.95 g) with S_2 (120 kg ha^{-1}) on D_3 (December-15). Shahazad et al. (2002) revealed that 1000-grain weight reduced with delay in planting.

Table 5: Effect of sowing time and seeding density on 1000-grain weight (gm) of spring wheat

Seed Rate	Date of Sowing			Mean
	25-Oct	20-Nov	15-Dec	
S_1 (100 kg ha^{-1})	45.96 bc	47.22 a	45.12 cd	46.10 a
S_2 (120 kg ha^{-1})	46.61 ab	46.74 ab	43.95 e	45.77 a
S_3 (150 kg ha^{-1})	44.82 de	44.21 de	41.06 f	43.63 b
Mean	45.80 a	46.06 a	43.37 b	

Various characters (S) are likely to be followed by the level of statistical significance of 5%.
LSD$_{0.05}$ for
Date of sowing = 0.625
Seed rate = 0.577
Interaction = 1.000

Grain yield (t ha^{-1})
Higher Grain yield of 7.42 (t ha^{-1}) on D_1 (October-25) sowing date with seeding density S_2 (120 kg ha^{-1}) was statistically at par with that on D_1 (Ocotober-25) with S_1 (100 kg ha^{-1}) producing 7.39 (t ha^{-1}) followed by 7.14 (t ha^{-1}) with S_3=150 (kg ha^{-1}) (Table 6). The lowest of 4.21 (t ha^{-1}) was obtained on D_3 (December-15) with recommended seeding density S_1= 100 (kg ha^{-1}) followed by D_3 (December-15) with S_2 120 (kg ha^{-1}) gave 5.21 (t ha^{-1}) grain yield. Early planted wheat on D_1 (October-25) gave the highest grain yield may be due to more growth duration which resulted longer spikes filled with heavier grain and higher number of fertile tillers. While in late planting, the higher seed rate increased the grain yield as compared to lowest seed rate might be attributed to highest tiller m^{-2} in highest seed rate. The decreasing trend in economic yield with delayed in sowing due to shorter growing period which resulted in low tiller m^{-2} plant height and grains spike^{-1}. Our results also confirmed the finding of Shahazad (2007) who reported that in late sowing the grain yield t ha^{-1} reduced due to short growing period.

Table 6: Effect of sowing time and seeding density on grain yield (t ha^{-1})of spring wheat

Seed Rate	Date of Sowing			Mean
	25-Oct	20-Nov	15-Dec	
S_1 (100 kg ha^{-1})	7.39 a	6.95 b	4.21 e	6.18 b
S_2 (120 kg ha^{-1})	7.42 a	7.04 b	5.21 d	6.56 a
S_3 (150 kg ha^{-1})	7.14 b	6.92 b	5.77 c	6.61 a
Mean	7.32 a	6.97 b	5.07 c	

Various characters (S) are likely to be followed by the level of statistical significance of 5% LSD

$LSD_{0.05}$ for
Date of sowing = 0.183
Seed rate = 0.128
Interaction = 0.22

Biological yield (t ha^{-1})

Biological yield (t ha^{-1}) was significantly affected by sowing date, seed rate and their interaction. The data showed that on date of sowing D_1 (October-25) with S_2 120 (kg ha^{-1}), highest biological yield of 19.58 t ha^{-1} was recorded followed by S_3-150 (kg ha^{-1}) and S_1-100 (kg ha^{-1}) seeding density by producing statistically at par biological yield of 19.51(t ha^{-1}) and 19.40 (t ha^{-1}) (Table 7). Minimum biological yield of 8.45 (t ha^{-1}) was found in delayed planting D_3 (December-15) with seed rate S_1 100 (kg ha^{-1}) followed by S_2-120 (kg ha^{-1}), S_3-150 (kg ha^{-1}) by producing biological yield of 9.68 (t ha^{-1}) and 9.46 (t ha^{-1}) respectively. The result supported the finding of Iqbal (2012) who stated that the highest biological yield was produced by using grain rate 150 kg ha^{-1}.

Table 7: Effect of sowing time and seeding density on biological yield (t ha^{-1}) of spring wheat

Seed Rate	Date of Sowing			Mean
	25-Oct	20-Nov	15-Dec	
S_1 (100 kg ha^{-1})	19.40 a	15.15 b	8.45 d	14.33 b
S_2 (120 kg ha^{-1})	19.58 a	15.51 b	9.68 c	14.93 a
S_3 (150 kg ha^{-1})	19.51 a	15.53 b	9.46 c	14.83 a
Mean	19.49 a	15.40 b	9.20 c	

Various characters (S) are likely to be followed by the level of statistical significance of 5% LSD.

$LSD_{0.05}$ for
Date of sowing = 0.362
Seed rate = 0.234
Interaction = 0.405

Harvest Index % (H.I)

Highest H.I (%) 61.10% was noted on D_3 (December-15) S_3 (150 kg ha^{-1}), followed by 53 (%) in S_2 (120 kg ha^{-1}) on same date D_3 (December-15) while minimum 38.10 (%) in S_1 (100 kg ha^{-1}) followed by 37.93(%) and 36.63(%) in S_2-120 (kg ha^{-1}), S_3-150 (kg ha^{-1}) on date D_1 (October-25) (Table 8). The lowest H.I in early planting might be due to higher biomass resulting from tall plants, more tillers m^{-2} and dense population which decreased the ratio of economic yield comparatively.

Table 8: Effect of sowing time and seeding density on harvest index (%) of spring wheat.

Seed Rate	Date of Sowing			Mean
	25-Oct	20-Nov	15-Dec	
S_1 (100 kg ha^{-1})	38.10 e	48.89 d	49.87 c	44.62 c
S_2 (120 kg ha^{-1})	37.93 e	45.38 d	53.88 b	45.73 b
S_3 (150 kg ha^{-1})	36.63 e	45.58 d	61.09 a	47.67 a
Mean	37.55 c	45.28 b	54.95 a	

Various characters (S) are likely to be followed by the level of statistical significance of 5% LSD.

$LSD_{0.05}$ for
Date of sowing = 1.69
Seed rate = 0.97
Interaction = 1.68

Early planting of wheat on (Oct-25) produced the highest economic yield of 7.39 and 7.42 t ha^{-1} with seed rates 100 kg ha^{-1} and 125 kg ha^{-1} indicating its suitability under the ecological environment of Dera Ismail Khan. It is suggested that in case of early planted of wheat (Oct-25 to Nov-10), the seed rate should be used from 100-120 kg ha^{-1} while in late planting, use the higher S_3 (150 kg ha^{-1}) for obtaining the reasonable grain yield.

REFERENCES

Ahmad Z, Kisana NS, Mujahid M, Majid A, Ahmad I and Mustafa S, 2000. Influence of plant density on yield parameters and yield of wheat. Pak. J. Bio. Sci. 3(9): 1389-1390.

Anonymous, 2008. Agricultural Statistics of Pakistan. Ministry of Food, Agriculture and Livestock. Govt. of Pakistan, Islamabad.

Baloch MS, Shah ITH, Nadim MA, Khan MI and Khakwani AA, 2010. Effect of planting time, seeding density on growth and yield parameter of wheat. J. Anim. Plant Sci. 20(4): 239-240.

Brown, 2011. World on the edge: How to prevent economic and environmental collapse. W.W. Norton and Company, New York. P.141–159.

Byerlee D, Aslam M, Hobbs RR and Sheikh AD, 1984. Wheat in rice based cultivation system of the Punjab. Implications for extension and research. NARC. Wheat Paper. page:49.

Chila AJ, 1993. Date of sowing and seed rate influence on spacing seeded small grain cultivars. Agron. J. 6 (1): 795-798.

Kristo I, Szel MH, Gyapjas J and Szekeres A, 2007. Effect of sowing dates and seeding rate on different winter wheat cultivars. Res. J. Agric. Sci. 39(1):61-64.

FAO, 2008. Statistical data base (FAO). Government of Pakistan. 2011-12. Agriculture statistics of Pakistan (Economic wing). Ministry of food, agriculture and livestock. Govt. of Pakistan, Islamabad. p. 22.

Gooding M.J and Davies, 1997. Wheat crop production and Utilization. Journal of Agriculture Sci. 138:317-331.

Harper IJ, 1977. Population biology of plants. Academic Press, London. pp: 41-54.

Iqbal MS, Yar A, Akram HM, Ali A, Iqbal J and Anser MR, 2001. Plant density and date of sowing influence on seed yield of wheat. J. Agric. Res. 39 (3-4): 217-220.

Iqbal J, Hayat K, Hussain S, Ali A, Ilyas M A and Bakhsh HA, 2012. Impact of plant population and nitrogen levels on yield parameters and yield of wheat. Pak. J. Nutr. 11 (7): 531-536.

Kabesh MO, Sary GA, El-Naggar HM and El-kramany MF, 2009. Impact of planting methods and some bio-organic fertilization treatments on economic yield and yield parameter of wheat. Res. J. Agric. Biol. Sci. 5: 97-102

Khan RU, 1993. Influence of plant density on the yield components and yield of three wheat cultivars. Sarhad J. Agric. 2 (1): 1-8.

Kirby EJM, 1993. Effect of sowing depth on seedling emergence, growth and development in wheat and barley. Field Crop Res. 35: 101-111.

Nakano H and Morita S, 2009. Impact of grain rate and nitrogen level on grain yield and protein content of the bread wheat variety. Int. J. Plant Production. Sci. 12: 109-115.

Razzaq A, Khan B, Mohammad D, Shah P and Saeed K, 1986. Impact of planting time on the growth development and straw yield of wheat varieties. Sarhad J. Agric. 2(2): 327-334.

Satorre EH, 1999. Plant density and distribution as modifiers of growth and yield. Wheat: Ecology and physiology of yield determination. Food Products Press, New York. p.144-159.

Shafiq HM, 2004. Radiation use efficiency, Modeling growth and yield of wheat at various planting date and nitrogen levels. Pak. J. Life Soc. Sci. 7(1):66-69.

Shah P, 1994. Cereal crop, Rabi. In cropprod. (EDS. Nazir, S., E. Bashir And R. Bantel), National Book Foundation, Islamabad. pp. 238 –245.

Shahazad K, Bakht J, Shah WA, Shafi M and Jabeen N, 2002. Yield and yield components of various wheat cultivars as affected by different sowing dates. Asian J. PI. Sci. 1(5): 522-525.

Shahazad MA, Khan MM, Sahi ST, Ahmad M, Din W and Ullah E, 2007. Influence of planting times, grain treatment on economic yield and feature of wheat. Pak. J. Agri. Sci. 44(4):581-583.

Singh H and Singh R, 1984. Effect of seed rates and nitrogen on wheat. Indi. J. Agron. 29(1): 129-130.

Soomro UA, Odhano EA, Rahman MU, Tareen AQ and Gul S, 2009. Effects of seed rate and sowing method on growth and yield of wheat. World J. Agric. Sci. 5 (2):159-162.

Stoskopf NC, 1981. Cereals in understanding crop production. Published by Reston Co. USA. 139p.

Tanveer SK, Hussain I, Kissana NS, Abbas SG and Sohail M, 2003. Influence of various planting methods on yield components and economic yield of wheat. Asian J. Plant Sci. 2: 811-813.

EFFECT OF SPLIT APPLICATION OF NITROGENON THE YIELD OF RICE CROP

Muhammad Nawaz[1], Anwar Javaid Wahla[1], Muhammad Saleem Kashif[1,] Masood Qadir Waqar[2] and Muhammad Anjum Ali[3]
[1]Adaptive Research Farm, Sheikhupura, [2]Directorate of Adaptive Research, Lahore, Punjab, [3]Directorate General of Agriculture Extension and Adaptive Research, Lahore, Punjab

ABSTRACT

A three year field study was conducted at Adaptive Research Farm, Sheikhupura to find the most suitable nitrogen application schedule for rice (Super Basmati) during Kharif 2012, 2013 and 2014. Randomized complete block design with 3 replications was employed to investigate 4 different treatments of nitrogen split application @ 140 kg ha^{-1} viz: T_1= Puddling and tillering, T_2= Puddling, tillering and panicle initiation, T_3= Tillering, panicle initiation and flowering stage and T_4= Puddling, tillering, panicle initiation and flowering stage. Nursery of rice was transplanted in 2nd week of July each year. Paddy yield and all yield associated parameter of rice super basmati were significantly affected by treatments and highest of all of them were recorded when nitrogen was applied in 3 equal splits *i.e.* at puddling, tillering and panicle formation stage. In conclusion, nitrogen @ 140 kg ha^{-1} in 3 equal splits (puddling, tillering and panicle formation), is the most suitable combination of nitrogen splits to obtain higher paddy yields under agro-ecological conditions of Sheikhupura, Pakistan.
Keywords: puddling, rice, super basmati, flowering, tillering

INTRODUCTION

Rice (*Oryza sativa* L.) is among the most widely used cereals across the globe, grown under diverse range of climatic conditions, to feed the mankind. It is a very important source of foreign exchange earning giving about US $ 1.667 billion annually through its export by Pakistan. Moreover, Pakistan ranks 4th among world rice producing countries and provides about 30% of global rice supplies. During 2013-14, about 2.8 million hectare land was under rice in Pakistan. Efforts are being made to enhance the rice yield all at levels including research, government and farmers, but average yield of rice is far less than the genetic potential in Pakistan. Its production depends upon several agronomic factors including irrigation management, selection of suitable cultivar, optimum sowing time, effective control of insects, diseases and weeds and balanced fertilization (Angus et al., 1994; Jing et al., 2008). Any of these factors if not managed well can cause significant yield losses in rice crop.

Fertilizers play key role in crop production as they provide essential plant nutrients, critical for meeting the world food supplies. Among different types of fertilizers, nitrogenous fertilizers are the most important. It is the integral part of different compounds, like amino acids, DNA, RNA, phytohormones, and nzymes. Moreover, nitrogen is also involved in carbon and amino acid metabolism and synthesis of proteins (Cai et al., 2012). Vegetative growth of plant is primarily dependent upon adequate supply of nitrogen, if not managed properly, a stunted growth with pale yellow leaves is observed for cereals.

In cereal crops, nitrogen use efficiency is about 29% and 42% in developing and developed agricultural countries of the world, respectively (Pilbeam, 1996; Raun and Johnson, 1999; Beatty et al., 2010). Moreover, among yield limiting factors for rice, nitrogen is the most important (Ladha and Reddy, 2003; Samonte et al., 2006). During vegetative growth stage of rice, nitrogen is required for promoting growth and tillering, that determines number of panicles and during early panicle formation stage, development of spikelets is also dependent much on nitrogen supply. Moreover, nitrogen is also involved in sink size at late panicle formation stage. It is also actively involved in photosynthesis, carbohydrate accumulation and grain filling in rice (Mae, 1997). So, proper nitrogen management is key factor in narrowing the yield gap of rice. Application of nitrogenous fertilizer at proper time with a suitable rate according to the soil conditions can cause significant increase in paddy yield by increasing nitrogen use efficiency of crop (Ganga Devi et al., 2012;

**Corresponding author's email:* kashif1637@gmail.com

Shakouri et al., 2012). Researchers have reported different outcomes of their studies for deciding the best time of nitrogen fertilization. Ehsanullah et al. (2001), reported that nitrogen split application at three growth stages (transplanting, tillering and panicle emergence) should be followed to achieve higher paddy yield whereas Sathiya and Ramesh (2009) reported that application of nitrogen at critical growth stages like tillering can increase number of panicle and number of spikelets per spike in rice. In addition to this, application of nitrogen during critical growth stages of rice avoids its losses and facilitates its increased translocation into plant parts (Pushpanathan et al., 2005). If nitrogen is applied in splits rather than single time application with consideration of crop growth stages, it can increase yield of rice crop by increasing leaf area per plant, decreasing spikelet sterility and increasing the number of grains per panicle (Ehsanullah et al., 2001). It is obvious from all these studies that time of nitrogen application matters a lot in gaining higher rice yield, whereas sufficient experimental evidence is not available about the proper nitrogen fertilizer management in Sheikhupura region, thus the unwise use of nitrogenous fertilizer causes huge yield losses in rice, so the present study was planned with the objective to find the best suitable nitrogen split application schedule for rice crop in Sheikhupura region, Punjab, Pakistan.

MATERIALS AND METHODS

Site description: The experiment was carried out at Adaptive Research Farm, Sheikhupura that lies between latitude 31° – 42° N and longitude 73° – 59° E on the globe at an altitude of 209.57 m. The climate is moist sub-humid with an annual rain fall of 250 – 500 mm. Rice-wheat cropping system is the major cropping system in this area. Composite soil sample to a depth of 30 cm was obtained from the experimental area with soil auger prior to sowing of crop. The sample was analyzed for its physio-chemical properties. Bouyoucos hydrometer method was employed to find the soil texture and 1% sodium hexametaphosphate was used as dispersing agent. The relative proportion of sand, silt and clay was 14%, 70% and 16%, respectively. The soil belongs to loamy type as described by the international textural triangle (Moodie et al., 1959). Moreover, a pH of 8.4 with 10.1% total soluble salts, 0.8% organic matter, 0.07% total nitrogen, 10.4 ppm available phosphorous and 204 ppm potash was recorded in chemical soil analysis for the soil of experimental site.

Crop husbandry: Nursery of rice cultivar, super basmati was raised in 2nd week of June each year and was transplanted manually on puddled soil during 2nd week of July at the age of about 30 days. Recommended dose of P and K, 80 and 62 kg ha^{-1}, respectively was applied to all plots at the time of puddling. The source of P and K was DAP and SOP for all the years in all treatments except in last treatment, where SSP was used at the time of puddling instead of DAP, as no nitrogen was required here at the time of puddling according to treatment structure. Zinc was applied @ 5 kg ha^{-1} in the form of zinc sulphate (33%) at active tillering stage. All other agronomic practices including weed management, irrigation scheduling and management of pests etc. were addressed uniformly for all the treatments. Each year, the crop was harvested and threshed manually from each plot at maturity.

Experimental Design and Treatments: Experiment was laid out in Randomized Complete Block Design (split plot arrangement) with 4 different combinations of nitrogenous fertilizer and years as the treatments and was replicated thrice. The source for nitrogenous fertilizer was Urea and DAP purchased from Fauji Fertilizer Limited, Pakistan. Nitrogen was applied at recommended rate (140 kg ha^{-1}) at different growth stages of rice in splits that are used as treatments and give below.

T_1= Puddling and tillering (2 equal splits)

T_2= Puddling, tillering and panicle initiation (3 equal splits)

T_3= Tillering, panicle initiation and flowering stage (3 equal splits)

T_4= Puddling, tillering, panicle initiation and flowering stage (4 equal splits)

The net plot size was 3.78 m × 3.12 m for each experimental unit.

Data Collection and Statistical analysis: Data regarding plant height at maturity, number of fertile tillers, number of grains per spike, 1000-grain weight and paddy yield were collected using the standard procedures. The collected data were analyzed statistically by employing the Fisher's analysis of variance technique. The significance of treatment means was tested using least significance difference (LSD) test at 5% probability level (Steel et al., 1997).

Table 1: Effect of different nitrogen split combinations on the yield and yield associated parameters of fine rice (Super basmati) during 2012, 2013 and 2014.

	2012	2013	2014	Mean
Plant height (cm)				
T_1= Puddling and tillering (2 equal splits)	100.0bc	99.0bcd	99.3bcd	99.4
T2= Puddling, tillering and panicle initiation (3 equal splits)	103.0a*	97.0d	100.3bc	100.1
T_3= Tillering, panicle initiation and flowering stage (3 equal splits)	98.0cd	102.0a	101.0ab	100.7
T_4= Puddling, tillering, panicle initiation and flowering stage (4 equal splits)	99.0bcd	101.3ab	99.0bcd	99.8
Mean	100.0	100.0	99.9	
LSD ($p \leq 0.05$) Year = Ns , Nitrogen splits= Ns , Year × Nitrogen splits= 2.39				
Number of Productive tillers (m^2)				
T_1= Puddling and tillering (2 equal splits)	116.7f	131.3bcd	135.0abc	127.7b
T2= Puddling, tillering and panicle initiation (3 equal splits)	131.0bcd	136.3ab	140.0a	135.8a
T_3= Tillering, panicle initiation and flowering stage (3 equal splits)	102.7g	127.3de	129.0cs	119.7c
T_4= Puddling, tillering, panicle initiation and flowering stage (4 equal splits)	121.0ef	135.0abc	137.0ab	131.0a
Mean	117.8b	132.5a	135.3a	
LSD ($p \leq 0.05$) Year = 4.28, Nitrogen splits= 3.45, Year × Nitrogen splits= 6.66				
Number of grains per spike				
T_1= Puddling and tillering (2 equal splits)	126.0	128.7	125.0	126.6c
T2= Puddling, tillering and panicle initiation (3 equal splits)	126.7	133.3	132.0	136.7a
T_3= Tillering, panicle initiation and flowering stage (3 equal splits)	124.7	126.7	122.3	124.6d
T_4= Puddling, tillering, panicle initiation and flowering stage (4 equal splits)	128.0	131.7	126.7	128.8b
Mean	127.1	130.1	126.5	
LSD ($p \leq 0.05$) Year = Ns , Nitrogen splits= 1.85, Year × Nitrogen splits= Ns				
1000-grain weight (g)				
T_1= Puddling and tillering (2 equal splits)	20.00ef	20.73cd	21.90b	20.88b
T2= Puddling, tillering and panicle initiation (3 equal splits)	21.10c	21.03cd	22.80a	21.64a
T_3= Tillering, panicle initiation and flowering stage (3 equal splits)	19.80f	20.60cde	20.40def	20.67c
T_4= Puddling, tillering, panicle initiation and flowering stage (4 equal splits)	20.90cd	20.97cd	20.90cd	20.92b
Mean	20.45b	20.83b	21.50a	
LSD ($p \leq 0.05$) Year = 0.540 , Nitrogen splits= 0.296, Year × Nitrogen splits= 0.692				
Paddy yield (t ha^{-1})				
T_1= Puddling and tillering (2 equal splits)	3.50fg	3.63f	4.62b	3.92c
T2= Puddling, tillering and panicle initiation (3 equal splits)	4.25c	3.73e	4.92a	4.30a
T_3= Tillering, panicle initiation and flowering stage (3 equal splits)	3.17h	3.33gh	4.03d	3.51d
T_4= Puddling, tillering, panicle initiation and flowering stage (4 equal splits)	3.63ef	3.70e	4.79ab	4.04b
Mean	3.64b	3.60b	4.59a	
LSD ($p \leq 0.05$) Year = 0.052 , Nitrogen splits= 0.108, Year × Nitrogen splits= 0.169				

*= Means not sharing the same letter for an individual factor or interaction for each parameter differ from each other at $p \leq 0.05$

RESULTS

Plant height (cm)

Results presented in Table 1 show that different nitrogen split combinations did not significantly affect the plant height in rice. Only the interaction of nitrogen splits and year was significant. Maximum plant height (103.0 cm) was recorded during 2013 with the application of nitrogen in 3 equal splits (puddling, tillering and panicle initiation) and minimum plant height (97.0 cm) was recorded in the same treatment during 2013 that may be the due to climatic effects. It is also obvious from the results that applying nitrogen at flowering stage does not cause significant increase in the plant height as in T_4 (puddling, tillering, panicle initiation and flowering stage) plant height (99.0 cm) was recorded which is statistically lesser than the maximum.

Number of fertile tillers (m^2)

Table-1 shows that different nitrogen split combinations significantly affected the number of fertile tillers in rice. Maximum number of fertile tillers (135.8) were recorded when nitrogen was applied at the time of puddling, tillering and panicle initiation (T_2) and minimum number of fertile tillers (119.7) was found when nitrogen was applied at the time of tillering, panicle initiation and flowering stage (T_3). Moreover, during 2013 and 2014, statistically higher numbers of productive tillers were recorded as given in Table 1. The interaction of nitrogen splits and year shows that maximum number of fertile tillers (140.0) were found during 2014 and the fertile tillers during 2013 (136.3) in the same treatment were statistically at par with it. Whereas, minimum number of fertile tillers (116.7) were found during 2012 in case of 2 splits of nitrogen (at the time of puddling and tillering).

Number of grains per spike

Different nitrogen split combinations significantly affected the number of grain per spike in rice. Maximum number of grains per spike (136.7) were found by the application of nitrogen @ 140 kg ha^{-1} at the time of puddling, tillering and panicle initiation (3 equal splits) and minimum number of grains per spike (124.6) were recorded for the treatment where nitrogen was not applied at the time of puddling (T_3). There was no significant effect of year and interaction was also non-significant (Table 1).

1000-grain weight (g)

Grain weight index or 1000-grain weight of rice was also significantly affected by the different treatments of nitrogen applications (Table 1). Application of nitrogen @ 140 kg ha^{-1} in 3 equal splits at the time of puddling, tillering and panicle initiation gave highest 1000-grain weight of rice (21.64 g) and minimum 1000-grain weight (20.67 g) was recorded when nitrogen was applied in 3 equal splits (tillering, panicle initiation an flowering stage). As for as interaction of nitrogen splits and year is concerned, maximum 1000-grain weigh (22.80 g) was found during 2014 with the application of nitrogen in 3 equal splits (puddling, tillering and panicle initiation) and minimum 1000-grain weight (19.80 g) was found by applying nitrogen in 3 equal splits (tillering, panicle initiation and flowering). A significant year effect was also observed.

Paddy yield (t ha^{-1})

Paddy yield of rice was also significantly affected by different nitrogen split combinations as maximum paddy yield (4.30 t ha^{-1}) was recorded with the application of nitrogen in 3 equal splits (puddling, tillering and panicle initiation) and minimum paddy yield (3.51 t ha^{-1}) was recorded in case of application of nitrogen in 3 equal splits (tillering, panicle initiation and flowering stage). The interaction of nitrogen splits and year shows that during all years, paddy yield was statistically better with the application of nitrogen in 3 equal splits (puddling, tillering and panicle initiation) than all other treatments, however, highest paddy yield (4.92 t ha^{-1}) was recorded during 2014 by applying nitrogen in 3 equal splits (puddling, tillering and panicle initiation) and minimum was recorded during 2012 when nitrogen was applied in 3 equal splits (tillering, panicle initiation and flowering stage). Year also had a significant effect on paddy yield.

DISCUSSION

A balanced and sufficient supply of nitrogen is quite necessary for proper vegetative growth and green colour of rice crop. Moreover, proper nitrogen balance in crop also promotes the proper utilization of other essential elements like P and K. Ha and Suh (1993), reported that if nitrogen is supplied in proper splits, it can make plant more responsive towards better utilization of other inputs and gives better plant height than that of applying all the nitrogen at the time of sowing. If nitrogen is applied in splits, it remains available to plant for the entire growth season as in case of 2nd treatment, higher plant height, more number of fertile tillers and more number of grains per spike were found in 2nd treatment as compared to other treatments and results also showed that omitting nitrogen at the time of puddling can cause severe reduction in all yield related parameters and subsequently lower paddy yield was observed, when no nitrogen was applied at the time of puddling. Maske et al (1997) also reported similar findings that proper management of nitrogenous fertilizer can affect the yield related parameters and if nitrogen application is not managed well, a declined yield will be observed. Increase in the number of grains per panicle was also observed, when nitrogen was applied in splits rather than applying whole nitrogen in single dose. The increased number of grains per spike in T_2 might be due to availability of nitrogen to crop at all critical growth stages. These results are in line with those of Ha and Suh (1993) who reported that when nitrogen is applied in splits, it increases number of grains per panicle in rice. Mathew et al (1990) also reported the similar findings that applying whole nitrogen at the time of puddling, decreased 1000-grain weight in rice. In conclusion, applying nitrogen in 3 splits @ 140 kg ha^{-1} at the time of puddling, tillering and panicle formation is the most suitable split combination of nitrogenous fertilizer for obtaining high yields of fine rice in agro-ecological conditions of Sheikhupura, Pakistan due to its availability at all critical growth stages.

REFERENCES

Angus JF, Ohnishi M Horie T and Williams L, 1994. A preliminary study to predict net nitrogen mineralization in a flooded rice soil using anaerobic incubation. Aust. J. Exp. Agric. 34:995-999.

Beatty PH, Anbessa Y, Juskiw P, Carroll RT, Wang J and Good AG, 2010. Nitrogen use efficiencies of spring barley grown under varying nitrogen conditions in the field and growth chamber. Ann. Bot. 105: 1171-1182.

Cai H, Lu Y, Xia W, Zhu T and Lian X, 2012. Transcriptome response to nitrogen starvation in rice. J. Bio. Sci. 37:731-747.

Ehsanullah M, Attaullah M, Cheema MS and Usman M, 2001. Rice Basmati-385 response to single and split application of nitrogen at different growth stages. Pak. J. Agri. Sci. 38:84-86.

Ganga DM, Tirumala RS, Sumatin V, Pratima T and John K, 2012. Nitrogen management to improve the nutrient uptake, yield and quality parameters of scented rice in aerobic culture. Int. J. App. Bio. Pharma. Tech. 3:340-344.

Ha WG and Suh HS, 1993. Collection and evaluation of Korean red rice nitrogen response on yield and its components. Kor. J. Crop Sci. 38:134-138.

Jing Q, Bouman B, Keulen HV, Hengsdijk H, Cao W and Dai T, 2008. Disentangling the effect of environmental factors on yield and nitrogen uptake of irrigated rice in Asia. Agric. Sys. 98:177-188.

Ladha JK and Reddy PM, 2003. Nitrogen fixation in rice system: State of knowledge and future prospects. Plant Soil. 252:151-167

Mae T, 1997. Physiological nitrogen efficiency in rice: Nitrogen utilization, photosynthesis, and yield potential. Plant Soil. 196: 201- 210.

Maske NS, Borkar SL and Rajgire HJ, 1997. Effects of nitrogen levels on growth, yield and grain quality of rice. J. Soils Crops. 7: 83-86.

Mathew J, Varghese K, Pillai GR and Santhakumari G, 1990. Response of winter rice to different water regimes and nitrogen levels. Ind. Kerala Agric. Res. J. 28: 26-30

Moodie GE, Smith WH and McGreery RA, 1959. Laboratory Manual for Soil Fertility. Dept. Agron., Washington State College, USA. pp: 31-39.

Pilbeam CJ, 1996. Effect of climate on recovery of crop and soil N^{15} labelled fertilizer applied to wheat. Fert. Res. 45: 209-215.

Pushpanathan KR, Vijayakumar M and Siddeswaran K, 2005. Effect of form of fertilizer nitrogen and timing of application on growth and yield of rice (*Oryza sativa* L.). Agric. Rev. 26:153-156.

Raun, WR and Johnson GV, 1999. Improving nitrogen use efficiency for cereal production. Agron. J. 91: 357-363.

Samonte SOPB, Wilson LT, Medley JC, Pinson SR, McClung AM, and Lales JS, 2006. Nitrogen utilization efficiency: relationships with grain yield, grain protein, and yield-related traits in rice. Agro. J. 98: 168-176

Sathiya K and Ramesh T, 2009. Effect of split application of nitrogen on growth and yield of aerobic rice. Asian J. Exp. Sci. 23:303-306.

Shakouri MJ, Vajargah AV, Gavabar MG, Mafakheri S and Zargar M, 2012. Rice vegetative response to different biological and chemical fertilizers. Adv. Env. Bio. 6:859-863.

Steel RGD, Torrie JH and Dickey D, 1997. Principles and Procedure of Statistics. A Biometrical Approach 3rd Ed. McGraw Hill Book Co. Inc. New York. pp. 352-358.

4

EFFECT OF POTASH LEVELS AND TIME OF APPLICATIONS ON SEED COTTON YIELD IN ECOLOGICAL ZONE OF RAHIM YAR KHAN

Muhammad Aslam[1], Ashiq Hussain Sanghi[2], Laila Khalid[3*] and Shamaun Javed[4]

[1] Senior Subject Matter Specialists (Agronomy, [2] Senior Subject Matter Specialists (Plant Protection)

[3, 4] Assistant Research Officers Adaptive Research Farm Rahim Yar Khan.

ABSTRACT

A study was conducted in three successive seasons during 2008-10 on sandy loam soil at Adaptive Research Farm, Rahim Yar Khan. The objective was to investigate the response of cotton (cv.MNH-786) to different levels and times of application of potash. The treatments comprised three potassium doses (0, 62.5 and 125 kg ha^{-1}) and applied at different times (full on sowing, full at 1st irrigation and ½ at sowing + ½ at 1st irrigation). Potassium doses applied at different times significantly affected seed cotton yield. On an average the highest seed cotton yield (3030.75 kg ha^{-1}) was obtained when crop was fertilized at the rate of 125 kg ha^{-1} K (full sowing) followed by 62.5 kg ha^{-1} K full at sowing (2875.25 kg ha^{-1}). So potassium fertilizer application at sowing proved more beneficial than its split application at sowing and 1st irrigation. The lowest yield was noted in control (2258.52 kg ha^{-1}) where no potassium fertilizer was applied. Hence application of K at economical doses of 62.5 kg ha^{-1} is recommended in ecological zone of Bahawalpur.

Keywords: potash, application time, dose rate, cotton yield, fertilizer doses

INTRODUCTION

Cotton (*Gossypium hirsutum* L.) pays a vital role in the economy of Pakistan. It contributes to the National economy by providing raw material to the local textile industry, such as cotton lint as an export item. It accounts for 7.8 percent of value added in Agriculture and 1.6 percent of GDP. During 2011-12 the crop was cultivated on an area of 2835 thousand hectares 5.4 percent more than last years (2689 thousand hectares). (Government of Pakistan, 2011-12). Fertilizers occupy pivotal position in raising seed cotton yield. Experiments have shown that an optimal yield could only be produced with the balanced application of all major nutrients in soil (Ahmad, 1998). Cotton growers in Pakistan use a desirable amount of N (125 kg ha^{-1}) but use of K fertilizer is negligible. The less use of K fertilizer in cotton may have serious consequences. Cotton appears to be more sensitive to K deficiencies than other crops, as root system of cotton is less dense than that of other crops (Mithaiwala et al., 1981). Colakoglu, (1980) recommended optimum dose of 80-120 kg ha^{-1} N, 60-90 kg ha^{-1} P and 100-200 kg ha^{-1} K for optimum seed cotton yield in Turkey. During the last two decades, cotton production scenario has changed. There is heavy drain of nutrients due to more demand by varieties at certain early maturing and high yielding cotton growth stages. Researchers (Makhdum, 2003 and Malik et al., 1989) reported that cotton crop could benefit from higher doses of potassium fertilizers when applied at different times after sowing. This may be attributed to equilibrium between various forms of potassium and degree of potassium fixation in soils (Mengel and Kirby 1982).

Potassium is one of the most important elements in plant nutrition. All living organisms require its large amounts for normal growth and development. Potassium increases the photosynthetic rates, CO_2 assimilation and facilitates carbon movement (Saleem, 2002). The experiments conducted in Indian Punjab showed that cotton crop absorbed a large quantity of potassium indicated that it was more than both nitrogen and phosphorous intake (Brar et al., 1987). The trend of potassium removal by cotton crop indicates its heavy drain from Pakistani soils. The exploitation of soils is likely to lead to severe depletion of would eventually limit the efficiency of other nutrients (Malavolta, 1985). Potassium is an essential macro element for all living organisms required in large amounts for normal plant growth and development.

Keeping in view the significant of cotton in Pakistan this study was conducted to see cotton

**Corresponding author:* e-mail: laila_kld@yahoo.com

response to varying levels of potash applied at different times.

MATERIALS AND METHODS

A trial was conducted for the three consecutive seasons (2008-10) at Adaptive Research Farm Rahim Yar Khan. Soil samples were collected from 0-15 cm and 15-30 cm depth before fertilizer application during each season. The physical and chemical characteristics of investigated soils were determined according to methods recommended by Ryan et al., (2001). Data on soil chemical analysis are given as PH 8.2; EC (ds/m) 1.1, Organic matter 0.42, available P (ppm) 8, K (ppm) 114-165 and soil was loamy in texture having low organic matter. The potassium level seemed to be in sufficient to obtain an economic yield according to limits set by WFDC (Saleem, 1989). Layout system was RCBD with three replications. The treatments comprised three K levels (0, 62.5 and 125 kg ha^{-1}) applied at different times (full on sowing, full at 1st irrigation and ½ at sowing and ½ with 1st irrigation). Cotton variety (MNH-786) was sown during second week of May having a seed rate of 30 kg ha^{-1}. P_2O_5 (57 kg ha^{-1}) as diammonium phosphate at planting and 150 kg ha^{-1} nitrogen as urea was applied in three splits at planting, flowering and peak flowering. Pre emergence weedicide stomp 330EC @ 2.5 L ha^{-1} was applied to control weeds. Crop was kept free from insect pest attack through regular sprays of insecticides. Normal irrigation was given to the crop and standard production practices for the area were adopted through out the season. Data on seed cotton yield and its yield parameter such as boll weight, no of boll $plant^{-1}$ and height were collected at maturity. Seed cotton yield was calculated on area basis. The yield data were analyzed statistically through computer as suggested by Steel *et al.*, (1997).

RESULTS AND DISCUSSION

The data on yield and its parameters differed significantly due to different levels of P_2O_5 applied at different times during 2008-10. During 2008 (Table-1) maximum seed cotton yield (3321 kg ha^{-1}) was produced when 125 kg ha^{-1} K was applied at sowing followed by (3005 kg ha^{-1}) yield of seed cotton when 62.5 kg ha^{-1} K was applied at sowing. The lowest yield of seed cotton (2491 kg ha^{-1}) was produced by control (no potash applied). The same trend of result was also seen for its parameters i.e., boll weight, no of boll per plant and height. There is immense scope of increasing productivity through adequate application of K (Bhattacharyya, 2000).

Table 1: Effect of potash levels applied at different times on yield of seed cotton and its parameters during 2008.

Treatments K (kg ha-1)	Plant height (cm)	No. of bolls $plant^{-1}$	Boll weight (g)	Yield (kg ha^{-1})
No. of K application	138e	21.40e	2.80f	2491f
62.5 full at sowing	152b	29.67ab	3.35ab	3005b
62.5 full at 1st irrigation	143.33d	23.27de	2.99e	2649e
62.5 half at sowing & half at 1st irrigation	143cd	25.20cd	3.18cd	2728d
125 full at sowing	158.67a	31.80a	3.47a	3321a
125 full at 1st irrigation	150bc	27.20bc	3.10de	2827cd
125 half at sowing & half at 1st irrigation	153.33b	29.00ab	3.30bc	2965bc
LSD(0.05)	**4.46**	**3.67**	**0.16**	**148.33**

During 2009 (Table-2) maximum seed cotton yield (3707 kg ha^{-1}) was produced when 62.5 kg ha^{-1} was applied at sowing followed by (3468 kg ha^{-1}) when 125 kg ha^{-1} K was applied at sowing followed by 3429 kg ha^{-1} yield of seed cotton 62.5 kg ha^{-1} was applied in splits ½ at sowing and ½ at 1st irrigation which are statistically at par. The same trend of results was also observed for other parameters.

Table 2: Effect of potash levels applied at different times on yield of seed cotton and its parameters during 2009.

Treatments K (kg ha-1)	Plant height (cm)	No.of bolls $plant^{-1}$	Boll weight (g)	Yield (kg ha^{-1})
No. of K application	142.33c	26.80b	2.49d	2671d
62.5 full at sowing	156ab	33.67a	3.34ab	3707a
62.5 full at 1st irrigation	149.33bc	27.27b	2.75cd	3269bc
62.5 half at sowing & half at 1st irrigation	151.67ab	29.20b	2.98bc	3429ab
125 full at sowing	158a	33.80a	3.46a	3468ab
125 full at 1st irrigation	152.33ab	29.20b	2.85cd	2910cd
125 half at sowing & half at 1st irrigation	153.67ab	29.67b	3.08bc	3110bc
LSD(0.05)	**7.02**	**3.31**	**0.37**	**392.16**

During 2010 (Table-3) maximum yield of seed cotton (1760 kg ha^{-1}) was produced when 125 kg ha^{-1}) K was applied at sowing followed by (1704 kg ha^{-1}) yield when same quantity of K was applied in splits ½ at sowing and ½ with 1st irrigation. The same trend of result was seen for boll weight but different trend of result for number of boll $plant^{-1}$ and height. Maximum height and number of boll $plant^{-1}$ were observed when 125 kg ha^{1} K was applied at sowing followed by 62.5 kg ha^{-1} K was applied at sowing.

Table 3: Effect of potash levels applied at different times on yield of seed cotton and its parameters during 2010.

Treatments K (kg ha-1)	Plant height (cm)	No.of bolls $plant^{-1}$	Boll weight (g)	Yield (kg ha^{-1})
No. of K application	135.6d	13.80c	2.76d	1155c
62.5 full at sowing	144.53b	20.53a	3.08bc	1430abc
62.5 full at 1st irrigation	135.73d	16.40bc	2.97c	1374bc
62.5 half at sowing & half at 1st irrigation	138.27cd	14.67bc	3.09bc	1374bc
125 full at sowing	153.8a	20.80a	3.38a	1760a
125 full at 1st irrigation	139.0cd	16.87b	2.98c	1595ab
125 half at sowing & half at 1st irrigation	142.27bc	16.87b	3.22ab	1704ab
LSD(0.05)	**4.95**	**2.72**	**0.18**	**347.88**

The trend of results of average of three years (Table-4) was also same for yield and other parameters as during 2008-2009. Maximum yield of seed cotton (3030.75 kg ha^{-1}) was produced when (125 kg ha^{-1}) K applied at sowing followed by (2875.25 kg ha^{-1}) yield of seed cotton when (62.5 kg ha^{-1}) K applied at sowing. The same trend of results for Number of boll plant^{-1}, boll weight and height were seen.

Table 4: Effect of potash levels applied at different times on yield of seed cotton and its parameters during 2008-2010 (average of three years).

Treatments K (kg ha-1)	Plant height (cm)	No.of bolls plant^{-1}	Boll weight (g)	Yield (kg ha^{-1})
No. of K application	138.64f	20.66d	2.68f	2258.52e
62.5 full at sowing	150.84b	27.96a	3.26b	2875.25b
62.5 full at 1st irrigation	142.80e	22.31c	2.92e	2583.28d
62.5 half at sowing & half at 1st irrigation	145.58d	24.13bc	3.08cd	2645.70cd
125 full at sowing	156.82a	28.80a	3.44a	3030.75a
125 full at 1st irrigation	147.11c	24.42b	2.98de	2598.09d
125 half at sowing & half at 1st irrigation	149.76b	25.29b	3.20bc	2738.79c
LSD (0.05)	**2.45**	**2.06**	**0.15**	**117.97**

An economic analysis showed that addition of K fertilizer beyond 62.5 kg ha^{-1} was not cost effective (Table-5). Moreover application of K fertilizer at sowing time was economical than other time of application.

Table 5: Seed cotton yield and economics of K_2O application to cotton crop (average of three years).

Treatments K (kg ha-1)	Seed Cotton Yield (kg ha^{-1})	Increase in yield over control kg ha^{-1}	Value seed cotton Rs ha^{-1}	Cost of fertilizer Rs ha^{-1}	Net profit Rs ha^{-1}	Cost benefit ratio
No. of K application	2258.52	-	-	-	-	-
62.5 full at sowing	2875.25	616.73	41043	2959	38084	1:12:87
62.5 full at 1st irrigation	2583.28	324.76	21663	2959	18654	1:6:30
62.5 half at sowing & half at 1st irrigation	2645.7	387.18	25767	3159	22808	1:7:71
125 full at sowing	3030.75	772.23	51352	5918	45474	1:7:68
125 full at 1st irrigation	2595.09	339.57	22598	5918	16680	1:2:82
125 half at sowing & half at 1st irrigation	2738.79	480.27	31962	6118	26044	1:4:40

Seed cotton @ 66.55 Rs. Kg^{-1} potassium fertilizers @ 47.34 Rs. Kg^{-1}

The positive response addition of potassium fertilizer on soils deficient in soil K has been reported earlier (Saleem, 1989). Other researchers also reported increased seed cotton yield with application of potassium fertilizer. Report of the earlier studies indicated that soils of Pakistan have low K selectivity and buffering capacity (Oosterhuis and Bednarz 1997). The desired amount of K unit time from soil reserves is not sufficient to meet the crop nutrient requirement. Potassium deficiency adversely affects the ability of cotton to fertilize soil nitrogen translocation of photosynthetic assimilates of leaves in cotton (Makhdum, 2003). The data suggested that addition of potassium fertilizer made the soil well buffered for sustained supply of K to cotton crop. As the cotton crop has an indeterminate growth habit which require regular supply of nutrients over a longer growth period. It is apparent ability of soil to supply K to the roots is more important in determining K uptake than ability of roots to absorb K in solution of most of the soils which are insufficient to meet total K requirement of the crop calculated on the basis of mass flow (Bhattacharyya, 2000). The application of potassium fertilizer at sowing time made the soil well buffered to increase K concentration in the solution. It is extended N absorption for vigorous vegetative growth during early part of the season. As K is associated with the transport of sugar (Oosterhuis and Bednarz 1997) application of 125 kg ha^{-1} K_2O at sowing produced significantly higher yield as compared to other treatments. Well balanced potassium nutrition is an important requirement for producing a high quality, high yielding cotton crop. Potassium plays an important role in translocation of sugar and activation of many of the enzyme required for various plant metabolic processes (Coker et al., 2000). Maximum yield was also reported at higher levels previously (Colakoglu, 1980, Makhdum, 2003 and Malik et al., 1989).

CONCLUSION

Cotton yield increased due to application of 125 kg ha^{-1} K_2O at sowing time showing positive response of seed cotton yield to the fertilizer. However application of 62.5 kg ha^{-1} K_2O proved economical in loamy soils of southern Punjab.

REFERENCES

Ahmad, N. 1998. Plant nutrition management for sustainable agriculture growth in Pakistan. Proc. Symp. Plant Nutrition Management for Sustainable Agriculture Growth. NFDC, Islamabad.pp.11-21.

Bhattacharyya, B. K. 2000.Response to applied potassium in important crops and cropping system in West Bengal. Workshop Proc. PPIC-IP, Calcutta. Pp.55-60.

Brar, MS, Brar, AS, Takkar, PN and Singh, TH. 1987. Effect of potassium supply on its concentration in plant and on yield parameters of American cotton (*Gossypium hirsutum* L.). J. Potassium Res. 3:1:149-154.

Colakoglu, H. 1980. Nutrient requirement and fertilization of cotton in Turkey. Potash Rev. 12:3.

Coker, D, Oosterhuis, D and Brown, R. 2000. Yield and physiological response of dryland and irrigated cotton to potassium fertilization: A four year summary. In: Oosterhuis, DM (Ed.). Summaries of Arkansas Cotton Research 2001. Univer-city of Arkansas Agriculture Experiment Station Research Series 101:100-110. Fayetteville, Ark.

Government of Pakistan (2011-12) "Economic Survey of Pakistan" Finance Division, Economic Wing, Islamabad, Pakistan pp-21.

Makhdum, MI. 2003. Response of some Cotton Cultivars to Sulphate of Potash (SOP) and Muriate of Potash (MOP). Ph.D. Thesis Bahauddin Zakaryia University, Multan. Pp. 206.

Malik, MNA., Makhdum, MI, Mirza, MB and Chaudhry, FI. 1989. Preliminary observations on potassium nutrition of cotton crop in silt and clay loam soils. The Pak. Cotton. 33 (3): 132-144.

Malavolta, E. 1985. Potassium status of tropical and sub-tropical region soils. Potassium in Agriculture. Menson, RD (ed.). ASA/CSS/. SSSA/, Madison, wt.USA. p. 163-200.

Mengel, K. and Kirkby, EA. 1982. Principles of plant nutrition. Worblanfen-Bern, Switzerland: International Potash Institute, pp. 435-441.

Mithaiwala, I. K., Mirbahar, MJ, Channa, AA and Arain, MH. 1981. Effect of fertilizers on the yield of long staple variety K-68/9 in

Guddu Barrage Area. The Pak. Cottons. 25: 73-79.

Oosterhuis, DM. and Bednarz, CW. 1997. Physiological changes during the development of potassium deficiency in cotton. pp 347-351. In: Ando, T et al. (eds). Plant Nutrition for Sustainable Food Production and Environment. Kluwer Academic Publishers, Japan.

Saleem, M.T. 2002. Fertilizer review 2002. Pakistan Situation. Farming Outlook. 1:24-27.

Saleem, MI. 1989. Potassium Research in Pakistan Present Status and Suggested Areas for Future Studies, Proc. Workshop on Potassium and Fertilizer Use Efficiency. NFDC Tech. Bull. 4/89: pp 229-242.

Steel, RGD, Torrie, JH and Dickey, DA. 1997. Principles and Procedures of Statistics. A biochemical approach. 3rd Ed. McGraw Hill Book. Int. Co. New York: pp172-177.

5

PRODUCTION POTENTIAL AND ECONOMICS OF WHEAT AS INFLUENCED BY LIMING IN NORTH EASTERN REGION OF BANGLADESH

Md. Rayhan Shaheb[1], Mahmudul Islam Nazrul[1] and M. Ataur Rahman[2]

[1]Senior Scientific Officer, On-Farm Research Division, Bangladesh Agricultural Research Institute (BARI), Sylhet-3100, Bangladesh and [2]Principal Scientific Officer, Regional Wheat Research Station, Bangladesh Agricultural Research Institute (BARI), Gazipur, Bangladesh

ABSTRACT

Soil acidification threatens dry land crop production in the soils of entire Sylhet region in Bangladesh. Our objective was to assess the efficacy of lime doses with agroeconomic performances of wheat varieties to lower soil acidity and increase wheat production in a direct-seeded system. Field trail was carried out at Eastern Surma Kushiara Flood Plain soil (AEZ 20) of Sylhet in the three consecutive years of 2010-11, 2011-12 and 2012-13. Treatments of wheat varieties (Bijoy, Prodip, Sufi, BARI Gom-25 and BARI Gom-26) and lime (0 and 1.5 tha^{-1}) arranged in a split-plot design. Lime was applied in main plots and varieties are distributed randomly in sub plots. Soil was sampled in November 2010 to assess pH and Low pH was evident at the depth of fertilizer placement (0-30 cm). Broadcast lime increased pH up to 6.0 in 2013. The result revealed that most of the yield component viz. spikes m^{-2}, thousand grain weight and grain yield of wheat was significantly influenced by liming. There were variations in lime response among the wheat varieties. Results observed that all the varieties performed well in the liming plot (3.39-4.09 tha^{-1}) and BARI Gom-26 produced the maximum yield (4.09 tha^{-1}) among them compare to non liming one (2.48-2.97 tha^{-1}). Cost benefit analysis indicated that the highest BCR was recorded in BARI Gom-26 in the liming plot that was 24% more compared to non liming plot. However, the index of relative adaptability of wheat was the maximum in BARI Gom-26 and it could be cultivated at acidic soil in Sylhet, Bangladesh.

Keywords: Soil acidity, liming, wheat, economic analysis, relative adaptability index.

INTRODUCTION

Global food demand is growing rapidly, and doubling food production and sustaining food production at this level, are major challenges for global food security (Tilman et al., 2011). Wheat (*Triticum aestivum* L.) is the second most important cereal crop in Bangladesh. It plays a vital role in national economy to close the gap between food production and import. Wheat has been cultivated in an area of 988 thousand acres with the total production of 0.99 mton in Bangladesh (KD, 2014). The average yield of wheat in Bangladesh is only 3.8 tha^{-1}, which is very low compared to other wheat growing countries like European Union and China where yields are 5.34 and 4.87 tha^{-1}, respectively (Taylor and Koo, 2012). In Sylhet, vast areas of lands remain fallow in each year during rabi season due to lack of irrigation facilities, labour shortages, absentee farmers, soil generally heavy, clay loams to clays and the top soil quickly becomes dry and hard after the harvest of transplanted aman (T. Aman) rice, soil are strongly acidic (pH 4-5.5) in nature. Rainfall prevails here from late October to early November usually in each year that offers the opportunity for the production of short duration crop by utilizing the residual moisture.

Soil acidity in the Sylhet region has become a serious problem for crop production, as both forage and grain yields have been reduced due to low soil pH. Soil acidity is harmful for plant growth due to nutritional disorders (e.g., deficiency of Ca and Mg, decreased availability of P and Mo) as well as the immediate toxicity of soluble Al, Mn, and H^+ (Carver and Ownby, 1995). The soils of northwest part of Bangladesh are light textured, low in organic matter and strongly acidic to moderately acidic in nature, pH ranges from 4.2 to 6.1 (FRG, 2012). The status of available P, Ca and Mg of these soils are low. The sandy soil has low cation exchange capacity. These soils have high content of Al, Fe and Mn (Breemen, 1993) and deficiencies of N, Ca, Mg, K, P and B are common. Al toxicity is responsible for poor yields in acid soils (Lierop et al., 1984).

**Corresponding author:* e-mail: smrayhan_bari@yahoo.com

Lime is applied to acid soils to neutralize excess acidity that causes reduced crop yields. Application of lime should be considered on all strongly acid soils and on many moderately acid soils to improve and maintain productivity. Application of lime at an appropriate rate brings several chemical and biological changes in the soils, which are beneficial or helpful in improving crop yields on acid soils (Fageria and Baligar, 2008). Several studies suggested lime recommendation for acidic soil in north-west Bangladesh under rice-wheat cropping system (Rahman et al., 2002; Rahman et al., 2004). Liming increases the availability of P and decreases the toxicity by reducing availability of Fe, Mn, and Al, thus ensures the balance of most of the plant nutrients resulting higher crop productivity (Stood and Bhardwaj, 1992; Rahman et al., 2004). The direct effect of liming are to increase nitrogen fixation and yield of most crops, reduce the soluble aluminum and manganese to nontoxic levels, improve availability of P and other nutrients (Agrifacts, 2002). Liming can eliminate the toxic effect of Al and Mn and increased the availability of several plant nutrients, such as Ca, Mg, P, N, and Mo (Mongia *et al.*, 1998; Rahman et al., 2005, and Fageria and Knupp, 2014). Zhang et al. (2004) reported that an application of lime 3.05 tha^{-1}, half of the rate recommended to raise soil pH from initial pH of 4.5 to 6.8, was found to be the most economical to lime low pH soils used for forage or dual-purpose winter wheat production.

Soil acidity limits wheat root growth and yield severely, probably as a result of extended water deficits during the vegetative stage. Grain yield were correlated positively with soil pH and exchangeable Ca^{2+}, and negatively with exchangeable Al^{3+} and Al^{3+} saturation, in the surface and subsurface layers (Caires et al., 2006). Soil pH and available P increased with the increase in the rate of lime addition and there was a high positive correlation between wheat yields and soil available P at both sites at harvest (Osundwa et al., 2013). Cifu et al. (2004) reported that lime application significantly increased the yield 4.67 times for barley, 2.24 times for mungbean, 57.3% for wheat, 44.1% for potato, 28.4% for corn, 11.0% for cowpea and 8.8 % for soybean. Tang et al. (2003) suggested that surface liming can ameliorate subsoil acidity. However, horizontal expansion is hardly possible therefore, vertical expansion such as varietals screening might be useful for the expansion of wheat cultivation in acidic soil as some of wheat varieties identified as relatively non-responsive to liming might be considered as adaptable in acidic soil.

Although numerous studies have investigated the effects of lime on grain yields of various crops, only a few studies have reported how the wheat is affected by soil pH in acidic soil of Sylhet. In view of these, the study was therefore, initiated to verify the liming effect and screen the wheat varieties suitable in acidic soil of Agroecological zone 20 (AEZ 20).

MATERIALS AND METHOD

A field trail was conducted in strongly acidic soil in farm level at Sylhet, Bangladesh during the year 2010-13 with the objectives to verify the liming effect and screen the wheat varieties suitable in acidic soil of north eastern region in Bangladesh (AEZ 20). Five wheat varieties viz. Bijoy, Prodip, Sufi, BARI Gom-25 and BARI Gom-26 were examined at two levels of lime viz. i) with lime (1.5 tha^{-1}) and ii) without lime (0 tha^{-1}) in split-plot design. Lime was applied in main plots and wheat varieties were grown in sub plots. The size of each main plot was 20m × 5m and subplot was 5m ×2m. The pH of initial soil was 4.2 -5.4. The wheat seeds were sown on 05th December in all the years. Lime-Dolomite $(CaMg).CO_3$ was applied in the main plots and incorporated into the soil 15 days before of wheat seed sowing. The five years of weather data is presented in the Figure 1. Initial soil nutrient status of the experimental field and chemical composition and neutralizing ability of the liming are presented in Table 1 and 2. Fertilizers @ of $N_{100}P_{30}K_{50}$ and S_{20} $Kgha^{-1}$ as urea, triple super phosphate, muriate of potash, and gypsum, respectively were applied in the soil. All fertilizers including two-third urea were applied in the field during final land preparation. The rest of urea was top dressed at the crown root initiation (CRI) stage at 21 days after sowing (DAS). The crop was irrigated thrice to bring the soil moisture near to field capacity during CRI, booting and grain-filling stages. Weeds were controlled once at 30 DAS manually. During maturity crops were harvested duly on 20th March and sun-dried to measured biomass. After harvest, the pH of the amended plot soil was ranging 6.2-6.3. The dried crop was threshed and after threshing, the grains were dried in the atmosphere and grain yields were converted to t ha^{-1} at 12 % moisture content.

To compare the varietals adaptability, an index of relative adaptability or performance (IRP) for yield and yield components were estimated following the equation below (Rahman et al., 2013) and adapted from Fisher and Maurer (1978).

$$IRP = \frac{\frac{Y}{Yp}}{\frac{X}{Xp}} \times 100$$

Where,
Y: Yield of a genotype under non-liming
Yp: Yield of same genotype under liming
X: Mean yield of all genotype under non-liming
Xp: Mean yield of all genotype under liming

Cost benefit analysis of wheat was calculated based on the current market price of the input and output costs of the specified year and expressed as US$ (1 US$=78 BDT). All data were statistically analyzed and the mean values were tested by the least significant difference (LSD) at 5% level of significance.

RESULTS AND DISCUSSION

Results revealed that yield related parameters and yield were significantly influenced except spike length of wheat by the application of lime (Table 3-6).

Plant height

Plant heights were improved due to the effect of liming (Table 3). Liming increases pH in acidic soil that favors root proliferation and by increasing nutrient availability that might contribute to higher plant height to wheat. The highest plant height was recorded in BARI Gom-26 (94.60cm) in liming while in the non liming plot, it was observed in BARI Gom-25 (90.95cm) followed by BARI Gom-26 (89.21cm).

Tiller per plant

Tiller per plant of wheat was also significantly influenced by application of liming (Table 3). Liming on acid soil increases the pH level, nutrient availability and microbial activities. These might be increased tillering of wheat crop in acid soils. The finding is in agreement with the findings of Sultana et al. (2009) who reported that liming increased tiller number per plant in wheat significantly. However, the highest number of tiller per plant was recorded in BARI Gom-26 variety while in non liming it was also observed in BARI Gom-26 variety.

Plants per m^2

Plants per m^2 was positively responded with the lime application (214.78) with few acceptations compared to without lime (198.36) condition (Table 4). It might be due to the application of liming as this increases soil pH in acidic soil, which favors root proliferation. Furthermore, liming also increase nutrient availability which influenced tillering that ultimately yielded higher spikes/m^2. The results partially agreed with Rahman et el. (2013). Similar results are also reported by Coventry et al. (1987) and reported that lime treatment increased more heads per metre of row of wheat.

Spike length

Research results suggested that spike length of wheat was non significant with due to the application of lime. Interestingly, it was almost same in both cases (Table 4). Further study may need to reach a definite conclusion in relation to spike length. However, among the variety, Prodip, Sufi, BARI Gom-25 and 26 showed the higher spike length except Bijoy in both liming and non liming cases.

Grain per spike

The result indicated that lime application resulted in relatively higher increment of grain per spikes in varieties than those under the control condition (Table 5). The main effect of liming showed that grain per spike was recorded 39% more in the liming plot (26.53) than that of without lime treatments plot (19.00). Among the combination of liming and variety, BARI Gom-26 was recorded the highest grain per spike (30.70) followed by BARI Gom-25 (28.28) and Sufi (25.45). The lowest grain per spike was noted at Bijoy variety in without lime condition. Plant nutrient uptake and their translocation from source to sink largely depend on chemical environment of the soil that generally favorable under lime application (Rahman et al. 2005) which might contribute to higher grain per spike in wheat. Similar result was also recorded by Sultana et al. (2009).

Thousand grain weight (TGW)

The study revealed that liming reflected positively sharp response in the thousand grain weight of wheat varieties (Table 5). The highest mean TGW was recorded in liming treatment (45.31g) compared to non liming treatment (41.90g). Among the combination of liming and wheat, liming with BARI Gom-26 produced the maximum TGW (47.46g) followed by liming with BARI Gom-25 (46.67g) and without liming with BARI Gom-26 (46.12g). Interestingly, it was observed that BARI Gom-26 in non liming plot produced almost similar TGW to liming condition. It might be due to the tolerance of the variety in acidic soil. However, it might also be due to the effect of liming that make the plant nutrients available to them and also create positive chemical environment of the soil that was also reported by Rahman et al. (2005) and Rahman et al. (2013).

Grain yield (GY)

Production potential of wheat in acidic soil in Sylhet revealed that grain yield of wheat was significantly varied due to the effect of liming as well as varietals difference (Table 5). The mean yield of wheat varieties was higher (3.70tha^{-1}) under lime condition than that of without liming plot (2.67tha^{-1}). The variety BARI Gom-26 gave the satisfactory yield (4.09tha^{-1}) among the wheat varieties followed by BARI Gom-25 (3.83tha^{-1}) and Pradip (3.61tha^{-1}) in liming condition. Studies suggest that liming has the both direct and indirect effects that improve the availability of nutrients including phosphorous fertilizers, increase microbial activity and release of plant nutrients (Agrifacts, 2002). Craig et al. (2013) reported that lime application resulted in significant increases in soil pH, phosphorus uptake and grain yield. Grain yield showed a linear relationship to phosphorus uptake. Grain yields of wheat were positively correlated with soil pH, available P, Ca and Mg contents of post harvest soils (Sultana et al., 2009). Similar results were also recorded from Rahman et al. (2005) and Rahman et al. (2013). However, the crop yield per unit area as well as farm productivity increased due to adoption of liming & better use of farm resources. Coventry et al. (1987) suggested that increased grain yield was mainly due to more heads per metre of row, although head size and grain weight were also increased by lime treatment.

Index of relative adaptability or performance (IRP)

The relative adaptability of different yield parameters and grain yield of wheat was described in table 3 to 5. Result indicated that the highest IRP showed in tiller per plant in the wheat variety BARI Gom-26 (111%) followed by Prodip variety (104%). IRP in plant per m^2 was observed in Bijoy (109%) followed by Sufi (107%) while in grain per spike, IRP% showed the highest in the Sufi variety (114%) followed by Prodip variety (110%). On the contrary, in case of grain yield the highest IRP (104%) was noted in Prodip variety followed by Sufi variety (102%). But the thousand grain weight and grain per spike was the highest in BARI Gom-26 compared to Sufi and others variety. Therefore, wheat variety BARI Gom-26, BARI Gom-25 and Prodip might have potentials in producing higher yield in acidic soil condition in Sylhet region of Bangladesh. The results are partially agreed with the findings of Rahman et al. (2013) who reported that the IRP % for yield of BARI Gom-26 and Bijoy was more than 100% and are relatively tolerant to low pH and could be adapted in acidic soil of Sylhet.

Cost benefit analysis

Details cost and benefit analysis was given in figure 2-4. Economic analysis were calculated base on the current price of the variable and non variable costs such as Urea- 0.26, TSP-0.28, MOP-0.19, Zypsum-0.13, Lime 0.13, and Cowdung-0.013 US$Kg^{-1}, respectively and labour- US$ 2.30 day^{-1}, wheat seed US$ 0.51, ploughing- US$ 28.80ha^{-1}, irrigation-US$1.28hr^{-1} and pesticides cost total of US$ 57.60 and output cost such as wheat grain- US$ 0.38Kg^{-1} and wheat straw US$0.032 5Kg^{-1}, respectively. Lime was applied once and the cost of lime was divided in each year. From the economic point of view, it was noted that the liming plot showed its superiority over the non liming one in all the cases of gross return, net return and also in benefit cost ration (BCR). The highest gross return (US$ 1580.48), gross margin (US$ 975.68) and BCR (2.61) was recorded in BARI Gom-26 (Fig 2-4). Interestingly, all the variety except BARI Gom-26 noted below 200% BCR in non liming condition. However, higher cost of production in liming plot over non liming one is due to the cultivation of wheat using liming in the field.

Table 1. Initial soil nutrient status of the experimental field for Wheat

Depth (cm)	pH	OM (%)	N (%)	P (µg/g soil)	K (m.eq /100g soil)	Ca(m.eq /100g soil)	Mg(m.eq /100g soil)	S (µg/g soil)	Zn (µg/g soil)	B (µg/g soil)	Fe
0-15											
Value	4.85	0.85	0.04	9.615	0.12	4.42	0.85	8.87	0.96	0.33	11.36
15-30											
Value	5.2	0.76	0.03	8.5	0.92	3.37	0.65	8.20	0.86	0.22	9.50

Table 2. Chemical composition and neutralizing ability of the liming material used in the experiment

CaO equivalent (%)	Ca	mg	K	Na	P	Zn	Mn	B
	%				Mg/kg			
48	20.10	7.20	0.18	0.32	0.00	0.00	44.00	62.40

(Source: Rahman et. el. 2013)

Table 3: Effect of lime on the plant height and tiller per plant of wheat (pooled of three years data).

Wheat Varieties	Plant height (cm)			Tiller per plant		
	With lime	Without lime	IRP (%)	With lime	Without lime	IRP (%)
Bijoy	92.83	89.47	100	4.04	3.97	102
Prodip	91.83	90.45	102	3.95	3.99	104
Sufi	90.56	88.73	101	4.10	3.57	90
BARI Gom-25	93.57	90.95	100	3.87	3.48	93
BARI Gom-26	94.60	89.21	97	3.81	4.09	111
Mean	92.68	89.76		3.95	3.82	
CV (%)	3.01			13.64		
LSD(0.05)	4.75			0.39		

Table 4: Effect of lime on the plants per m^2 and spike length of wheat (pooled of three years data).

Wheat Varieties	Plants per m^2			Spike length (cm)		
	With lime	Without lime	IRP (%)	With lime	Without lime	IRP (%)
Bijoy	233.11	233.86	109	14.9	13.95	93
Prodip	217.75	192.09	96	16.57	17.34	104
Sufi	202.79	200.14	107	14.65	15.73	107
BARI Gom-25	208.62	182.57	95	15.34	15.02	97
BARI Gom-26	211.65	183.15	94	14.93	14.79	98
Mean	214.78	198.36		15.28	15.37	
CV (%)	11.23			6.78		
LSD (0.05)	40.14			NS		

Table 5: Effect of lime on the grain per spike, thousand grain weight (TGW) and grain yield of wheat (pooled of three years data).

Wheat Varieties	Grain per spike			TWG (g)			Grain yield (tha^{-1})		
	With lime	Without lime	IRP (%)	With lime	Without lime	IRP (%)	With lime	Without lime	IRP (%)
Bijoy	23.13	16.38	99	43.9	40.47	1.00	3.6	2.60	100
Prodip	25.11	19.77	110	44.15	37.18	0.91	3.61	2.71	04
Sufi	25.45	20.82	114	44.37	41.96	1.02	3.39	2.48	02
BARI Gom-25	28.28	18.94	94	46.67	43.76	1.01	3.83	2.58	94
BARI Gom-26	30.70	19.11	87	47.46	46.12	1.05	4.09	2.97	101
Mean	26.53	19.00		45.31	41.90	1.00	3.70	2.67	
CV (%)	6.41			6.24			5.67		
LSD (0.05)	2.53			4.71			0.31		

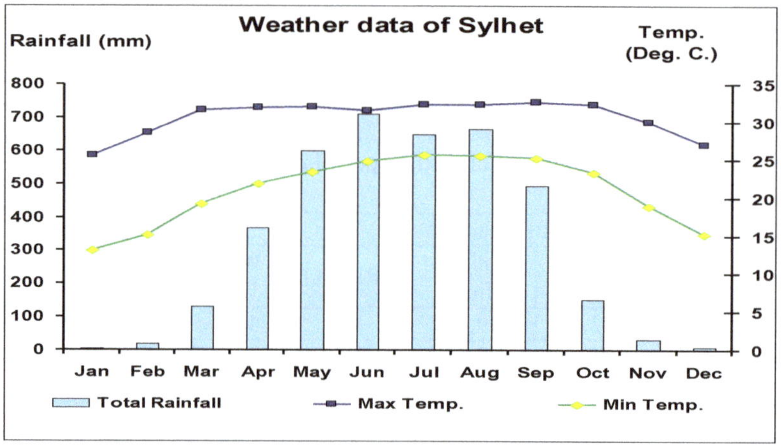

Fig. 1. Annual monthly total rainfall (mm), maximum and minimum temperatures during the period 2007-2013 in Sylhet

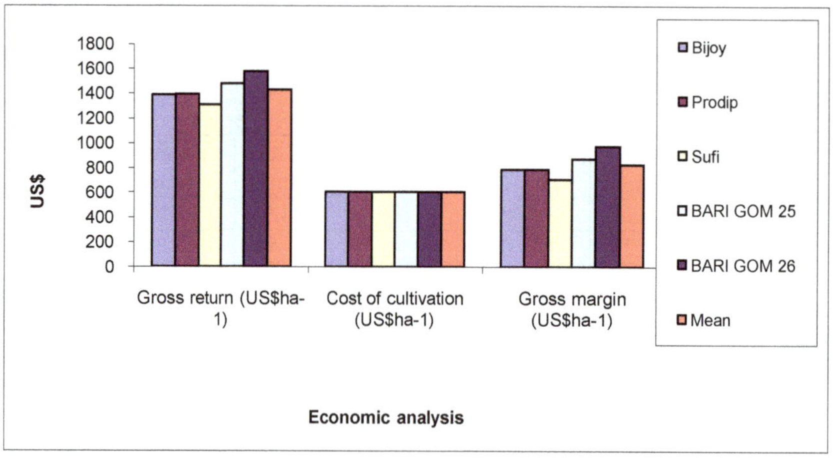

Fig. 2. Economic analysis of wheat as affected by lime

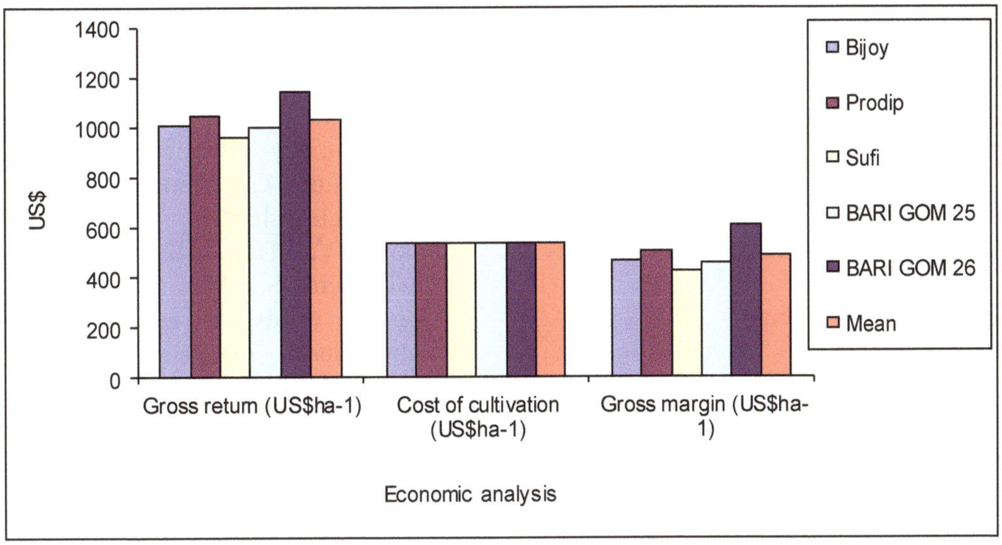

Fig. 3. Economic analysis of wheat without lime

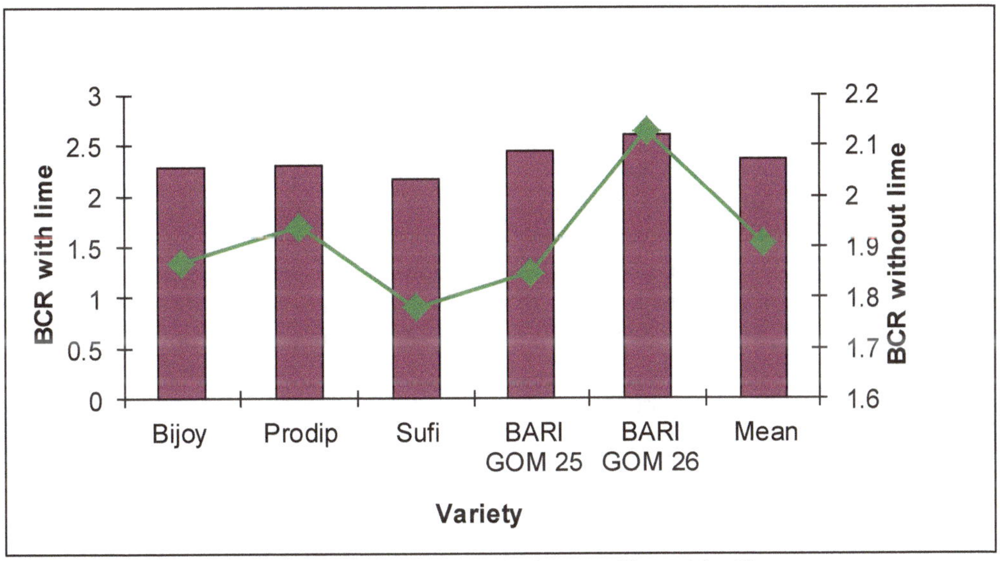

Fig. 4. BCR of wheat varieties as affected by lime

CONCLUSION

Soil acidity in the Sylhet region has become a serious problem for crop production, as both forage and grain yields have been reduced due to low soil pH. Result revealed that all the wheat varieties performed well in the liming plot (3.39-4.09 tha^{-1}) and BARI Gom-26 produced the maximum yield (4.09 tha^{-1}) among them compare to non liming one (2.48-2.97 tha^{-1}). BCR was also recorded higher in the liming plot (2.37) that was 24% more compared to non liming plot. However, the index of relative adaptability or performance of wheat was showed maximum in BARI Gom-26 followed by Prodip and BARI Gom-25. So, these wheat varieties could be cultivated at acidic soil in the agroecological zone 20 (Eastern Surma Kushiara Flood Plain soil) in Sylhet, i.e. northern eastern part of Bangladesh successfully.

ACKNOWLEDGEMENTS

The authors are offered their cordial thanks and acknowledgement to metrological department, Sylhet office for providing weather data timely. The authors are also thanked to the scientists of Regional Soil Testing Laboratory, Soil Resources Development institute, Sylhet for soil nutrient analysis of the present study.

REFERENCES

Agrifacts, 2002. Liming Acid Soils. Practical information for Alberta's agriculture industry. http://www1.agric.gov.ab.ca/department/deptdocs.nsf/all/agdex3684. (Source: AgDex 534-1 and website assessed on 15 February 2014).

Breemen NV, 1973. Genesis and Solution Chemistry of Acid Sulphate Soils in Thailand. Wagneningen, Centre for Agricultural Publishing and Documentation.

Caires EF, Corrêa, JC, Churka LS, Barth G and Garbuio FJ, 2006. Surface application of lime ameliorates subsoil acidity and improves root growth and yield of wheat in an acid soil under no-till system. Sci. agric. (Piracicaba, Braz.), 63(5): 502-509.

Carver BE and Ownby JD, 1995. Acid soil tolerance in wheat. Adv. Agron. 54:118-173.

Cifu M, Xiaonan L, Zhihong C, Zhengyi H and Wanzhu M, 2004. Long-term effects of lime application on soil acidity and crop yields on a red soil in Central Zhejiang. Plant and Soil, 265(1-2): 101-109.

Coventry DR, Reeves TG, Brooke HD, Ellington A. and Slattery WJ, 1987. Increasing wheat yields in northeastern Victoria by liming and deep ripping. Australian J. Exp. Agric. 27:679-685.

Craig S, Brennan R and Sarre G, 2013. Crop rotation and liming history affect the availability of soil phosphorus. Grain research and Development Corporation. Department of Agriculture and Food, Government of Western Australia. 2013 WA Crop Updates (1-4).

Fageria NK and Knupp AM, 2014. Influence of Lime and Gypsum on Growth and Yield of Upland Rice and Changes in Soil Chemical Properties. DOI:10.1080/01904167.2014.890219.

Fageria, N.K., Baligar, V.C., 2008. Ameliorating soil acidity of tropical oxisols by liming for sustainable crop production. In: D. L. SPARKS, editor Advances in Agronomy. Acad. Press. Brazil., p. 345-389.

Fisher RA and Maurer R, 1978. Drought resistance in spring wheat cultivars. I. Grain yield responses. Australian J. Agric. Res. 29: 897-907.

FRG, 2012. Fertilizer Recommendation Guide, Bangladesh Agricultural Research Council (BARC), Farmgate, Dhaka-1215. 274p

KD, 2014. Krishi Diary, Agricultural Information Services, Ministry of Agriculture, Government Peoples Republic of Bangladesh. P. 2.

Lierop WV, Tran TS, Banville, G and Morissette S, 1984. Effect of liming on potato yields as related to soil pH, aluminium, manganese and calcium. Agron. J. 74: 1050-1055.

Mongia AD, Singh, NT, Mandal LN and Guha A, 1998. Effect of liming, super phosphate and rock phosphate application to rice on the yield and uptake of nutrients on acid sulphate soils. J. Indian Soc. Soil Sci. 46: 61-66

Osundwa MA., Okalebo JR, Ngetich WK, Ochuodho JO, Othieno CO, Langat B and Omenyo VS, 2013. Influence of Agricultural Lime on Soil Properties and Wheat (*Triticum aestivum L.*) Yield on Acidic Soils of Uasin Gishu County, Kenya. American J. Experimental Agric. 3(4): 806-823.

Rahman MA, Barma NCD, Sarker MH, Sarker MM. and Nazrul MMI, 2013. Adaptability of wheat varieties in strongly acidic Soils of sylhet. Bangladesh J. Agril. Res. 38(1): 97-104.

Rahman MA, Chikushi J, Lauren JG, Duxbury JM, Meisner CA. and Yasunaga E, 2005. Chemical control of soil environment by lime and nutrients to improve the productivity of acidic alluvial soils under rice-wheat cropping system. Environmental Control in Biology 43(4): 259-266.

Rahman MA, Chikushi J, Lauren, JG, Duxbury JM and Meisner CA, 2004. Liming and nutrient management for sustainable productivity of acidic alluvial soils under rice-wheat cropping system in Bangladesh. Proc. 6th International symposium on plant-soil interaction at low pH. pp. 304-305. August 1-5, 2004. Sendai, Japan. Organized by International steering committee on Soil-Plant Interaction at low pH.

Rahman MA, Meisner CA, Duxbury JM, Lauren J and Hossain ABS, 2002. Integrated approach of liming and nutrient management to improve the productivity of acidic soil within a rice-wheat cropping system. Poster no. 773, Symposium no. 05,

17th World Congress of Soil Science. 14-21 August 2002. Bangkok, Thailand.

Stood RD and Bhardwaj RK, 1992. Effect of liming and phosphorus fertilization on the P fractions in an acid Alfisol. J. Indian Soc. Soil Sci. 40: 299-301.

Sultana B, Mian MMH, Islam MR, Rahman MM, Sarker BC and Zoha MS, 2009. Effect of liming on soil properties, yield and nutrient uptake by wheat. Curr. World Environ 2009;4(1) :39-47.

Tang C, Rengel Z, Diatloff, E and Gazey C. 2003. Response wheat and barley to liming on sandy soil with subsoil acidity. Field Crops Res. 80: 235-244.

Taylor RD and Koo WW, 2012. Outlook of the U.S. and World Wheat Industries. Agribusiness & Applied Economics, June 2012-2021. Center for Agricultural Policy and Trade Studies, Department of Agribusiness and Applied Economics 696, North Dakota State University, Fargo, North Dakota 58108-6050, 235p.

Tilman D, Balzer C, Hill J and Befort BL, 2011. Global food demand and the sustainable intensification of agriculture. Proceedings of the National Academy of Sciences of the United States of America108, 20260–20264.

Zhang H, Schroder JL, Krenzer EG, Kachurina OM and Payton ME, 2004. Yield and quality of winter wheat forage as affected by lime. Online. Forage and Grazing lands doi:10.1094/FG-2004-1020-01-RS.

VARIABILITY IN HCN CONTENT OF WHITE CLOVER GROWING IN DIFFERENT ALTITUDINAL ZONES OF NEPAL

A. M. Shrestha Malla*
Botany Department, Tri-Chandra College, Tribhuvan University, Kathmandu, Nepal

ABSTRACT

Plants contain cyanogenetic glucoside and also the appropriate enzyme which hydrolyses this glucoside and produces HCN. There are different types of plants which contain: the glucoside but not the enzyme; the enzyme but not the glucoside; and neither the glucoside nor the appropriate enzyme. *Trifolium repens* L. (White clover) is polymorphic for cyanogenesis, with both cyanogenic and acyanogenic plants occurring in nature. The study was done in eight different varieties of *Trifolium repens* L. collected from different altitudinal zones such as subtropical representing Kirtipur (1245-1300 masl), subtropical to temperate representing Dhunche (1962 -2012masl) and temperate to sub-alpine representing Chandanbari (3400-3500masl). Plants collected from these areas were tested for HCN contents by "Feigl spot test" method. The results showed that highest quantity of HCN was present in WNML, medium in WPML and minimum in LPML variety growing at Kirtipur. Medium quantity of HCN was present in WNML, WPML, WPNML and PNML varieties growing at Dhunche. PML and LPNML varieties have no HCN and were suggested as the best fodder plants for livestock. HCN contents in plants growing in Chandanbari were very low or absent. Similarly, the transplant from Chandanbari to Kirtipur also did not show HCN except a medium amount in PNML variety. The study showed a marked variation in HCN content among the varieties of white clover. It also revealed that the higher amount of HCN was present in white coloured flower than pink coloured flower; HCN content of plant is lesser in colder and higher altitudinal zone and is higher in plant with larger size of the leaves as well as in plants growing in lower moisture and higher phosphorous containing soil.

Keywords: HCN, White clover, Flower colour, Altitude, Moisture, Phosphorous

INTRODUCTION

Varieties of white clover in Nepal *Trifolium repens* L. (white clover), is an allotetraploid, from the cross of putative parents T. pallescenes and T. occidentale (Richards, 2011). *Trifolium repens* L. is a legume of moist temperate pastures and one of the most important forage crops containing high protein and minerals than any other grass and other species of *Trifolium*. It is commonly used in renovation of permanent pastures lacking legumes in temperate regions, Mediterranean areas and some cool subtropical parts of the world (Frame et al., 1998). It fixes atmospheric nitrogen due to the presence of nitrogen fixing bacteria in the root nodules and improves feed quality. *Trifolium repens* was first introduced in Nepal during 1860 AD. Subsequently it was introduced in different parts of the country as a ground fodder for pasture land in government as well as in private dairy farms due to its value as a good fodder. Different white clover varieties from eight different countries have so far been introduced and believed that they had freely crossed with one another. Eight varieties of white clover were presently identified in Nepal (Shrestha, 2013; Table1).

HCN content of plants

The release of HCN from damaged leaves is known as Cyanogenesis. White clover leaves contain two cyanogenic glucosides e.g. linamarin and lotaustralin. It also contains the enzyme known as linamerase. When the leaves are damaged e.g. on grazing by animals, the glucosides are hydrolysed by linamarase to produce glucose and an unstable aglycone. The aglycone dissociates to release HCN and acetone or 2-butanone (Hart, 1987). The level of glucosides in the foliage may be sufficiently high for grazing animals to suffer from cyanide poisoning. Cyanogenesis is regarded as a protection against predators, herbivores, especially insect pests, slugs and snails. Cyanogenic glucosides in plants act as a defensive mechanism against feeding by some herbivores (Richard, 1973). Acyanogenic white

**Corresponding author:* e-mail: rajeshanjali_hanoi@yahoo.com clover may be favoured under certain climatic or edaphic conditions. Cyanogenesis is controlled by two independently segregating genes: Ac/ac controls the presence/absence of cyanogenic glucosides; and Li/li controls the presence/absence of their hydrolysing enzyme, linamarase. Whereas Li is well characterized at the molecular level, Ac has remained unidentified. The nature of these polymorphisms may reflect white clover's evolutionary origin as an allotetraploid derived from cyanogenic and acyanogenic diploid progenitors (Kenneth et. al., 2008; Kenneth et al., 2013).

MATERIALS AND METHODS

Study areas

Different varieties of white clover were collected from different altitudinal areas of Nepal such as Kirtipur in Kathmandu district, Dhunche and Chandanbari in Rasuwa district. Kirtipur represents the sub-tropical zone at 1245-1300 metre at sea level (masl) altitude. Dhunche represents sub-tropical to sub-temperate zone at 1962-2012 masl altitude. Chandanbari represents temperate to sub-alpine zone at 3400-3500 masl altitude.

Plant collection

Fresh and healthy plants were collected for HCN test from Kirtipur. Different varieties of white clover were brought from Dhunche and Chandanbari and grown in pots containing soil from Kirtipur at 1300 m altitude. The growth and flowering of every transplant were monitored for three consecutive years. So, fresh and healthy transplants from Dhunche and Chandanbari were collected for HCN test. HCN test was carried out in Kirtipur as well as in Dhunche.

Method of HCN test

Hydrocyanic acid (HCN) test was carried out in different varieties of white clover by "Feigl spot test" method (Feigl and Anger, 1966).

Preparation of test paper

100gm. each of Copper (II) ethyl acetoacetate and 4, 4'- Methylenebis (N, N-dimethylaniline), were dissolved separately in 10 ml of chloroform in two beakers. These two solutions were mixed together. The test paper was made by saturation of a sheet of filter paper (Whatman No.1) with the mixed (1:1) solution and was dried at room temperature. If the colour of the paper did not change, it was cut into small strips measuring 6 cm in length and half cm width. The test papers were preserved in a clean and dry amber bottle with lid for few months.

Testing plants for HCN

Three healthy and similar sized leaflets of each variety of white clover were collected for the test. It was kept at the centre of the sterilized dry vial. The leaflets were crushed and squeezed by a sterilized dry glass rod. Immediately a strip of test paper was inserted into the vial, without touching the leaf sample and kept hanging with the help of rubber lid of the vial. The vial was then kept in a dark place or away from the sunlight and the time was noted down. The procedure was repeated five times for each variety of white clover. The time when the colour of the test paper changed was noted. If the colour of test paper is changed into blue, it indicated a positive result for HCN content i.e. the plant was cynophoric or cyanogenic, while the test paper with no colour change indicated a negative result or absence of HCN content i.e. the plant was non-cynophoric or acyanogenic (Photograph 1).

Calculations of HCN

The HCN content calculation is shown in table 2.

RESULTS AND DISCUSSION

The presence or absence of hydrocyanic acid (HCN) content was tested in different varieties of *Trifolium repens* growing in different altitudinal zones of Nepal such as Kirtipur , Dhunche and Chandanbari (Table 3, Photograph 1). The results based on calculations as on table 2 (Table 3, figure 1) are shown below:

- WML variety of white clover growing at Kirtipur and Dhunche do not contain HCN while it showed maximum quantity of HCN when the plant transplanted from Dhunche to Kirtipur were tested. However, the specimen transplanted at Kirtipur from Chandanbari also did not show HCN.
- WNML variety of white clover from Kirtipur and the same variety transplanted at Kirtipur from Dhunche showed maximum quantity of HCN content while it is at a medium range in the same type growing at

Dhunche. However, the transplants from Chandanbari did not contain HCN.

- WPML variety of white clover growing at Kirtipur and Dhunche contained a medium quantity of HCN while those varieties transplanted at Kirtipur from Dhunche and Chandanbari did not contain HCN.
- WPNML variety of white clover growing at Kirtipur did not contain HCN while it contained medium quantities of HCN at Dhunche. The same variety when transplanted at Kirtipur from Dhunche and Chandanbari did not show HCN content.
- PML variety of white clover growing at Kirtipur and Dhunche as well as those varieties transplanted at Kirtipur from Dhunche and Chandanbari did not contain HCN at all.
- PNML variety of white clover from Kirtipur did not contain HCN while this variety in Dhunche showed a medium amount. PNML variety transplanted at Kritipur from Dhunche also did not show HCN while those transplanted from Chandanbari contained a medium amount of HCN.
- LPML and LPNML varieties of white clover, only from Kirtipur were analysed. LPML contained a minimum amount of HCN while LPNML did not contain it at all.

HCN content of different varieties of white clover at Kirtipur and Dhunche shows significant variations among themselves. WML and PML varieties of white clover growing at Kirtipur and Dhunche do not show HCN content. So these varieties are referred to be good fodder plants for livestock. Acyanogenic plants have advantage of being less injured by frost than cyanogenic plants (Daday, 1965) and therefore this could be an adaptation of plants growing in colder and higher areas where frost is frequent.

The transplant of WML variety from Dhunche to Kirtipur showed HCN content. It might be due to water stress because Kirtipur is drier than Dhunche while this variety was growing in the soil of low phosphorous (Shrestha Malla, 2002). HCN content in white clover were found to increase by moisture stress, low light intensity, cool grazing condition and low soil-phosphorous supply (Vickery et al. 1987). The transplant of PML from Dhunche and Chandanbari to Kirtipur has no HCN content. PML variety does not contain HCN at all. So PML variety is one of the best fodder plants for livestock. It might be due to smaller size of leaflets. There is a positive association between increasing leaf size and cyanogenesis levels for most of the New Zealand white clover cultivars (Caradus et al., 1996).

Maximum quantity of HCN was present in WNML variety growing at Kirtipur and medium quantity at Dhunche. The transplant from Dhunche to Kirtipur showed maximum HCN, however the transplant from Chandanbari to Kirtipur does not show HCN. Maximum HCN present in WNML growing at Kirtipur was due to moisture stress, low soil phosphorous on which it grows and due to large size of leaflets. Similarly medium content of HCN present in WNML variety at Dhunche was due to medium content of phosphorous (Table 4). Cyanogenic white clover populations were larger leaved, medium productive, and more persistent than acyanogenic populations (Doak, 1933; Foy and Hyde, 1937). On the other hand, an American study of diverse germplasm showed cyanogenic plants to be larger in diameter but shorter, earlier flowering and less persistent than acyanogenic plants (Paim and Dean, 1976).

WPML variety of white clover has medium content of HCN at Kirtipur and Dhunche areas due to large size of leaflets. WPNML variety actually is a good fodder plant for Kirtipur, as it has no HCN content. However WPNML variety has medium content of HCN at Dhunche. It might be due to cool grazing condition of the area. PNML variety also seems good due to absence of HCN at Kirtipur because it has small leaflets size and growing in soil with low phosphorous and high moisture content. However it has medium content of HCN at Dhunche, it might be due to larger leaflets compared to Kirtipur, growing in soil with low moisture and medium amount of phosphorous (Table 4). HCN content in LPML is negligible or low and LPNML variety has no HCN content at all. So, it showed that these varieties were good for cattle feedings.

HCN study in White clover at Kirtipur and Dhunche revealed that maximum HCN was found at Kirtipur and medium at Dhunche and to be low or absent at Chandanbari, but HCN study could not be conducted at Chandanbari. Meanwhile, the transplant of white clover from Chandanbari to Kirtipur also did not show HCN except for the PNML variety. The study also showed that HCN content in plants was higher in lower altitude zone while it was lower in higher altitude zone. HCN content of white clover showed that it also depends upon altitude i.e. there was a negative correlation between HCN content of plants and altitude. The gene frequencies for the genes for glucoside and enzyme were 100% in warm regions and decreased to 0 % in cold northern latitudes or at high elevations (Leffel and Gibson, 1973). De Araujo, and Aldo (1976); Brighton and Horne (1977) also showed that the frequency of cyanogenic plants decreased at high altitudes. The proportion of cyanogenics showed a striking reduction with increasing altitude. Plants growing below 100 m, north-facing sites contained more cyanogenics than south-facing sites, but aspect did not effect on higher sites. Grassland management had no effect on the proportion of cyanogenics. Grazing by large herbivores does not favour cyanogenesis, but some invertebrate herbivore may do so (Richards et al., 2001).

TABLES

Table 1. : Different varieties of *Trifolium repens* in Nepal (Shrestha, 2013)

Flower colour	Leaf marking	Name of varieties	Symbol
White	Marked	White flower with marked leaf	WML
White	Non-marked	White flower with non-marked leaf	WNML
White with pink tinged	Marked	White with pink tinged flower and marked leaf	WPML
White with pink tinged	Non-marked	White with pink tinged flower and non-marked leaf	WPNML
Pink	Marked	Pink flower with marked leaf	PML
Pink	Non-marked	Pink flower with non-marked leaf	PNML
Light pink	Marked	Light pink flower with marked leaf	LPML
Light pink	Non-marked	Light pink flower with non-marked leaf	LPNML

Table 2: Calculation for HCN content

	Colour change	Time taken	Symbol	HCN Content
1	Plants whose colour changes	Within 5 mins.	+4	Maximum
2	Plants whose colour changes	Between 5 mins. to 2 hrs.	+3	Medium
3	Plants whose colour changes	Between 2 to 6 hrs.	+2	Medium
4	Plants whose colour changes	Between 8 to 24 hrs.	+1	Minimum
5	Plants whose colour changes	After 24 hrs.	0	Negative results

Table.3: HCN test on different varieties of *Trifolium repens* L.

Varieties	Kirtipur	Dhunche	Dhunche / Kirtipur	Chandanbari/Kirtipur
WML	0	0	+4 (Max)	0
WNML	+4 (Max)	+3 (Med)	+4 (Max)	0
WPML	+3 (Med)	+2 (Med)	0	0
WPNML	0	+3 (Med)	0	0
PML	0	0	0	0
PNML	0	+3 (Med)	0	+3 (Medium)
LPML	+1 (Min)			
LPNML	0			

Table 4: Comparative study of leaflet size, soil moisture and soil phosphorous among different varieties of *Trifolium repens* L. (Shrestha, 2002)

Parameters	Study Areas	WML Var.	WNML Var.	WPML Var.	WPNML Var.	PML Var.	PNML Var.	LPML Var.	LPNML Var.
Leaflet size (sq. mm)	Kirtipur (1245-1300masl)	1.40	2.04	1.46	0.76	0.99	0.24	1.22	1.20
	Dhunche (1962-2012 masl)	0.62	0.54	0.58	0.55	0.49	0.49		
	Chandanbari (3400-3500 masl)	1.07	0.32	0.19	0.53	0.91	0.44		
Soil moisture % on which the plant grows	Kirtipur	14.82	14.96	12.99	14.17	14.26	13.44	13.44	12.47
	Dhunche	30.72	28.73	29.89	6.19	32.5	6.40		
	Chandan Bari	46.4	24.08	40.31	19.66	25.85	23.17		
Soil phosphorous % on which the plant grows	Kirtipur	7.41	7.99	1.09	1.19	5.07	1.36	1.09	6.04
	Dhunche	0.003	0.002	0.002	0.002	0.001	0.002		
	Chandan Bari	0.001	0.002	0.000	0.001	0.000	0.001		

Photograph 1: HCN testing with strips in sealed bottles

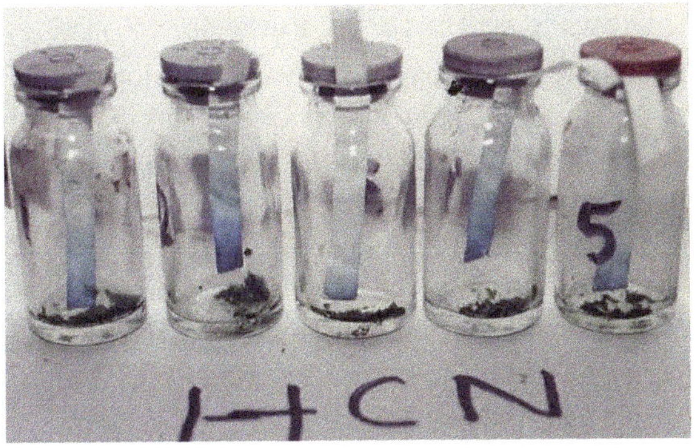

Figure 1: HCN content in different varieties of *Trifolium repens* L. growing on different altitude.

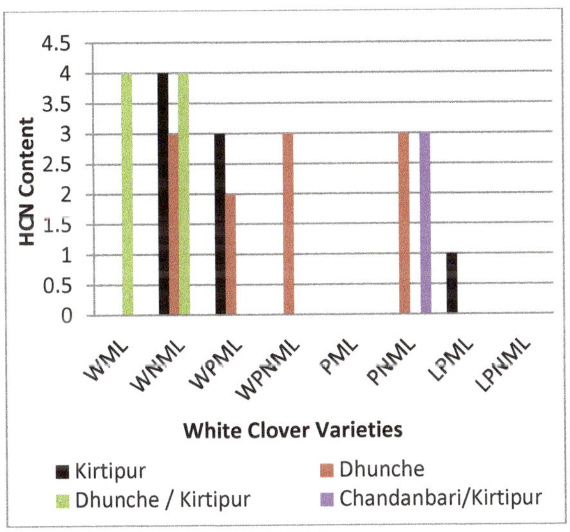

CONCLUSIONS

In Kirtipur, the maximum quantity of HCN was present in WNML variety of white clover, medium in WPML variety and minimum in LPML variety. In Dhunche, medium quantity of HCN was present in WNML, WPML, WPNML and PNML varieties of white clover. In Chandanbari, the content of HCN in plants might be very low or absent because the transplant of plants from Chandanbari to Kirtipur did not show HCN except a medium amount in PNML variety. It can be concluded that the maximum amount of HCN was present in plants with white coloured flower than pink coloured flower.

Furthermore, the amount of HCN was lesser in colder regions and higher altitude than in plants of warmer regions and lower altitude i.e. the frequency of cyanogenic plants decreased. Acyanogenic WML and PML from Kirtipur and Dhunche have an advantage of being less injured by frost than cyanogenic plants. However, the slugs and snails consume these acyanogenic plant types rather than cyanogenic ones (Crawford- Sidebotham, 1972). The ability to produce free cyanide would therefore be more advantageous to plants in warmer habitats, where slugs and snails remain active during the winter. Cyanogenesis would be less valuable to plants in areas where winter weather controls herbivore populations.

HCN content of plants also depended on moisture and phosphorous content of soil; low light intensity; cool grazing condition of areas and size of the leaf.

No HCN content was found in LPNML growing at Kirtipur and PML variety growing at Kirtipur, Dhunche , transplant from Dhunche and transplant from Chandanbari at all, so these varieties are recommended as a best fodder plant for livestock.

ACKNOWLEDGEMENT

My sincere thanks go to Dr. Dilli P Poudel and Mr. Rajesh Bhakta Shrestha for their valuable help and suggestions during my research.

REFERENCES

Brigton F and Horne MT, 1977. Influence of temperature on cyanogenic polymorphisms. Nature. UK. 265:437-438.

Cahn MG and Harper JL, 1976. The biology of leaf mark polymorphism in *Trifolium repens* L. 2. Evidence for the selection of leaf marks by rumen fistulated sheep. Heredity. 37: 327-333.

Caradus JR, Woodfield DR and Hay RJM, 1996. The poisoning of White clover cultivars in New Zealand. In: Woodfield, D. R. (Ed). White clover. New Zealand's competitive Edge. Grassland Research and Practice series No. 6, New Zealand Grassland Association, Palmerston North, pp. 45-49.

Crawford- Sidebotham TJ, 1972. The role slugs and snails in the maintenance of the cyanogenesis polymorphisms of Lotus corniculatus and *Trifolium* repens. Heredity.28: 405-411.

Daday H, 1965. Gene frequencies in wild populations of *Trifolium repens* L. IV. Mechanism of natural selection, Heredity. 20: 355-365.

De Araujo AM, 1976. The relationship between altitude and cyanogenesis in white clover (*Trifolium repens* L.). Heredity. 37: 293.

Doak BW, 1933. A chemical method for determination of type in white clover. New Zealand Journal of science and Techn. 14: 359-365.

Feigl and Anger, 1966. "The Analyst", American Journal. 91: 282-284.

FoyNR and Hyde EOC, 1937. Investigation of the reliability of the picric-acid test for distinguishing strains of white clover in Zealand. New Zealand Journal of Agriculture. 55: 219-224.

Frame J, Charlton JFL and Laidlaw AS, 1998. White clover. Temperate Forage Legumes, CAB International, Willingford, Oxon ox10 8DE, UK.198.

Hart AL, 1987. Physiology In: White clover. M.J. Baker and W.M. Williams (Ed). CAB international. UK.pp.125-152.

Kenneth MO, Nicholas JK, Lindal S and Shih-chung H, 2008. Evidence on the Molecular Basis of the AC/ac Adaptive Cyanogenesis Polymorphism in White clover *Trifolium repens* L. Genetics. 179 (1):517-526.

Kenneth MO, Nicholas JK and Lindal S, 2013. Recurrent gene deletions and the evolution of adaptive cyanogenesis polymorphism in white clover (*Trifolium repens* L.). Molecular Ecology.22 (3):724-738.

Leffel RC and Gibson PB, 1973. White clover. Forage grasses and legumes. 3 rd. Ed. Agricultural Research Service, USDA. Iowa State University Press. pp. 167-176.

Paim NR and Dean CE, 1976. Evaluation of cyanogenic and acyanogenic clones and progenies of *Trifolium repens* L. Agronomy Abstracts. 59.

Richard AJ and Fletcher A, 2001. The effects of altitude, aspect, grazing and time on the proportion of cyanogenics in neighbouring populations of *Trifolium repens* L. (white clover). Heridity.

Richard JW, 1973. Herbivore feeding and Cyanogenesis in *Trifolium repens* L. Heridity. 30(2): 241-245.

Richards R, 2011. A survey of the genetic diversity in populations of white clover, *Trifolium repens* with a focus on South-western Europe: M Sc. thesis in Plant Molecular Biology at Massey University, Palmerston North, NewZealand.

Shrestha AM (Malla), 2002. The Study on The Diversity of *Trifolium repens* L. (white clover) In Nepal. Ph. D Dissertation. Central Department of Botany. T.U. Kirtipur. Kathmandu, Nepal.

Shrestha AM (Malla), 2013. Identification of Different Varieties of *Trifolium repens* L. In Nepal. Asian J. Agri Biol.1 (2):51-62.

Vickery PJ, Wheeler JL and Mulcahy C. 1987. Factors affecting the hydrogen cyanide potential of white clover (*Trifolium repens* L.). Australian J. of Agriculture Research.38:1053-1059.

7

EFFECT OF TILLAGE AND INTEGRATED NUTRIENT MANAGEMENT PRACTICES ON YIELD AND WATER USE EFFICIENCY OF WHEAT UNDER SUB-HUMID CONDITIONS

Shaheen*, N. Sabir and M. Zafar

Department of Soil and Environmental Sciences. The University of Poonch, Rawalakot, Azad Jammu & Kashmir-Pakistan

ABSTRACT

Conventional practices of farming in mountainous and hilly topographic features of Rawalakot Azad Kashmir-Pakistan accelerate runoff and erosion. To study effect of tillage and integrated nutrient management practices field study was conducted in 2009. Layout of experiment was split plot with four main plots of shallow (0-15 cm) and deep tillage (15-30 cm) with residues (wheat straw; 4 Mg ha^{-1}) and without residues incorporation. Four tillage levels were: S (Shallow tillage), SR (Shallow tillage with residues; wheat straw @ 4 Mg ha^{-1}), D (Deep tillage) and DR (Deep tillage with residues; wheat straw @ 4 Mg ha^{-1}). Sub plots were having three fertilizer levels viz. F0 (control), F1 (Recommended rate of inorganic fertilizers at per hectare rate; 120 Kg N-90 Kg P_2O_5-60 Kg K_2O) and F3 (application of farm yard manure (FYM @ 7692 kg ha^{-1}) + 60 Kg N- 90 Kg P_2O_5-60 Kg K_2O ha^{-1}). Significantly ($p \leq 0.05$) higher wheat grain yield of 2.9 kg ha^{-1} was obtained under D while wheat straw yield was statistically similar under all tillage practices. Fertilizer's level F2 and F3 increased grain yield and straw yield compare to F0 but between F2 and F3 statistically no difference was observed. Statistically higher Water Use efficiency (WUE) was obtained for D (4.56 kg ha^{-1} mm^{-1}) and lower with S (2.79 kg ha^{-1} mm^{-1}). Among fertilizers levels F2 gave higher WUE (4.24 kg ha^{-1} mm^{-1}) which is similar to F1 (3.78 kg ha^{-1} mm^{-1}).

Keywords: Farm yard manure, residues incorporation, Tillage practices, water storage, wheat

INTRODUCTION

In sub-humid to humid mountainous region sustainable crop production demands that soil should be in good biological, physical and chemical condition. A mountainous land form causes more rain water losses as run-off and less infiltration. There is a need of extensive experimentation on soil water conservation practices that could increase infiltration and productivity to encourage farmers of these areas for adoption of these measures. Due to global change in rainfall pattern the humid areas having uncertain conditions of rainfall, winter rains are becoming less during wheat germination period. For efficient utilization of rain water focus on more infiltration and less runoff is required. Research on tillage system is necessary in sub-humid region of Rawalakot (Latitude 33° 51' to 33° 85' N and longitude 73° 48' to 73° 80' E) Azad Kashmir-Pakistan to lessen land degradation and to get potential crop yields as average yield production of wheat is 1 Mg ha^{-1} to 1.5 Mg ha^{-1}.

Soil water is important factor that is influenced by tillage practices because of changes produced in infiltration and surface run-off (Zhai *et al.*, 1990). Implementation of tillage practices and incorporation of crop residues varies according to climatic conditions and soil types. Deep tillage (30 cm) disrupts water impeding layers down in soil profile, conserve rainwater and increase grain yield and water use efficiency (Shaheen *et al.*, 2010a). Tillage and residue management variations form a complex association of soil and surface condition and influence on crop yield. For example, crop residues incorporation has been observed to increase yield and in some studies decreases yield. As on the plus side, residues have protected soil from rain drop crusting (Blevins *et al.*, 1983) and have conserved soil and water (Doran *et al.*, 1984). Crop residues incorporation improves soil condition and enhances crop productivity. Residues management affects on soil moisture storage, bulk density, build up soil structure and reduce raindrop impact (Rasmussen and Collins, 1991). Residues may be incorporated or left on the soil surface as mulch. Few studies showed lower wheat yield during first 2 to 3 years of straw incorporation due to wide C:N ratio of

**Corresponding author:* e-mail: aqeela.shaheen@gmail.com

added straw, but in later years straw incorporation did not affect adversely (Mandal *et al.*, 2004). Swan *et al.* (1994) reported poor crop yield with crop residues because of poor weed control, excessively wet and cold soils (Irena, 2011), and poor seed placement and stand. During winter wheat, as low temperature prevails and slow decomposition of residues takes place, so short term studies shows contrasting results and reasons for contrasting results are also related to factors such as existing soil organic carbon, climatic conditions, physico-chemical conditions of soils and crop management practices (Wilhelm *et al.*, 2004).

With adoption of water conservation practices, fertility build up is equally important. In sub-humid to humid conditions losses of inorganic fertilizers with leaching can be minimized by application of integrated nutrients from both inorganic and organic source. Integrated nutrient management is a rational strategy (Palm *et al.*, 1997) that reduces losses by converting inorganic N into organic form (Kramer *et al.*, 2002), increases the efficiency of the fertilizers and reduces the environmental problems that may arise due to use of inorganic source. Farm yard manure has the potential to improve soil physical conditions (El-Shakweer *et al.*, 1998) and provision of nutrients (Schjonning *et al.*, 1994). Unlike inorganic fertilizers, nutrients from organic source release slowly and have long residual effects (Sharma and Mittra, 1991). Many studies conducted on integrated nutrient management showed variable results due to the climatic, soil type quality effects on the performance of the organic component of the system (Palm *et al.*, 2001).

Objectives of study were to determine (i) Effect of straw incorporation and tillage depths on grain yield and water use efficiency. (ii) Effect of integrated nutrient management practices on yield and water use efficiency.

MATERIALS AND METHODS

Field experiment was carried out in 2009 on clay loam soil at experimental research fields of Rawalakot university and wheat was used as test crop. Soil characteristics of experimental sites are given in Table 1. The experiment was conducted as a split plot arranged in randomized complete block design with three replications. The treatments of main plots consist of: shallow tillage (0-15 cm; S), Shallow tillage with wheat straw incorporation at the rate of 4 Mg ha^{-1} (SS), Deep tillage (0-30 cm; D) and deep tillage with wheat straw incorporation at the rate of 4 Mg ha^{-1} (DS). The sub-plots were having three fertilizers levels viz. Control (no application of fertilizers; F0), recommended rate of fertilizer application (at per hectare rate of 120 kg N, 60 Kg P_2O_5 and 90 Kg K_2O; F1) and integrated application of inorganic and organic fertilizers (at per hectare rate of 7692 kg Farm yard manure (it provided 60 kg ha^{-1} N), 60 Kg N, 60 Kg P_2O_5 and 90 Kg K_2O; F2). Urea was applied as source of nitrogen and Diammonium phosphate as a source of phosphorus and sulphate of potash as a source of potassium. All inorganic fertilizers were broadcasted at the time of sowing and nitrogen was applied as split dose, half at the time of sowing and half after germination. The Farm yard manure (FYM) was collected from neighboring house's pens and its analysis showed total nitrogen percent was 0.78%. FYM was mixed thoroughly and spread evenly and incorporated into the experimental plots at the time of sowing.

Wheat straw in specified treatments according to layout plan was incorporated and properly mixed up to 15 cm depth at the time of sowing. The sub-plot size of experimental field was 3.6 m^2. The rainfall data during growing season of wheat is given in Figure 1. Monthly minimum and maximum temperature distribution during growing season of wheat is given in Figure 2.

Soil water contents and bulk density was measured at the time of sowing and after harvest of crop by gravimetric and core method respectively up to depth of 30 cm. The water contents were converted into volumetric water contents by multiplying with bulk density. Threshing was done manually and yield data was recorded at 12 % water contents after harvesting middle rows of each plot. Water use efficiency was calculated as: economic yield / amount of water applied. Where amount of water applied= water at sowing – water at harvesting + growing season precipitation Schlegel *et al.*, 2002. Leaf area was measured with leaf area meter at the flowering stage. Harvest index was calculated as the ratio of grain yield to above ground biomass. Partial factor productivity (Pfp) was calculated (Cassman *et al.*, 1996) as ratio of economic yield to all nutrients applied and stated as:

$$Pfp = Y/Nr$$

Where, Y is the yield in kg ha^{-1} and Nr is the amount of fertilizer nutrients (N+P +K) in kg ha^{-1}.

Analysis of variance was performed by using MSTAT C software, whenever treatment differences were found significant based on F-test, critical differences were calculated at the 5% level of probability by using least significant difference (LSD) test.

Table 1. Soil properties of experimental site

Soil characteristics	0-15 cm	15-30 cm
pH $_{1:1}$	7.38	7.67
EC $_{1:1}$ (dSm^{-1})	0.99	0.64
Organic matter (%)	1.03	----
Bulk density	1.12	1.17
Sand (%)	39.12	----
Silt (%)	30.02	----
Clay (%)	30.86	----
Textural class	Clay loam	----
Average annual precipitation	1481 mm; 33 % rainfall is received in winter (Dec. to March)	
Mean soil temperature	Winter 15 ◦C	
Parent material	Residuium-colluvium from shales	

Table 2. Effect of tillage practices and fertilizers levels on wheat yield harvest index, leaf area and partial factor productivity

S= Shallow tillage, 0-15 cm depth, SR= Shallow tillage with incorporation of wheat straw, D= deep

	Wheat grain yield (Mg ha^{-1})	Wheat straw yield (Mg ha^{-1})	Harvest Index	Leaf area (cm^2)	Partial factor productivity (Pfp) (kg grain yield / kg N+P+K applied)
Tillage practices					
S	1.79 C	3.99 NS	0.33 NS	26 C	7.53 B
SR	2.08 BC	3.76	0.37	29 BC	9.01 B
D	2.86 A	5.10	0.34	34 AB	11.97 A
DR	2.21 B	4.23	0.34	36 A	8.95 B
LSD value	0.41	--	--	5.47	2.67
Fertilizers levels					
F0	1.64 B	2.68 B	0.37 NS	27 B	------
F1	2.39 A	4.88 A	0.33	32 A	8.86 NS
F2	2.68 A	5.28 A	0.33	34 A	9.88
LSD value	0.53	0.79	--	3.76	--

tillage, 0-30 cm depth, DR= Deep tillage with incorporation of wheat straw. F0= control, F1= recommended rate of fertilizer application, F2= Farm yard manure application at the rate of 7.6 Mg ha^{-1} with inorganic fertilizer application

Columns sharing same letters are statistically non-significant $P \leq 0.05$; NS= non-significant

Table 3. ANOVA for the effect of tillage practices and fertilizers levels on yield, harvest index, leaf area and partial factor productivity

	Mean Squares					
	df	Grain yield	Straw yield	Harvest index	Leaf area	Partial factor productivity
Replications	2	0.16	0.48	0.001	29.19	0.73
Tillage (T)	3	1.83 **	3.09 ns	0.002 ns	195 *	21 *
Error	6	0.13	1.77	0.004	22.51	3.56
Fertilizers (F)	2	3.38 **	23.52 **	0.006 ns	183 **	6.23 ns
T X F	6	0.40 ns	1.61 ns	0.007 ns	9.65 ns	6.19 ns
Error	16	0.39	0.83	0.008	18.86	6.99
Total	35					

*; ** significant at $p \leq 0.01$ and $P \leq 0.05$ respectively
ns: non-significant

Table 4. Effect of tillage practices and fertilizers levels on bulk density, water contents and water use efficiency (kg ha^{-1} mm^{-1}).

	Bulk Density Mg m^{-3}		Water contents mm		Water use efficiency kg ha^{-1} mm^{-1}
Tillage practices (T)	0-15 cm	15-30 cm	0-15 cm	15-30cm	
S	1.12 C	1.07 B	24.8 B	21.15 A	2.79 B
SR	1.16 BC	1.08 B	34.4 A	26.70 AB	3.31 B
D	1.22 AB	1.17 A	34.3 A	29.3 A	4.56 A
DR	1.24 A	1.10 AB	29.6 AB	26.05 AB	3.45 B
Fertilizers levels (F)					
F0	1.22 A	1.14 A	29.4 A	24.6 A	2.57 B
F1	1.19 AB	1.11 AB	31.2 A	25.7 A	3.78 A
F2	1.14 B	1.08 B	31.8 A	27.2 A	4.24 A
LSD value					
T	0.06	0.08	6.88	5.69	
F	0.06	0.03	5.04	4.18	0.84
TxF	Ns	ns	ns	ns	ns

S= Shallow tillage, 0-15 cm depth, SR= shallow tillage with incorporation of wheat straw, D= deep tillage, 0-30 cm depth, DR= Deep tillage with incorporation of wheat straw. F0= control, F1= recommended rate of fertilizer application, F2= Farm yard manure application at the rate of 7.6 Mg ha^{-1} with inorganic fertilizer application
Columns sharing same letters are statistically non-significant $P \leq 0.05$; ns= non-significant

Mean Squares						
	df	Bulk Density (0-15 cm)	Bulk Density (15-30 cm)	Water contents (0-15 cm)	Water contents 15-30 cm	Water use efficiency
Replications	2	0.001	0.015	0.001	29.19	0.73
Tillage (T)	3	0.028**	0.018*	0.002 ns	195 *	21 *
Error	6	0.0026	0.004	0.004	22.51	3.56
Fertilizers (F)	2	0.021*	0.009*	0.006 ns	183 **	6.23 ns
T X F	6	0.002 ns	0.001	0.007 ns	9.65 ns	6.19 ns
Error	16	0.005	0.002	0.008	18.86	6.99
Total	35					

Table 5. ANOVA for the effect of tillage practices and fertilizers levels on bulk density, water contents and water use efficiency

*; ** significant at $p \leq 0.01$ and $P \leq 0.05$ respectively

ns: non-significant

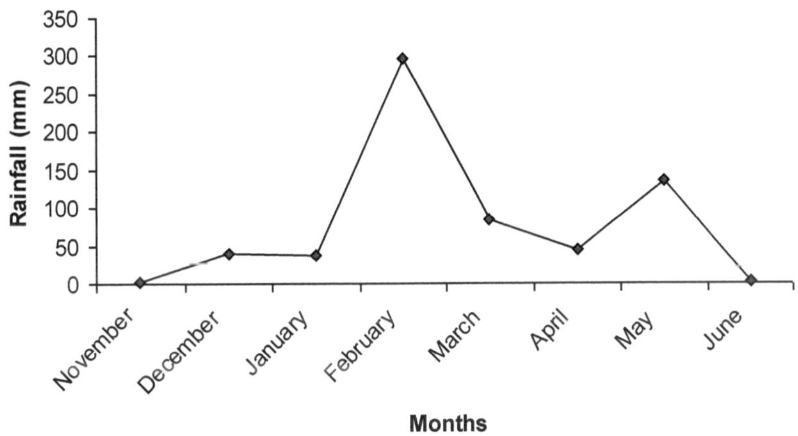

Fig. 1 Rainfall during growing season of wheat 2009-2010.

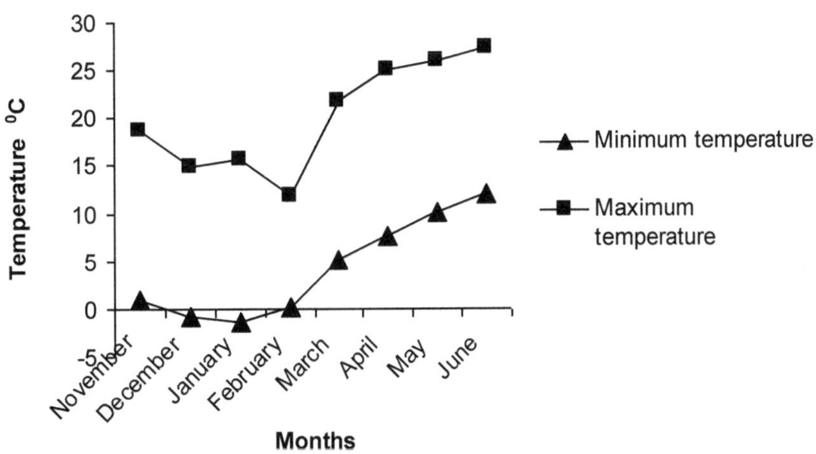

Fig. 2 Monthly minimum and maximum temperature during growing season of wheat 2009-2010

RESULTS AND DISCUSSION

Tillage practices and fertilizers levels (Table 2) have showed statistically ($p \leq 0.05$) significant effect on grain yield. Among tillage treatments D gave higher yield (2.86 Mg ha^{-1}) while the lowest yield (1.79 Mg ha^{-1}) in S. Among fertilizers levels, F1 and F2 gave statistically similar grain yield relative to control (F0). Higher straw yield was observed under D that is statistically non-significant with DR, SR and S. Among fertilizers levels same trend of increase in straw yield was studied as in grain yield. Interactive effects of tillage and fertilizers are non-significant for all parameters (Table 3). Both tillage practices and fertilizers levels showed statistically ($p \leq 0.05$) similar harvest index values.

Crop residues management with different tillage practices and integrated nutrient application does not show same results for all soil types and climate but it varies. In this study residue incorporation with shallow and deep tillage (Table 2) did not show significant impact that may be due to lower temperature (Cook and Haguland., 1991) or application of residues at the time of sowing and nutrient contents of residues that have not been converted into readily available form. Few studies has shown decreased grain yield of wheat due to poor germination with crop residues incorporation (Hejazi *et al.*, 2010). Higher yield under D compare to S (Table 2) may be attributed to more infiltration of water and higher proliferation of roots due to breaking of hard layer up to 30 cm depth (Shaheen *et al.*, 2010a). Data of grain yield (Table 2) suggests that integrated nutrient management was about as efficient as inorganic fertilizer source. The combined use of FYM with inorganic fertilizers improve soil permeability to air and water, water stable aggregates, checks N losses by forming organic mineral complexes and uptake of nutrients witch in turn increase plant growth and yield (Satyanarayana *et al.*, 2002). In long-term studies (Bayu *et al.*, 2006; Yadav, 2001) application of 5 Mg FYM ha^{-1} in combination with 50 % of the recommended inorganic fertilizer rate has increased grain yield comparable to 100 % of the inorganic recommended rate of fertilizer. The results propose that by application of 50 % the recommended fertilizers with FYM can save cost on inorganic fertilizers. Wilhelm and Wortmann (2004) found that tillage methods had no effect on crop yield.

Leaf area response for both tillage and fertilizers practices (Table 2) is statistically non-significant. D and DR showed higher leaf area than S and SR. Both F1 and F2 showed higher leaf area than F0. Interaction of fertilizers and tillage practice was found non-significant (Table. 3). Higher leaf area under D and DR may be attributed to higher uptake of both nutrients and water under these treatments and shows higher growth under these treatments.

Higher values of Pfp (11.97) were obtained under D while S, SR and DR showed statistically similar values (Table 2). Fertilizers levels showed more or less similar values while interaction of fertilizers and tillage (Table. 3) practices was non-significant statistically.

There are many measures of productivity (Dawe and Dobberman, 1999) one example is the partial factor productivity (PFP) of fertilizer, which is the average productivity, measured by grain output divided by quantities of fertilizer. This is relatively easy to measure, but its interpretation is not clear. Some studies have calculated trends in the partial factor productivity of fertilizer (PFP-F) over time. Invariably, these calculations show sharply declining trends, and they are cited as a cause for concern.

Partial factor productivity (Pfp) is a useful measure of nutrient use efficiency as it provides an integrative index that quantifies total economic output relative to the utilization of all nutrient resources in the system (Yadav, 2003). Decline in partial factor productivity for tillage with residues application may be attributed to nutrients immobilization in short term experimental trial. Non-significant effect of fertilizers levels (Table 2) may be due to slow release of nutrients from integrated nutrient application in one growth season of crop. According to Cassman *et al.* (1996), Pfp can be increased by increasing the efficiency with which applied nutrients are taken up by the crop and utilized to produce grain.

Data of water contents at the time of harvest in given in Table 4. Statistically non-significant water contents for depth of 0-15 cm were calculated for SR, D and DR tillage treatments but significantly higher than S. Water contents at depth of 15-30 cm was calculated higher for D at the time of harvest. Fertilizers levels showed statistically non-significant difference but overall non-significant higher water contents were observed under F2. Effect of tillage practices shows (Table 4) higher bulk density values with DR and lower with S for depth of 0-15 cm. Similar trend was found for 15-30 cm depth. However, among fertilizers levels lower bulk density values are obtained with F2 and higher with F0 for both soil depths. Table 4 shows that statistically higher WUE values were obtained for D (4.56 kg ha^{-1} mm^{-1}) and lower with S (2.79 kg ha^{-1} mm^{-1}). Among fertilizers levels, F2 gave 4.24 kg ha^{-1} mm^{-1} WUE which is statistically higher than F0 (2.57 kg ha^{-1} mm^{-1}) and similar to F1 (3.78 kg ha^{-1} mm^{-1}). Interaction of TxF (Table 5) for water contents, bulk density and WUE was statistically non-significant.

Higher water contents and WUE under deep tillage may be due to disruption of hard layers or naturally occurring dense pans that develops due to ploughing at same depths year after year (Campbell *et al.*, 1994). Similarly crop residue management, especially in sub-humid and semiarid regions, is to improve soil water conservation (Steiner, 1994) due to reduced runoff of surface water and improved soil surface condition that allows more time for and permits greater water infiltration. Improved water conservation with suitable tillage practices is an important factor in humid regions where the short-term drought can severely limit crop yields on soils that have low water-holding capacities. In short-term studies combined use of organic and mineral fertilizers gave comparable amount of water used by the crop and similar water use efficiency values (Shaheen *et al.*, 2010b). Apart from increasing yield, fertilizers reduce the rate of evaporation by rapid growth of canopy that shades the soil surface (Cooper *et al.*, 1987) so water contents under F_1 and F_2 were similar statistically. Residue management practices affect soil physical properties such as soil moisture content and bulk density. In some studies however, researchers failed to obtain a significant difference in soil water content between conservation tillage and conventional tillage (Hill *et al.*, 1985; Nuttall *et al.*, 1986). Such results occur because of extended dry periods or because of very low residue rates. Soil water contents affected by tillage practices because of changes produced in infiltration, surface runoff, evaporation and protection from rainfall impact (Zhai *et al.*, 1990; Triplett *et al.*, 1968). The effect of residue management on soil bulk density is variable as some workers reported no effect (Skidmore *et al.*, 1986; Hill, 1990), whereas others found lower soil BD in a conservation tillage-residue management system (Edwards *et al.*, 1992) and residue incorporation (Sidhu and Sur, 1993). Several authors found greater bulk density under conservation tillage than conventional tillage (Hammel, 1989; Ferreras *et al.*, 2000). Water-use efficiency depends upon physiological processes in plants, and these processes are affected by soil fertility. Any practice that leads to an increase in soil water in the root zone may have a positive impact on WUE, due to increased water availability and nutrient uptake (Hatfield *et al.*, 2001). As during moisture-stress conditions the shortage of assimilates and the reduced availability of nutrients cause reduction of grain growth and duration of grain filling (Ren *et al.*, 2003) which results lower WUE.

CONCLUSIONS

The higher yield was obtained with deep tillage than shallow tillage and it is attributed to breaking of hard pans that develops with year to year ploughing of same soil depth (0-15 cm). Crop residues incorporation with shallow and deep tillage practices did not give higher yield as deep tillage without residues incorporation but showed higher yield than shallow tillage. Farm yard manure application as integrated approach with inorganic fertilizers gave comparable yield, pfp as inorganic fertilizers application.

REFERENCES

Bayu W, Rethman NFG, Hammes PS. Alemu G, 2006. Effect of farmyard manure and inorganic fertilizers on sorghum growth, yield and nitrogen use in semi-arid area of Ethiopia. J. Plant Nutr., 29: 391-407.

Blevins RL, Thomas GW, Smith MS, Frye WW, Cornelius PL, 1983. Changes in soil properties after 10 years continuous non-tilled and conventionally tilled corn. Soil Till. Res., 3: 135-146.

Campbell JA, Akhtar ME, Robertson GW, 1994. Water balance and deep tillage in rainfed agriculture of the potowar plateau of Pakistan. In: Proc. Efficient Use of Plant Nutrients. 4th National Congress of Soil

Science. May 24–26, 1992 at Islamabad. pp. 439–50.

Cassman KG, Gines GC, Dizon MA, Samson MI, Alcantara JM, 1996. Nitrogen use efficiency in tropical low land rice systems: Contributions from indigenous and applied nitrogen. Field Crops Res., 47: 1-12.

Cook RJ, Haguland WA, 1991. Wheat yield depression associated with conservation tillage caused by root pathogens in the soil, not phytotoxins from the straw. Soil Biol. Biochem., 23: 1125-1132.

Cooper PJM, Gregory PJ, Keatinge JDH, Brown SC, 1987. Effect of fertilizer variety and location on barley production under rain-fed conditions in northern Syria. Field Crops Res., 16: 67-84.

Dawe D, Dobermann A, 1999. Defining productivity and yield. IRRI Discussion Paper Series No. 33. Manila (Philippines): International Rice Research Institute. 13 p.

Doran JW, Wilhelm WW, Power JF, 1984. Crop residue removal and soil productivity with no-till corn, sorghum and soybean. Soil Sci. Soci. Am. J., 48:640-645.

Edwards JH, Wood CW, Thurlow DL, Ruf ME, 1992. Tillage and crop rotation effects on fertility status of a Hapludult soil. Soil Sci. Soc. Am. J., 56:1577-1582.

EI-Shakweer MHA., EI-Sayad EA, Ewees MSA, 1998. Soil and plant analysis as a guide for interpretation of the improvement efficiency of organic conditioners added to different soils in Egypt. Comm Soil Sci. Plant Anal,. 29: 2067-2088.

Ferreras LA, Costa JL, Garcia FO, Pecorari C, 2000. Effect of no tillage on some soil physica properties of a structural degraded Petrocalcic Paleudoll of the southern "pamps" of Argentina. Soil Till. Res,. 54:31-39.

Irena JUG, Danijel JUG, Mirjana SABO, Bojan IPEŠEVIĆ, Miro STOŠIĆ, 2011. Winter wheat yield and yield components as affected by soil tillage systems. Turkish J. Agri. Forestry. , 35:1-7.

Hammel JE, 1989. Long term tillage and crop rotation effects on bulk density and soil impedance in northern Idhao. Soil Sci. Soc. Am. J., 53: 1515-1519.

Hatfield JL, Thomas JS, Prueger JH, 2001. Managing soils to achieve greater water use efficiency. Agron. J., 93: 271-280.

Hejazi A, Bahrani MJ, Kazemeini SA, 2010. Yield and yield components of irrigated rapeseed-wheat rotation as influenced by crop residues and nitrogen levels in a reduced tillage method. Am-Eur. J. Agric. Environ. Sci., 8 (5): 502-507.

Hill RL, Horton R, Cruse RM, 1985. Tillage effects on soil water retention and pore size distribution of two mollisols. Soil Sci. Soc. Am. J., 49:1264-1270.

Hill RL, 1990. Long-term conventional and no-till effects on selected soil physical properties. Soil Sci. Soc. Am. J., 54:161-166.

Kramer AW, Doane TA, Horwath WR, Kessel CV, 2002. Combining fertilizer and organic inputs to synchronize N supply in alternative cropping systems in California. Agri. Eco. Environ., 91: 233-243.

Mandal KG, Misra AK, Hati KM, Bandyopadhyay KK, Gosh PK, Mohanty M, 2004. Rice-residue management options and effects on soil properties and crop productivity. Food Agri. Environ., 2(1): 224-231.

Nuttall, WF, Bowren, KE, Campbell, CA, 1986. Crop residue management practices and nitrogen and phosphorus fertilizer effects on crop response and some physical and chemical properties of a black chernozem over 25 years in a continuous wheat rotation. Can. J. Soil Sci., 66:159-171.

Palm CA, Gachengo CH, Delve RJ, Cadisch G, Giller KE, 2001. Organic inputs for soil fertility management in tropical agro-ecosystems: Application of an organic sources database. Agri. Eco. Environ., 83: 27-42.

Palm CA, Myers RJK, Nandwa SM, 1997. Combined use of organic and inorganic nutrient sources for soil fertility management in tropical agro-ecosystems: Application of an organic resources database. Agri. Eco. Environ., 83: 27-42.

Rasmussen, PE, Collins, HP. 1991. Long-term impacts of till-age, fertilizer, and crop residues on soil organic matter in temperate semi-arid regions. Adv. Agron., 45:93-134.

Ren JL, Li SQ, Wang J, Ling L, Li FM, 2003. Effects of plastic film mulching and fertilization on water consumption and water use efficiency of spring wheat in semiarid agro-ecosystem. Journal of Northwest Sci-Tech University of Agriculture and Forestry, 31: 1-3.

Satyanarayana V, Prasad VV, Murthy VRK, Boote KJ, 2002. Influence of integrated use of farmyard manure and inorganic fertilizers on yield and yield components of irrigated lowland rice. J. Plant Nutr., 25: 2081-2090.

Schjonning P, Christensen BT, Carstensen B, 1994. Physical and chemical properties of a sandy loam receiving animal manure, mineral fertilizer or no fertilizer for 90 years. Eur J Soil Sci., 45: 257-268.

Schlegel, A. J., T. J. Dumler and C. R. Thompson. 2002. Feasibility of four-year crop rotations in the central high plains. Agron. J., 94: 509-517.

Shaheen A, Naeem MA, Jilani G, Shafiq M, 2010a. Integrated soil management in eroded land augments the crop yield and water-use efficiency. Acta Agriculturae Scandinavica Section B - Plant Soil Sci., 60 (3): 274-282.

Shaheen A, Shamsher A, Stewart BA, Naeem MA, Jilani G, 2010b. Mulching and synergistic use of organic and chemical fertilizers enhances the yield, nutrient uptake and water use efficiency of sorghum. Afr. J. Agric. Res., 5(16): 2178-2183.

Sharma AR, Mittra BN, 1991. Effect of different rates of application of organic and nitrogen fertilizers in a rice-based cropping system. J. Agri Sci., 117: 313-318.

Sidhu A S, Sur HS, 1993. Effect of incorporation of legume straw on soil properties and crop yield in a maize-wheat sequence. Trop. Agric., 70: 226-229.

Skidmore EL, Layton JB, Armbrust DV, Hooker ML, 1986. Soil physical properties as affected by cropping and residue management. Soil Sci. Soc. Am. J., 50: 415-419.

Steiner JL, 1994. Crop residue effects on water conservation. In. Unger P (ed), "Managing Agricultural Residues," CRC Press/Lewis Publishers, Boca Raton, FL: 41-76.

Triplett JR, Van Doren DM, Schmidt BL, 1968. Effect of corn stover mulch on no-tillage corn yield and water infiltration. Agron J., 60:236-239.

Swan JB, Higgs RL, Bailey TB, Wollenhaupt NC, Paulson WH, Peterson AE, 1994. Surface residue and in row treatment on long term no tillage continuous corn. Agron. J., 86: 711-718.

Wilhelm WW, Johnson JMF, Hatfield JL, Voorhees WB, Linden DR, 2004. Crop and soil productivity response to corn residue removal: A Literatue review. Agron. J., 96:1-17.

Wilhelm WW, Wortmann CS, 2004. Tillage and rotation interactions for corn and soybean grain yield as affected by precipitation and air temperature. Agron J., 96: 425-432.

Yadav RL, 2001. On-farm experiments on integrated nutrient management in rice-wheat cropping systems. Exp Agr., 37: 99-113.

Yadav RL, 2003. Assessing on-farm efficiency and economics of fertilizer N, P and K in rice wheat systems of India. Field Crop Res., 18: 39-51.

Zhai R, Kachanoski RG, Voroney RP, 1990. Tillage effects on the spatial and temporal variation of soil water. Soil Sci. Soc. Am. J., 54: 186-192.

ASSESSMENT OF YIELD LOSSESS CAUSED BY BACTERIAL BLIGHT OF RICE IN UPPER DIR, KHYBER PAKHTUNKHWA PROVINCE

Mehran Khan[1], Abdur Rafi[1], Aqleem Abbas[*1], Tauheed Ali[1] and Akhtar Hassan[2]

[1] *Department of Plant Pathology, The University of Agriculture, Peshawar Pakistan*

2Department of Agricultural Chemistry, The University of Agriculture, Peshawar Pakistan

ABSTRACT

The present study was conducted to assess the yield losses caused by bacterial brlight of Rice (*Xanthomonas oryzae pv oryza*) in Upper Dir a district of Khyber Pakhtunkhwa province of Pakistan during cropping season 2013-2014. Four rice growing locations of district namely Barawal Bandi, Satara, Maidan khwar and Thikar Kot were selected. In Upper Dir disease incidence (%) ranged between 57.14%-75% while disease severity (%) was between 36.04 to 66.12%. Highest disease incidence (75%) and disease severity (66.12%) was recorded from Barawal Bandi location whereas lowest disease incidence and disease severity was recorded from Thikar kot. Similarly maximum 1000 grain weight losses was also recorded in Barawal Bandi while minimum 1000 grain weight loss was recorded in Thikar kot. It can be concluded that Bacterial blight of rice was more severe in Barawal location and less severe in Thikar kot.

Keywords: Rice, bacterial blight, % Incidence, % Severity

INTRODUCTION

Rice (*Oryzae sativa* L.), is an annual plant which belongs to family *Poaceae* and is mostly cultivated in tropical and subtropical regions of the world (Ezuka and Kaku, 2000). Rice is an important staple food crop of the world especially of the Asian countries such as Bangladesh, Pakistan, China, Vietnam and Korea. It ranks as the third largest crop after wheat and maize and has an integral position in the economy of Pakistan as a staple food as well as an export item (Salim et al., 2003; Zahid et al., 2005). There are several varieties of rice grown in Pakistan but basmati and kernel are the major varieties grown in rice growing areas of Pakistan (Ahmad et al., 2007). These two varieties are famous for their taste and aroma. Unfortunately, yield of rice obtained from per unit area in Pakistan is very low from neighboring countries and the world. Rice production is severely affected by various diseases such as viral, bacterial and fungal which are primarily responsible for the low yield of rice. Among the bacterial diseases of rice, bacterial leaf blight (BLB) i.e., *Xanthomonas oryzae* pv. *oryzae*, is a major bacterial disease of rice which frequently appeared in rain fed, irrigated and deep water rice growing areas (Mew, 1987). The *Xanthomonas oryzae* pv. *oryzae* is a gram negative bacterium belongs to Gracillicutes and is known to be one of the most disparaging disease of rice throughout the world and especially in South Asia (Swings et al., 1990; Ou, 1985). Leaf blight disease was first recorded in Japan and later on, its occurrence has been reported from Asian and African countries (Mew et al., 1993; Ezuka and Kaku, 2000). In Pakistan, bacterial leaf blight (BLB) occurrence was confirmed from all rice growing areas in 1977 except the Tribal belt and Dir valley (Mew and Majid, 1977). Recently its incidence has augmented in "Kaller" belt which is prominent for rice farming (Khan et al., 2000). BLB infects almost all the growth stages of rice and reveals either leaf blight "kresek" symptoms. The pathogen enters into plants through water pores and wounds. Since the water pores are located at the margins of upper surface of the leaves. Therefore, the lesion caused by bacterial leaf blight frequently emerges from the leaf margins near its tips (Mizukami, 1956; Mizukami, 1961; Tabei and Mukoo, 1960). Symptoms of this disease includes, tiny water soaked lesion, which then turn yellow as the disease progresses and finally develops into an elongated irregular lesion with wavy margins. Sign of the causal agent as bacterial ooze can be seen on the margins or veins of the freshly infected leaf under moist conditions (Tagami

**Corresponding author:* e-mail: aqlpath@gmail.com

and Mizukami, 1962; Octa, 1970; Ou, 1985). If the plant is infected with the bacterial leaf blight disease, the panicles will only contain sterile immature grains, which are easily fractured during milling process. In case of severe infection, 50% reduction in yield can be expected from bacterial leaf blight disease (BLB) (Mew et al., 1993) whereas in case of mild infection 10-12% yield reduction has been recorded (Ou, 1985).

Breeding resistance varieties has become an essential objective for the management of leaf blight disease (Mew et al., 1987). Actually, no approach is considered as an effective and economic practice to control this disease (Devadath, 1989). Keeping in view the importance of bacterial leaf blight disease, this study was initiated to assess the incidence and severity of BLB and to determine yield losses caused by BLB in Upper Dir district of Khyber Pakhtunkhwa province of Pakistan.

MATERIALS AND METHODS

The present research work was carried out in different rice growing areas of Upper Dir, located in the Khyber Pakhtunkhwa.

Survey: A comprehensive survey was conducted for various locations of Upper Dir during 2013-14 cropping season at the booting stage of the crop. Leaves showing typical symptoms of Bacterial Blight of rice were observed. Four different locations of upper dir including Barawal Bandi, Satara, Maidan Khwar and Thikar Kot were visited during the survey. The Data on disease incidence (%), severity (%) and 1000 grain weight loss were recorded as follows:

Disease incidence (%): In order to calculate disease incidence (%) five random spots of 1m^2 area were selected in the field and the infected plants were counted and percentage of incidence was calculated from the total number of number of plants from that area.

Disease severity (%): Five plants in each location were randomly selected. The data regarding lesion length and total leaf area were recorded on five leaves of the selected plants and percent disease severity was worked out.

For disease rating, scale of Chaudhry (1996) was used (Table 1).

Table 1: Disease severity scale for bacterial leaf blight disease (Chaudhry, 1996).

Disease Score	Lesion area (%)	Disease Reaction
0	0	Highly Resistant (HR)
1	1-10	Resistant (R)
3	11-30	Moderately Resistant (MR)
5	31-50	Moderately Susceptible (MS)
7	51-75	Susceptible (S)
9	76-100	Highly Susceptible (HS)

Loss in grain weight (GW)

For calculating loss in grain weight, samples were dried and their grains were separated from both healthy and diseased plants. Loss (%) in grain weight was calculated using the following formula:

$$\% \text{ loss in 100 GW} = \frac{W_1 - W_2}{W_1} \times 100$$

Where, GW= Grain weight, W_1= Weight of 1000-grain obtain from healthy plants and W_2= Weigh of 1000-grain obtain from the disease plants.

Statistical analysis

Descriptive statistics was used for the analysis of obtained data i.e. disease severity and disease incidence of bacterial leaf blight disease of rice were expressed in percentages.

RESULTS AND DISCUSSION

Disease incidence: Data regarding disease incidence of bacterial leaf blight (BLB) in Upper Dir during year 2011-2012 are shown in Table 2. The peak disease incidence (75%) was recorded for Barawal Banda, followed by Satara 68.18%. However, lowest (57.14%) disease incidence was recorded for Thikar Kot. This difference may be attributed due to high relative humidity in Barawal Bandi. Same experiments of bacterial blight of leaf were also conducted by Akhtar et al. (2003) who found that BLB disease of rice prompted by *Xanthomonas oryzae* pv. *oryzae* has created a serious situation in all provinces of Pakistan viz., Punjab, Khyber pakhtukhwa, Sindh, Baluchistan including Azad Kashmir. In Punjab the Disease Incidence of BLB ranges from 16-100%, 11-71%, 11-91%, 16-65%, 1-50%, 2-100%, 32-80%, 47%, 50-71% and 40-50% in Sargodha, Hafizabad, Sheikhupura, Sialkot, Narowal, Gujranwala, Gujrat, Lahore, Kasurand and Okara district. In Sindh the Disease Incidence (%) was 0% in Larkana, 0-5% in Shikarpur, 5% in Dadu, 0% in Nawab Shah, 5% in Thatta, 0-5% in Badin, 0-5% in Jacob Abad. In Khyber Pakhtunkhwa province the disease incidence (%) range was 3-96, 2-100 and 0 in Lower Dir, Swat and Malakand Agency respectively. In Baluchistan the Disease Incidence (%) range was 0 and 1-5 in Nasirabad and Usta Muhammad respectively while in Azad Jammu and Kashmir (AJK) no disease was observed. There were no reports of bacterial blight diseases from Gilgit Baltistan. Khan et al. (2008) carried out a survey on disease incidence and losses caused by BLB disease of rice in Pakistan. It was found that 80% rice crops in some fields were affected near Manawala and almost 100% in Feroz watwan, areas of district Sheikupura. In Mureedkey and Sheikhupura, 70-80% rice fields were found to be infected with rice blight whereas in Narung and Wahndo, 90-100% rice fields were found to be infected with bacterial leaf blight disease.

Table 2. Disease Incidence (%) for bacterial leaf blight disease recorded in four locations of Upper Dir District.

Location	Disease Incidence (%)
Barawal Bandi	75
Satara	68.18
Maidan Khwar	66.66
Thikar Kot	57.14

Disease severity (%): Maximum 66.12% disease severity was recorded for Barawal Bandi location of Upper Dir Khyber Pakhtunkhwa, followed by Maidan Khwar (55.9%) and Satara (41.02%) while in Thikar Kot minimum disease severity (36.04%) was recorded (Table 3). Therefore in Barawal Bandi location rice crop was categorized as more susceptible to the BLB pathogen and this might be due to high relative humidity (>70%). These results are in accordance with previous observations of Akhtar et al. (2003) who recorded disease severity in Swat, Malakand Agency and lower Dir.

Table 3: Disease Severity (%) of bacterial leaf blight disease recorded in four locations of Upper Dir District.

Location	Average severity (%)	Category
Barawal Bandi	66.12	Susceptible (S)
Satara	41.02	Moderately Susceptible (MS)
Maidan Khwar	55.9	Susceptible
Thikar Kot	36.04	Moderately Susceptible (MS)

1000 Grain Weight loss of five locations: The highest loss in 1000 grain weight was 15.59% at Barawal Bandi, followed by Satara, Maidan Khwar and Thikar Kot (12.13, 13.37 and 11.94% respectively). Increase in disease severity is the main cause of the decrease in 1000 grain weight (Table 4). Thousand grain weight losses due to BLB are varied depending on location, season, crop growth stage, and cultivar (Ahmad and Singh, 1975; Mohiuddin et. al., 1977; Muralidharan and Rao, 1979; Srivastava and Kapoor, 1982).

Table 4: Thousand Grain Weight loss due to bacterial leaf blight at four locations of Upper Dir District.

Locations	1000g Grain Weight (Healthy Plants)	1000g Grain Weight (Disease Plants)	Loss in Weight (%)
Barawal Bandi	32.89g	27.76g	15.59
Satara	30.91g	27.16g	12.13
Maidan Khwar	28.12g	24.36g	13.37
Thikar Kot	28.80g	25.36g	11.94

CONCLUSION

It was concluded that among the four locations of Upper Dir, highest disease incidence(%) and disease severity(%) was recorded in Barawal Bandi and Maidan Khwar whereas the other locations i.e. Thikar Kot and Satara were not severely infected with bacterial leaf blight (BLB) disease of rice. It is recommended to raise awareness in the farming community of Uper Dir about gradual increase bacterial leaf blight (BLB) disease of rice. Further farming community must be trained on management of bacterial leaf blight (BLB) disease of rice to minimize yield losses.

ACKNOWLEDGEMNTS

The authors are very grateful to Department of Plant Pathology, The University of Agriculture Peshawar for providing facilities to carry out this research.

REFERENCES

Ahmad K, Zafar Y and Arif M, 2007. Path analysis of some leaf and panicle traits affecting grain yield in double haploid lines of rice (*O. sativa*). J. Agri. Res. 15: 245-252.

Ahmad KM and Singh RA, 1975. Disease development and yield losses in rice varieties by bacterial leaf blight. Indian Phytopathol. 28: 502-507.

Ahmad KM and Singh RA, 1975. Disease development and yield losses in rice variety by BLB. Indian Phytopathol. 28: 502-507.

Akhtar MA, Zakria M, Abbasi FM and Masood MA, 2003. Incidence of bacterial blight of rice in Pakistan during 2002. Pak. J. Bot. 35: 993-997.

Chaudhary RC, 1996. Internationalization of elite germplasm for farmers: Collaborative mechanisms to enhance evaluation of rice genetic resources. In: New Approaches for Improved use of Plant Genetic Resources; Fukuyi, Japan; pp. 26.

Devadath S and Dath AP, 1983. Role of inoculum in irrigation water and soil in the incidence of bacterial blight of rice. Indian Phytopathol. 36: 142-144.

Ezuka A and Kaku H, 2000. A historical Review of Bacterial Blight of rice. Nat. Insti. of Agrobio. Resourc. Bull. Jap. 207.

Khan JA, Jamil AFF and Gill MA, 2000. Screening of rice varieties/lines against Bakanae and Bacterial Leaf Blight (BLB). Pak. J. Phytopathol. 12: 6-11.

Khan H, Hidayat-ur-Rahman, Ahmad H, Ali H, Inamullah and Mukhtar A, 2008. Magnitude of combining ability of sunflower genotypes in different environments. Pak. J. Bot. 40: 151-160.

Mew TW and Majid A, 1977. Bacterial Blight of rice in Pakistan. International Rice Research Newsletter. 5: 5.

Mew TW, 1987. Current status and future prospects of research on bacterial blight of rice. Annu. Rev. Phytopathol. 25: 359-382

Mew TW, Alvarez AM, Leach JE and Swings J, 1993. Focus on bacterial blight of rice. Plant Diseases 77: 5-12.

Mizukami T, 1956. Studies on the bacterial leaf blight of rice plan on the entrance and multiplying portion of the pathogen upon the rice plant leaves (preliminary report). Agric. Bull. Saga. Univ. 4: 169-175.

Mizukami T, 1961. Studies on the ecological properties of *Xanthomonas oryzae* (Uedya et Ishiyama) Dowson, the causal organism of bacterial leaf blight of rice plant. Agri. Bull. Saga. Univ. 13: 1-85.

Mohiuddin MS, Verma JP and Rao YP, 1977. Losses due to bacterial blight of rice. Indian J. Agric. Sci. 47: 221-223.

Mohiuddin MS, Rao YP, Mohan and Verma JP, 1976. Role of *Leptocorrisa acuta* Thun, in the spread of bacterial blight of rice. Curr. Sci. 45: 426-427.

Muralidharan K, Venkata R, 1979. Bacterial leaf blight on rice in Nellore district, A.P. Indian Phytopathol. 32: 483-485.

Octa, 1970. Bacterial leaf blight of rice in Southern Asia. *Overseas Technical Cooperation Agency*. Tokyo, Japan. 71p.

Ou SH, 1985. Rice Diseases. 2nd ed. Commonwealth. Instt. Kew, Surrey. England. pp 61-96.

Salim M, Akram M, Akhtar MH and Ashraf M, 2003. Rice - a production hand book. Pak. Agric. Res. Council, Islamabad.70p.

Singh GP, Srivasttava KM, Singh VR and Singh MR, 2000. Variation in quantitative and qualitative losses caused by bacterial blight in different rice varieties. Indian Phytopathol. 30: 180-185.

Srivastava MP and Kapoor TR, 1982. Yield losses due to bactrial leaf bligh. Int. Rice. Res. Newsl. 7(3): 7-10

Swings J, Van den Mooter M, Vauterin L, Hoste B, Gillis M, Mew TW and Kersters K, 1990. Reclassification of the causal agents of bacterial blight *(Xanthomonas campestris* pv. *oryzae)* and bacterial leaf streak *(Xanthornonas campestris* pv. *oryzicola)* of rice as pathovars of *Xanthomonas oryzae.*

Tabei H and Mukoo H, 1960. Anatomical studies of rice plant leave affected with bacterial leaf blight, in particular reference to the structure of water exudation system. Bull. Nat. Agric. Sci. 11: 37-43.

Tagami Y and Mizukami T, 1962. Historical review of researches on bacterial leaf blight of rice caused by *Xanthomonas oryzae* (Uyeda et Ushiyama) Dowson. Special report of the plant diseases and insect pest forecasting service No. 10. Plant protection Division, Ministry of Agriculture and Forestry, Tokyo, Japan. 112 p.

Zahid MA, Akhtar M, Sabir M, Anwar M and Jamal A, 2005. Genotypic and phenotypic correlation and path analysis in coarse grain rice. Proceeding of the International seminar in rice crop. Oct 2-3. Rice Research Institute, Kala Shah Kaku. 29.

RESPONSE OF CORIANDER (*Coriandrum sativum* L.) TO DIFFERENT NITROGEN LEVELS AND SOWING DATES

Haidar Ali[1*], Gohar Ayub[1], Ehsan Elahi[1] Muhammad Shahab[1], Saeed Ahmed[2], Nazeer Ahmed[3]

[1]*Department of Horticulture, The University of Agriculture Peshawar-Pakistan.*

[2]*State University of Londrina Centre of Agriculture Sciences Londrina, Parana (PR) Brazil.*

[3]*State Key Laboratory of Crop Stress Biology for Arid Areas, Northwest A&F University. Yangling, China*

ABSTRACT

Field experiment for different levels of nitrogen and sowing dates for coriander crop was performed under field conditions during year 2011-12. Sowing of Coriander was practiced on three dates with fortnight interval from 20th October to 20th November). Nitrogen fertilizer was applied at four levels of 0, 20, 40 and 60 kg ha^{-1}. Maximum days to germination, plant height, number of days to first umbel maturity and seed yield kg ha^{-1} were observed in plants provided with at 60 kg ha^{-1} nitrogen at 5th November sown crop, hence recommended for better production of coriander under the agro climatic conditions of Swat.

Keywords: Coriander (*Coriandrum sativum* L.), Nitrogen levels and sowing dates.

INTRODUCTION

Regarding low yield of coriander in Pakistan, several factors are responsible. Appropriate sowing time and judicious use of fertilizers are of supreme importance in this regard. This crop has its own requirement for low temperature and light to precede germination, growth and maturity. Its need of low temperature initially could be observed in early sowing as temperature is above optimum which adversely affects germination due to the action of several bacteria and fungi on embryo and endosperm. This mainly happens by the activities of fungi and bacteria encouraged by temperature above optimum (Naeem et al., 2002). Late crop sowing results in less number of branches per plant and poor or stunted crop growth due to severe low temperature. In order to obtained higher yield of coriander, there is need to improve the efficiency of available land by judicious use of fertilizers and selection of proper sowing time. Soil fertility and crop productivity could be improved by proper nitrogenous fertilization (Akanbi et al., 2001).

Application of nitrogen fertilizer improved seed yield and biomass of coriander by 43-68% and 25-42%, respectively (Heng et al., 2001). Various production factors are responsible for improving crop yield, however, nutrient management the most important significantly affecting yield in crops. Low crop productivity in recent years is mainly attributed to decline in fertility status of the soil. Major problems faced by farmers in maintaining soil fertility are, shortage in production of fertilizer and availability at required times. Due to variation in yield potential of different soils, ensuring proper nutrient supply throughout the growing season is the greatest challenge of the day for agriculture scientist. For sustainability in crop production and improvement in soil health, balanced fertilization is very important (Sharif et al., 2004). Nitrogen fertilization plays major role in crop yield improvement with better plant health (Kizil and Ipek, 2004). Regarding various nutrients, nitrogen is the most limiting nutrient and coriander requires approximately 50 Kg N ha^{-1}. Efficient application of nitrogen from organic and inorganic sources effectively enhances yield and quality of coriander and soil health. Nitrogen as a fertilizer is consider the most limiting factor affecting marketable yields as well as nitrate accumulation in coriander. Application of excess nitrogen, however, is often much more than utilized by the plants. Higher nitrate build up due to arbitrary incorporation of nitrogen is getting importance as a public health concern. Considering the importance of nitrogen dose rate and effective sowing date for improved coriander crop yield, present study was planned under field conditions to find out the best sowing time and nitrogen application rate.

MATERIALS AND METHODS

The experimental plots were ploughed well and the soil sample was taken for analysis. Soil

**Corresponding author:* e-mail: haidarali212@gmail.com

texture class was silt loam which contained (0.057) percent nitrogen, 41 mg kg^{-1} phosphorus (P_2O_5) and 1.59% organic matter. The pH and lime content ($CaCO_3$) were 6.4 and 2.40, respectively. Two factors (sowing dates and N application rates) were investigated in the course of experiment. Randomized complete block (RCBD) design was used with split plot arrangement with each treatment replicated three times. Corriander cultivar "Irani" was used with seed rate of 25 kg ha^{-1} on all three particular dates of sowing. Two factors were combined in various combinations and twelve treatments were formed. Field was divided in main and sub plots having sub-plot size of 1.2 m x 1m. Row to row distance was 30 cm. Urea was used as a source of nitrogen and diammonium phosphate (DAP) for phosphorus.. Following are detail of the treatment used in the course of study.

Sowing dates	*Nitrogen levels*
D_1:20th October	N_1: 0 kg ha^{-1}
D_2:5th November	N_2: 20 kg ha^{-1}
D_3:20th November	N_3: 40 kg ha^{-1}
	N_4: 60 kg ha^{-1}

All other cultural practices were kept constant for the treatments. Field was irrigated when needed. In order to avoid loses due to weeds infestation, hand weeding was done after twenty and fifty days of sowing.

The data recorded from different parameters were analyzed statistically using analysis of variance techniques appropriate for randomized complete block design with split plot arrangement. Means were compared using LSD test at 0.05 level of probability, when the F-values were significant. The statistical software GenStat release 8.1 was used for analysis of the data.

RESULTS AND DISCUSSION

Days to germination

Days to germination of coriander was affected by different sowing dates and nitrogen application (Table 1). There existed non-significant difference for different sowing dates but significant for nitrogen application. The interaction was, however, found non-significant. Maximum days to germination recorded in control treatment followed by treatments of 40 kg ha^{-1} and 60 kg ha^{-1} nitrogen levels and minimum in 20 kg ha^{-1}. Emergence of coriander was statistically similar in different sowing dates. Environmental conditions at the end of October is not that much warmer as compared to the temperature in mid of November. However, 20 kg ha^{-1} nitrogen application showed early emergence which might be the optimum range of nutrition for coriander.

Table I: ***Days to germination as affected by different Sowing dates and nitrogen levels***

Nitrogen	**Sowing Dates**			**Mean**
	20th Oct	**5th Nov**	**20th Nov**	
0	9	9	9	9a
20	7	8	7	7b
40	8	7	8	8ab
60	7	8	9	8ab
Mean	8	8	8	

LSD for nitrogen = 1.1316
LSD for sowing dates= 0.4627
CV (REP*SOW DATE) = 4.90
CV (REP*SOW DATE*NITROGEN) = 13.71

Plant height (cm)

Nitrogen application at 60 kg ha^{-1} resulted in tallest plants (132 cm) as compared to other treatments followed by that of 40 kg ha^{-1} (121 cm) (Table 2) while control plants resulted in short stature (96 cm). Higher plant height (118 cm) was recorded in plots sown on 5th November similar to 20th November (115 cm) but short stature (113 cm) in early sown plots (20th October). Height of plant affects crop growth in several ways and directly influences biological yield and total biomass. Higher plant height in N treated plots could be attributed to synergetic effect of N fertilization on vegetative growth of the plant (Tiwari et al., 2002). Possible reason for this increase is the fact that higher amounts of nitrogen trigger vigorous vegetative growth of crops (Carrubba et al., 2006). The superiority of higher N application over low N may be due to the fact that N incorporated plots utilized the water and nutrients in a better way resulted in optimum

plant growth and attributed to deeper and denser rooting or vigorous vegetative growth.
As planting were delayed, coriander plant height were increased and may be attributed to day length. Delay planted plants were more exposed to longer day which promoted more vegetative growth and maximum plant height was obtained.

Table II: ***Plant height as affected by different Sowing dates and nitrogen levels.***

Nitrogen	Sowing Dates			Mean
	20th Oct	5th Nov	20th Nov	
0	92	103.66	95	96.88d
20	111	111	117.33	113.11c
40	121.33	125.33	117.33	121.33b
60	129.33	133.66	133	132a
Mean	113.41b	118.41a	115.66b	

LSD for Nitrogen= 2.3768 CV (REP*SOW DATE) = 1.88
LSD for Sowing date= 2.4704 CV (REP*SOW DATE*NITROGEN) = 2.07

Number of days to first umbel maturity
Different N application rates and sowing dates significantly affected days to first umbel maturity of coriander with significant interaction (table 3). Days to first umbel maturity increased as N application rate increased from 0 to 60 kg ha^{-1}. Early maturity of first umbel was observed in plots where no N was applied. Similarly, N application at the rate of 60 kg ha^{-1} resulted in delay maturity and it took more days (70days) to first umbel maturity followed by (68 days) of N application at 40 kg ha^{-1} statistically at par (67 days) with N application at the rate of 20 kg ha-^{1}. Regarding sowing dates, early maturity of first umbel was observed in 20th October sowing plots (69 days) followed by (68 days) sowing on 5th November. Sowing of coriander on 20th November took (67 days) to first umbel maturity. The delay in maturity of first umbel in coriander might be attributed to vigorous growth and development of coriander because more nutrients were supplied and available in high nitrogen fertilizer plots. Nitrogen fertilization improved coriander and enhanced growing season. Later planting may delay progress of the crop which shortened the phonological developmental (Zheljazkov et al., 2008) and thus assumed earlier planting with greater time for phonological development than later plantations.

Table III: Days to first umbel maturity as affected by different Sowing dates and nitrogen levels

Nitrogen	Sowing dates			Mean
	20th Oct	5th Nov	20th Nov	
0	66.33	68.33	65.33	66.67c
20	68.33	67.66	67.33	67.77bc
40	70.33	68.33	66.33	68.33b
60	72.33	69.33	70.33	70.66a
Mean	69.33a	68.41ab	67.33b	

LSD for Sowing dates=1.4010 CV(REP*SOW DATE)= 1.81
LSD for Nitrogen= 1.6589 CV(REP*SOW DATE*NITROGEN= 2.45

Seed yield (kg ha^{-1})
Higher seed yield (162 kg ha^{-1}) were recorded in plots where N was applied at rate of 60 kg ha^{-1} followed by (152 kg ha^{-1}) in plots with 40 kg ha^{-1}. Control plots resulted in lower (126 kg ha^{-1}) seed yield. Sowing of coriander on 5th November resulted in maximum (148.5 kg ha^{-1}) seed yield followed by (146.3 kg ha^{-1}) 20th November while, sowing on 20th October produced minimum (141.41 kg ha^{-1}). Nitrogen application increased seed yield in linear manner and 60 kg ha^{-1} improved seed yield by 40% as compared to control. Reduction in yield due to lower fertilization might be attributed to nutritional imbalance and deficiency of certain important plant growth nutrients at various important growth stages like flowering, seed formation and seed maturity (Dierchesen,

1996). These results are in accordance with Singh et al. (2010) who suggested that timely availability of N could be insured and productivity can be positively increased.

Table IV: ***Seed yield (kgha^{-1}) as affected by different sowing dates and Nitrogen levels.***

Nitrogen	Sowing dates			Mean
	20th Oct	5th Nov	20th Nov	
0	124.67	127.66	126	126.11d
20	133	143.33	147.66	141.33 c
40	152	153	151	152b
60	156	170	160.66	162.22a
Mean	141.41 b	148.5 a	146.33 ab	

LSD for Sowing dates=5.2455
LSD for Nitrogen = 4.9930
CV(REP*SOW DATE)= 3.18
CV(REP*SOW DATE*NITROGEN)= 3.47

CONCLUSION AND RECOMMENDATIONS

Nitrogen application at the rate of 60 kg ha^{-1} with a sowing date of 5th November improved growth and yield of coriander. Considering all the factors of good crop production and yield, this dose rate with sowing date is recommended for the coriander growers of Swat valley.

REFERENCES

Akanbi WB, Adediran JA, Togun AO and Soul RA, 2001. Effect of organic base fertilizer on the growth, yield and storage life of tomato (*Lycopersicon esculoutum*). Biosci. Res. Comm. 12(4): 14-18.

Ayanoglu F, Mert A, Aslan N and Gürbüz B, 2002. Seed yields, yield components and essential oil of selected corriander (*Coriandrum sativum* L.) lines as affected by sowing dates and N fertilizer. J, Herbs, Spices Med, Pl. 9(2/3): 71-76.

Carrubba A, Torre A and Calabrese R. 2002. Cultivation trials of corriander (*Coriandrum sativum* L.) in a semi-arid Mediterranean environment. Acta Hort. 576: 237-242.

Dierchesen A, 1996. Promoting the conservation and use of underutilized and neglected crop, coriander. Int. Pl. Gen. Res. Inst. (IPGRI). Italy. 82pp.

Heng LK, White RE, Helyar KR, Fisher R and Chen D, 2001. Seasonal differences in the soil water balance under perennial and annual pastures on an acid Sodosol in southeastern Australia. Eur, J, Soil Sci. 52: 227-236.

Kizil S and Ipek A, 2004. The effects of different row spacing and nitrogen on yield, yield components and essential oil content of some coriander (*Coriandrum sativum* L.) lines. J Agric. 10(3): 237-240.

Naeem M, Chohan MSM, Khan AH and Salahuddin S, 2002. Evaluations of different varieties of pearl millet for green fodder yield potential under different sowing dates. Asian J. Pl. Sci. 1: 326–331.

Sharif M, Ahmed M, Sharir MS and Khattak RA, 2004. Effect of organic and inorganic fertilizers on the yield and yield components of maize. Pak. J. Agric. Engg. Vet. Sci. 20(1): 11-15.

Singh UB, Tomar SP and Rathi RS, 2010. Effect of NPK and varying irrigation levels on the production of small seeded coriander (*Coriandrum sativum* L.). Indian J. Agron. 16: 313-315.

Tiwari RS, Agarwal A and Sengar SC, 2002. Effect of dates of sowing and number of cutting on growth, seed yield and economics of coriander. Crop Res (Hisar). 23(2): 324-329.

Zheljazkov VD, Pickett KM, Caldwell CD and Pincock JA, 2008. Cultivar and sowing date effects on seed yield and oil composition of coriander in Atlantic Canada. Indust. Crop Prod. 28: 88-94.

10

HETEROSIS AND COMBINING ABILITY ESTIMATES FOR EAR TRAITS AND GRAIN YIELD IN MAIZE HYBRIDS

Sultan Mahmood[1], Saad Imran Malik[1*] and Mozamil Hussain[2]

[1]*Department of Plant Breeding and Genetics, PMAS Arid Agriculture University Rawalpindi.*

[2]*Maize, Sorghum and Millet Programme, NARC, Park Road, Islamabad.*

ABSTRACT

To increase maize yields, exploitation of heterosis has been proven as highly effective and is widely used over the world to produce new and improved hybrids. Present study was undertaken to estimate combining ability and heterosis in maize hybrid combinations for ear traits, plant height and leaf area. Five inbred parents were crossed in a diallel fashion excluding reciprocals and fifteen genotypes including 10 hybrid combinations and 5 parents were tested during summer 2013. Statistical analysis revealed significant difference for kernel rows ear^{-1} ($P \leq 0.05$), and highly significant difference ($P \leq 0.01$) for days to silking, plant height, leaf area, kernels row^{-1}, 1000-kernel weight and grain yield. Hybrid $P_3 \times P_5$ and $P_3 \times P_4$ was marked as suitable for breeding early maturing hybrids due to negative heterosis values. High heterosis for plant height was recorded for $P_2 \times P_3$ with significant SCA effects. P_2 in combination with P_3, P_4 and P_5 showed significant SCA for greater leaf area, however, a significant positive heterosis was obtained only for $P_2 \times P_3$. Maximum numbers of kernel rows ear^{-1} were counted in hybrid $P_3 \times P_4$; nevertheless, hybrid $P_1 \times P_3$ exhibited the highest MPH while $P_3 \times P_4$ was the only combination to have significant BPH. On the other hand $P_1 \times P_4$ had the highest number of kernels row^{-1}. The highest 1000-kernel weight was obtained for the hybrid $P_2 \times P_3$ with highly significant heterosis and SCA. The grain yield remained highest in $P_2 \times P_3$ (4719 kg/ha) followed by $P_4 \times P_5$ (4616 kg/ha) and $P_3 \times P_4$ (4532 kg/ha). Hence the hybrid combination $P_2 \times P_3$ was marked as superior for the development of high yielding hybrids among 10 hybrids tested considering its significant and positive SCA for yield related traits and highest grain yield.

Keywords: Maize, *Zea mays*, heterosis, combining ability, diallel.

INTRODUCTION

Maize (*Zea mays* L.) is the most produced cereal grain in the world with a total global production of 1038 M-t in 2013-14, followed by rice (740 M-t) and wheat (720 M-t) (FAOSTAT, 2014). Maize is a source of calories in many African countries serving as a staple food for over 300 million people (La Rovere *et al*., 2011). In Pakistan it is third most important cereal crop with the annual grain production of 4.7 M-t (FAO, 2014). Maize is primarily used for animal and poultry feed in Pakistan which accounts for about 60% of the total production. Remaining 25% is used in the industry and a small fraction is used for human consumption (Pioneer, 2012). Increasing industrial demands and growing poultry and livestock sector in the country has raised maize demands more than ever today.

Historically, maize hybrids were first developed in 1920s and commercial cultivation was started in 1930s (Hayes, 1946; Crow, 1998; Duvick, 2005). To increase maize yield, exploiting the phenomenon of heterosis proved to be very effective and widely used over the world to significantly boost maize yields (Shull, 1948; Guo et al., 2014; Jiang et al., 2015; Oliveira et al., 2015). Useful heterosis is the superior performance of hybrids when compared to their parents (Elisa, 2001; Ding et al., 2014) and usually economic heterosis is commercial gain in yield over existing check hybrid cultivars. Since grain yield is a complex quantitative trait, this gain is indirectly achieved by improving yield-related traits with improves ear parameters, photosynthesis efficiency and response to fertilizer application along with better resistance to insect pests and diseases (Hirel, 2001; Jiang et al., 2015). The manifestation of heterosis is linked to specific combining ability (SCA) of inbred lines described by the performance of an inbred in a specific combination and characterized by non-additive gene action. On the other hand, general combining ability (GCA) is the overall performance of an inbred line in multiple crosses with other inbreds and is related to the

Corresponding author: e-mail: saad.malik@uaar.edu.pk

additive genetic effects (Bauman, 1981; Rodrigans et al., 2001; Moterle et al., 2011; Zhang et al., 2015). Sprague and Tatum (1942) elaborated for the first time the concept of general and specific combining ability. Both GCA and SCA are important in determining the improvement or decline in hybrid traits (Vieira et al., 2009).

For developing hybrids with desirable traits information about combining ability of parents and crosses is very important (Mohammad et al., 2006; Amir-uz-zaman et al., 2010; Khan et al., 2012; Kage et al., 2013). There are a number of studies with heterosis estimates and evaluation of the performance of diverse breeding materials *per se*. Temperate inbred lines have shown more significant effects of GCA for grain yield and yield related traits (da Silva et al., 2010). Due to negative effects of combining ability for days to pollen shedding, subtropical inbred lines happened to be usefulness in developing earliness in hybrids while tropical inbred lines are reported to have significant effects of GCA for days to pollen shedding (Malik et al., 2004a and b). Diallel analysis is a reliable and well known method of determining GCA and SCA effects for inbred lines in different crossing schemes (Griffing, 1956; Garcia et al., 1999; Larish and Brewbaker, 1999; Lee and Kannenberdg, 2003).

Hybrid maize has much higher yield potential as compared to open pollinated varieties (OPVs). As in the USA and other developed countries, hybrid maize seed industry is very much developed making maize the most productive cereal per unit area of land. Such efforts by breeders in Pakistan could be very fruitful to improve maize yield in the country by testing promising inbreds in hybrid combinations, their evaluation and identification of best crosses to develop and release commercial hybrid maize varieties.

MATERIALS AND METHODS

Experimental site and plant material

Experiment was conducted at the research area of Maize, Sorghum and Millet Program, National Agricultural Research Centre, Islamabad where the seed for inbred lines was also obtained. The seed for five inbred lines was sown in the March of the year 2013 and tested in *Kharif* season during summer of the same year.

Crossing plan

Inbred lines were grown under irrigation using standard cultural practices during spring 2013. The plants were spaced 25 cm while plant rows were 75 cm apart and 5 m long. The ears were bagged before the emergence of silks to prevent cross pollination and the pollens were collected in the Kraft paper bags for crossing. The inbreds and their hybrids were sown in *Kharif* of the same year in a randomized complete block design (RCBD) with three replications for data recording on several parameters. Similar row spacing and uniform cultural practices were undertaken.

Harvesting and data recording

The ears obtained after cross fertilization were manually harvested and dried under shade for two weeks before shelling. Seeds were harvested and grown next year for evaluation including five self-pollinated parental lines and 10 hybrids obtained by diallel fashion crossing of inbred lines excluding reciprocals. Data were recorded for days to 50% silking by counting the number of days from the date of germination when silks in 50% plant emerged out of the ears for every single entry in each replication. The plant height was measure in centimeters from the base of the stem to the lowest tassel branch. The leaf area was measured according to Payne et al. (1991). Kernel rows ear^{-1} were counted by selecting five random ears from each row. Kernels row^{-1} were counted by randomly selecting five ears and counting the numbers of kernels per row. Thousand-kernel weight (g) was obtained by counting 1000-seeds and allowing them to get dry to 15% moisture content. The seeds were weighted with the help of table top weighing balance. Grain yield was calculated by the following formula:

$$\text{Grain yield (GY)} = \frac{\text{FW} \times (100 - \text{MC}) \times 0.8 \times 10{,}000}{\text{Plot area} \times 85}$$

Where grain yield is in kg/ha, FW = Fresh weight of ears/plot, MC = Percent moisture content, 0.8 is the shelling percentage, 10,000 is the area in m^2 for one hectare and 85 is the factor for grains stored at 15% moisture content.

Heterosis and combining ability analysis

Combining ability was estimated according to Griffing's Method-II (Fixed Model) (Griffing,

1956) while reciprocal crosses were excluded in this study. The data were subjected to statistical analysis using Statistix 8.1 and diallel analysis package in MSTAT-C.
Heterosis was calculated as described by Haullauer and Miranda, 1988.

$$\text{Midparent (MP)heterosis (\%)} = \frac{(\text{F1} - \text{MP})}{\text{MP}} \times 100$$

$$\text{Better parent (BP) heterosis (\%)} = \frac{\text{F1} - \text{BP}}{\text{BP}} \times 100$$

RESULTS AND DISCUSSION

Analyses of variance showed significant difference ($P\leq0.05$) for kernel rows ear^{-1} and highly significant difference ($P\leq0.01$) for days to 50% silking, plant height, leaf area, kernels row^{-1}, 1000-kernel weight and grain yield (Table 1). The difference between replications was recorded as non-significant for all parameters under study.
Similarly, mean square values from the ANOVA performed for GCA effects showed highly significant difference ($P\leq0.01$) among inbred lines for plant height, leaf area, kernels row^{-1} and grain yield while days to 50% silking and 1000-kernel weight exhibited significant difference ($P\leq0.05$) and difference for kernel rows ear^{-1} was found as non-significant (Table 2). Considering SCA effects, highly significant difference ($P\leq0.01$) was observed among hybrids for days to 50% silking, plant height, kernels row^{-1}, 1000-kernel weight and grain yield (Table 2), while the difference for leaf area and kernel rows ear^{-1} was significant at $P\leq0.05$ (Table 2)

Heterosis and combining ability effects for days to 50% silking

Days required to 50% silking by all the genotypes were significantly different (Table 1). The results revealed that among the F_1 from the crosses, $P_3{\times}P_5$ showed highly significant SCA of -4.67 for days to 50% silking and took 58 days to reach silking stage compared to its parental lines P_3 (64 days) and P_5 (67 days). This suggests that $P_3{\times}P_5$ cross is a good combination for early maturity hybrid breeding. P_3 showed negative and significant value of 0.93 for GCA effect which means this parent had a diminishing effect on days taken to 50% silking. This suggests that P_3 can be a good choice for developing early maturing hybrids. Estimation of heterosis showed that hybrid $P_3{\times}P_5$ has -17.53% MPH and -12.77% BPH values (Table 5). Hence this hybrid combination is marked as suitable for early maturity hybrid breeding. Likewise, hybrid $P_3{\times}P_4$ also exhibited -12.32% MPH and -5.23% BPH (Table 5). Considering highly significant negative SCA effects of -4.67 for $P_3{\times}P_5$ (Table 4), the combination is identified as suitable for early maturity hybrid breeding.

Heterosis and combining ability effects for plant height

Plant height varied significantly ($P\leq0.01$) among all hybrids and their parental lines (Table 1). The highest value for plant height was recorded in hybrid $P_2{\times}P_3$ (136.47 cm), while the lowest value was measured for $P_2{\times}P_2$ (74.09 cm). Statistical analysis for combining ability (both GCA and SCA) showed highly significant difference for plant height (Table 2). P_1 and P_2 gave negative values suggesting that these parents are not good combiner for increased plant height (Table 3). P_3 and P_5 both had positive and highly significant value for GCA suggesting that these two inbred parents are good combiners for increasing plant height (Table 3) which may be good to increase fodder production but might have reduced lodging resistance. Like GCA, the effects of SCA were also highly significant for most of the hybrid combinations (Table 4). Hybrid $P_2{\times}P_3$ showed the highest and significant SCA (34.53) for greater plant height which proves this combination to be the best for this trait (Table 4). Likewise, highly significant heterotic effects were seen for $P_2{\times}P_3$ with 67.37% MPH and 53.36% BPH (Table 5). Similarly, significant and positive heterosis was recorded for plant height in $P_3{\times}P_4$ (Table 5) with highly significant SCA effects of 10.88 (Table 4).

Heterosis and combining ability effects for leaf area

All genotypes showed statistically significant difference ($P\leq0.01$) for leaf area. Hybrid obtained by the cross $P_2{\times}P_3$ had the highest leaf area (244.47 cm^2) despite parental line P_2 had a lowest value of 143.44 cm^2. The GCA effects were significant for P_1 and P_5 only (Table 3) while P_2 in combination with P_3, P_4 and P_5 showed significant SCA (Table 4). Overall, a significant and positive heterosis was obtained only for $P_2{\times}P_3$ regarding leaf area with 51.87% MPH and 36.83% BPH, hence, regarded as the best hybrid for greater leaf area in the crosses tested (Table 5). Since, leaf area is directly correlated with increased biomass and grain filling with more photosynthates, the inbred combination identified could be a valuable source for yield increase in hybrid combinations.

Heterosis and combining ability effects for number of kernel rows ear^{-1}

Statistical analysis showed significant difference regarding the number of kernel rows per ear ($P \leq 0.05$) (Table 1). Maximum numbers of kernel rows (14.8) were observed in hybrid $P_3 \times P_4$. None of the parents showed significant effect of GCA (Table 2). Among all specific combinations, four hybrids had significant SCA effects for this trait. Based on these estimates, $P_1 \times P_2$ was highlighted as the best combination with significant positive SCA effects (Table 4) and highly significant 15.26% MPH (Table 5). Hybrid $P_1 \times P_3$ showed the highest value for MPH and $P_3 \times P_4$ was the only combination to have significant BPH (Table 5). These data suggest that P_1 may be a good parent choice for increasing kernel rows in maize in certain specific combinations mentioned.

Heterosis and combining ability effects for number of kernels row^{-1}

The data for number of kernels row^{-1} showed significant statistical difference when hybrids and their parents were compared (Table 1). Highest average number of kernels produced on a row was measured in the hybrid combination $P_1 \times P_4$ and was calculated as 29.0 kernels row^{-1}. Statistical analysis of GCA and SCA revealed highly significant difference for number of kernels row^{-1} (Table 2), while significant and positive GCA effects were only recorded for P_4, hence regarded as a good general combiner for this trait (Table 3). SCA effects for number of kernels row^{-1} remained positive and highly significant for four hybrids evaluated (Table 4). Hybrid $P_1 \times P_2$ gave negative value showing an undesired combination, while $P_2 \times P_3$ yielded significant and positive SCA effects (Table 4). Among 10 hybrids tested, six hybrids showed highly significant and positive MPH and only two hybrids showed significant and positive BPH (Table 5). Hybrid $P_1 \times P_3$ was marked as the best with 52.81% MPH and 49.45% BPH for this parameter (Table 5).

Heterosis and combining ability effects for 1000-kernel weight

Highly significant difference was recorded for this trait when means were compared by the ANOVA. The highest 1000-kernel weight was recorded for the hybrid $P_2 \times P_3$ (230 g) while inbred P_1 had the lowest value of 126.3 g. Highly significant but negative value for GCA was obtained only in P_1 while rest of the parental lines showed non-significant GCA effects (Table 3). Hence, P_1 was identified as a poor general combiner for 1000-kernel weight (Table 3). Results of SCA also varied for all hybrid combinations tested. Hybrid $P_2 \times P_3$ showed highly significant and positive SCA for 1000-kernel weight suggesting it as the best combination amongst 10 hybrids for this trait. On the other hand, hybrid $P_2 \times P_4$ showed negative SCA value suggesting this combination results in reduced kernel weight and ultimately grain yield (Table 4). The results for heterosis remained highly significant in certain crosses (Table 5). Out of the 10 hybrids tested, eight hybrids showed significant and positive heterosis for 1000-kernel weight. Hybrid $P_4 \times P_5$ had the highest and highly significant BPH of 41.21% (Table 5). Hybrid $P_2 \times P_3$ was designated as a cross with highly significant mid-parent and better-parent heterotic effects with 46.34% MPH and 36.09% BPH (Table 5).

Heterosis and combining ability effects for grain yield (kg/ha)

A highly significant difference for grain yield was observed in hybrid combinations as well as the inbreds under investigation ($P \leq 0.01$) (Table 1). Hybrid $P_2 \times P_3$ gave the highest estimated grain yield of 4719 kg/ha followed by $P_4 \times P_5$ (4616 kg/ha), $P_3 \times P_4$ (4532 kg/ha) and $P_1 \times P_4$ (4506. GCA for grain yield remained highly significant for P_4 and significant but negative for P_1 while other inbred lines did not show significant GCA effects (Table 3). These data suggests that P_4 is a good general combiner for higher grain yield while P_1 is a poor general combiner.

SCA effects were positive and highly significant for $P_1 \times P_4$, $P_2 \times P_3$, $P_3 \times P_4$, $P_4 \times P_5$ (Table 4). Hybrid $P_2 \times P_3$ was regarded as the best combination for high grain yield with highest 61.24% MPH and 34.31% BPH (Table 5). Specifically, this combination showed the highest and highly significant SCA (1184.72) increasing the worth of this combination for higher grain yield in hybrid combination. Significant negative SCA was exhibited by the hybrid $P_2 \times P_4$ which suggests it unsuitability for hybrid development for high grain yield (Table 4). Considering MPH, four hybrid combinations showed high heterosis, while in view of better parent heterosis, only two hybrids showed high heterotic values (Table 5). Hybrid $P_4 \times P_5$ showed the highest and significant BPH of 35.2% and 52.32% MPH (Table 5). This hybrid combination also had highly significant and positive SCA (915.98). As mentioned, hybrid $P_2 \times P_3$ due to its high heterosis, can be further used for the development of high yielding hybrids. Rest of the hybrids did not show significant heterosis so the lines in specific combinations for these hybrids cannot be regarded as useful.

The inbred line development is a tedious job with 6-7 cycles of self-pollination followed by an extensive and rigorous selection. Majority of the lines are discarded in 3rd or 4th selection cycle and only a few are carried further for SCA and GCA estimates along with the measurement of heterotic effects in cross combinations. Hence, the hybrid breeding program completely relies on the quality and performance of inbred lines *per se* which is evaluated by manual crossing and field trials for agronomic performance. There are a number of studies carried out for such purposes with different results obtained for diverse breeding materials and crosses (Larish and Brewbaker, 1999; Katna et al., 2005; Uddin et al., 2006; Devi et al., 2007; Kage et al., 2013). Different scientists obtained diversified results for the material tested, hence such studies are important for the evaluation of inbred lines in specific hybrid breeding programs and identification of those inbreds which have high and significant SCA in specific hybrid combinations and give significant heterosis over the better parent to increase maize yield in the growing demands for poultry and livestock sector of the country.

Table 1: Mean square values from the analyses of variance for grain yield and related traits of maize hybrids.

Source of variation	DF	Days to silking	Plant Height (cm)	Leaf area (cm^2)	Kernel rows ear^{-1}	Kernels rows^{-1}	1000-kernel weight (g)	Grain yield (t-ha^{-1})
Genotypes	14	18.01**	10.31**	2564**	1.81*	5.92**	2632.4**	1981.1**
Replications	2	4.5^{ns}	155.62^{ns}	6078^{ns}	0.152^{ns}	2.14^{ns}	846.0^{ns}	245.4^{ns}
Error		125.7	78.8	900.1	0.90	7.02	632.34	474.8
CV (%)		3.4	8.2	15.4	7.0	11.1	13.8	20.0

** Significant at $P \leq 0.01$, *Significant at $P \leq 0.05$, ^{ns}Non-significant

Table 2: Mean square values for GCA and SCA effects in maize inbred lines and their hybrid combinations for grain yield and related traits.

Source of variation	DF	Days to 50% silking	Plant height (cm)	Leaf area (cm^2)	Kernel rows ear^{-1}	Kernels rows^{-1}	1000 kernel weight (g)	Grain yield (kg-ha^{-1})
GCA	4	5.7*	267.2**	1321.3**	0.33^{ns}	10.52**	629.7*	653061**
SCA	9	6.1**	272.4**	667.9*	0.71*	15.16**	976.6**	663269**

** Significant at $P \leq 0.01$, *Significant at $P \leq 0.05$, ^{ns}Non-significant

Table 3: Estimated GCA effects for grain yield and related traits for each parental line.

Parents	Days to 50% silking	Plant height (cm)	Leaf area (cm^2)	Kernel rows ear^{-1}	Kernels rows^{-1}	1000-kernel weight (g)	Grain yield (t-ha^{-1})
P1	-0.44^{ns}	-6.6**	-17.96**	-0.3^{ns}	-1.41*	-16.5**	-352.41*
P2	1.04*	-6.17**	-11.31^{ns}	0.14^{ns}	0.03^{ns}	7.27^{ns}	119.96^{ns}
P3	-0.96*	5.14**	8.37^{ns}	0.14^{ns}	0.1^{ns}	1.98^{ns}	-26.41^{ns}
P4	-0.53^{ns}	1.04^{ns}	7.63^{ns}	0.18^{ns}	1.9 *	5.17^{ns}	445.97**
P5	0.9*	6.6**	13.28*	-0.16^{ns}	-0.62^{ns}	2.08^{ns}	-187.1^{ns}

** Significant at $P \leq 0.01$, *Significant at $P \leq 0.05$, ^{ns}Non-significant

Table 4: Estimated SCA effects for grain yield and related traits in hybrid combinations.

Hybrids	Days to 50% silking	Plant height (cm)	Leaf area (cm^2)	Kernel rows ear^{-1}	Kernels rows^{-1}	1000 kernel weight (g)	Grain yield (t-ha^{-1})
1 × 2	-1.33*	-6.18**	4.0^{ns}	1.17**	-1.58**	12.54^{ns}	-244.42^{ns}
1 × 3	0.67^{ns}	3.98^{ns}	4.19^{ns}	1.05**	4.54**	7.83^{ns}	254.27^{ns}
1 × 4	0.9^{ns}	12.52**	13.0^{ns}	0.47^{ns}	4.54^{ns}	19.3*	971.39**
1 × 5	0.81^{ns}	9.02**	-11.5^{ns}	-0.46^{ns}	0.0^{ns}	5.73^{ns}	-176.55^{ns}
2 × 3	-1.81**	34.53**	52.67**	-0.13^{ns}	4.83**	38.73**	1184.72**
2 × 4	0.1^{ns}	0.22^{ns}	-26.0**	-0.05^{ns}	-3.17**	-13.13*	-719.57**
2 × 5	-1.33*	4.51^{ns}	27.6**	-0.17^{ns}	-0.17^{ns}	16.97*	113.9^{ns}
3 × 4	-1.24	10.88**	10.8^{ns}	0.9**	1.29^{ns}	24.16**	671.25**
3 × 5	-4.67**	-0.84^{ns}	-1.9^{ns}	-0.23^{ns}	1.28^{ns}	10.59^{ns}	-14.47^{ns}
4 × 5	-1.1*	-2.44^{ns}	1.1^{ns}	0.59*	3.02**	35.73**	915.98**

** Significant at $P \leq 0.01$, *Significant at $P \leq 0.05$, ^{ns}Non-significant

Table 5: Mid-parent heterosis (MPH) and better parent heterosis (BPH) for grain yield and related traits.

Hybrids	Days to 50% silking		Plant height		Leaf area		Kernel rows ear^{-1}		Kernels rows^{-1}		1000-kernel weight		Grain yield	
	MPH%	BPH%	MPH%	BPH%	MPH%	BPH%	MPH%	BPH%	MPH%	BPH%	MPH%	BPH%	MPH%	BPH%
1 × 2	1.38^{ns}	-5.64^{ns}	8.98^{ns}	4.88^{ns}	14.23^{ns}	10.39^{ns}	15.26**	8.42^{ns}	1.29^{ns}	-12.74^{ns}	25.51*	9.66^{ns}	1.38^{ns}	-15.64^{ns}
1 × 3	4.9^{ns}	1.71^{ns}	24.78**	18.56*	14.01^{ns}	6.01^{ns}	16.04**	10.71^{ns}	52.81**	49.45**	29.08*	20.64^{ns}	41.91^{ns}	41.71^{ns}
1 × 4	6.81^{ns}	2.01^{ns}	25.95**	16.39*	9.13^{ns}	-5.55^{ns}	12.06*	7.18^{ns}	37.01**	16.31^{ns}	33.02*	19.25^{ns}	56.81**	32.01^{ns}
1 × 5	9.41^{ns}	2.93^{ns}	17.17*	0.85^{ns}	-2.77^{ns}	-16.47^{ns}	0.26^{ns}	-5.47^{ns}	15.24^{ns}	6.13^{ns}	24.70^{ns}	14.29^{ns}	9.41^{ns}	2.93^{ns}
2 × 3	-11.24**	-4.31*	67.37**	53.36**	51.87**	36.83*	3.52^{ns}	1.98^{ns}	36.91**	20.22*	46.34**	36.09**	61.24**	34.31*
2 × 4	-5.09^{ns}	-6.44^{ns}	16.36*	3.82^{ns}	-6.53^{ns}	-21.39^{ns}	4.79^{ns}	2.97^{ns}	-7.21^{ns}	-8.82^{ns}	10.46^{ns}	7.30^{ns}	-5.09^{ns}	-6.44^{ns}
2 × 5	3.22^{ns}	-0.74^{ns}	16.57*	-2.83^{ns}	25.43*	4.73^{ns}	-0.25^{ns}	-0.50^{ns}	3.73^{ns}	-3.60^{ns}	29.94*	23.27^{ns}	13.22^{ns}	-0.74^{ns}
3× 4	-12.32**	-5.23^{ns}	30.86**	27.07**	14.00^{ns}	5.48^{ns}	13.55*	13.27*	26.43**	9.36^{ns}	40.04**	33.89*	57.53**	32.77^{ns}
3 × 5	-17.53**	-12.77*	13.85*	2.54^{ns}	9.30^{ns}	0.30^{ns}	0.76^{ns}	-0.50^{ns}	27.27**	19.68^{ns}	32.44*	29.67*	28.84^{ns}	21.37^{ns}
4 × 5	8.84^{ns}	1.37^{ns}	5.27^{ns}	-2.59^{ns}	2.26^{ns}	1.36^{ns}	7.58^{ns}	5.97^{ns}	23.98**	13.37^{ns}	44.69**	41.21**	52.32**	35.2 *

** Significant at $P \leq 0.01$, *Significant at $P \leq 0.05$, ^{ns}Non-significant

CONCLUSION

Our results highlighted that none of a single hybrid was superior for all of traits and that hybrid superiority varied for each individual trait evaluated. The hybrid combination $P_2 \times P_3$ was marked as the best with highest heterotic effects regarding high yield among 10 hybrids tested considering its significant and positive SCA for yield related traits and highest grain yield. The cross combinations $P_3 \times P_5$ and $P_3 \times P_4$ were found suitable for the development of early maturity varieties while highest kernel rows ear^{-1} were obtained in the hybrid $P_3 \times P_4$ which is an important character having a positive correlation with grain yield. The hybrids $P_1 \times P_4$ yielded the highest number of kernels row^{-1} while $P_1 \times P_3$ combination was marked as the best for high heterotic effects for the said trait. The highest heterosis for 1000-kernel weight was obtained for $P_2 \times P_3$ along with a highly significant and positive SCA effects for this trait. The cross combinations presented in this study can be used for the improvement of specific traits highlighted in this study and further development of hybrid cultivars according to the needs of the farmers.

REFERENCES

Amir-uz-zaman M, Islam MA, Hassan L, and Rohman MM, 2010. Combining ability and heterosis for yield and yield components in maize. Acad. J. Plant Sci. 3(2): 79-84.

Bauman LF, 1981. Review of methods used by breeders to develop superior corn inbred. Proc. Corn Sorghum Ind. Res. Conf. 36: 199-209.

Crow J F, 1998. 90 years ago: The beginning of hybrid maize. Genetics 148: 923-928.

da Silva VQ, do Amaral Junior AT, Goncalves LS, Freitas Junior SP, Candido LS, Vittorazzi C, Moterle LM, Vieira RA, and Scapim CA, 2010. Combining ability of tropical and temperate inbred lines of popcorn. Genet. Mol. Res. 9(3):1742-1750.

Devi B, Barua NS, Barua PK, Talukar P, 2007. Analysis of mid-parent heterosis in a variety diallel in rainfed maize. Indian J. Genet. Plant Breed. 67(2):67-70.

Ding H, Qin C, Luo X, Li L, Chen Z, Liu H, Gao J, Lin H, Shen Y, Zhao M et al., 2014. Heterosis in early maize ear inflorescence development: a genome-wide transcription analysis for two maize inbred lines and their hybrid. Int. J. Mol. Sci. 15(8):13892-13915.

Duvick DN, 1977. Genetic rate of gain in hybrid maize yields during the past 40 years. Maydica 22: 187-196

FAO, 2014. Food and Agriculture Organization of the United Nations. (http://www.faostat3. fao.org/faostat-gateway/go/to/browse/Q/*/E]

Garcia A, Lasa JM, Alvarez A and Gaacia P, 1999. Combining ability effects for seed vigour and traits related to emergence at low temperature in maize and grain sorghum. Invest. Agr. Prod. Prot. Veg. 14(3): 383-391.

Griffing B, 1956. Concept of general and specific combining ability in relation to diallel crossing systems. Aust. J. Bio. Sci. 9: 463-456.

Guo T, Yang N, Tong H, Pan Q, Yang X, Tang J, Wang J, Li J and Yan J, 2014. Genetic basis of grain yield heterosis in an "immortalized F(2)" maize population. Theor. Appl. Genet. 127(10):2149-2158.

Haullauer AR and Miranda JB, 1988. Quantitative genetics in maize breeding. 2nd ed. Iowa State University Press, Ames, Iowa, USA. 468 pp.

Hayes, HK, 1946. Yield genes, heterosis and combining ability. Am. Nat. 80: 430-445.

Hirel B, Bertin P, Quillere I, Bourdoncle W, Attagnant C, Dellay C, et al., 2001. Towards a better understanding of the genetic and physiological basis for nitrogen use efficiency in maize. Plant Physiol. 125: 1258–1270

Jiang L, Ge M, Zhao H, and Zhang T, 2015. Analysis of heterosis and quantitative trait loci for kernel shape related traits using triple testcross population in maize. PLoS One 10(4):e0124779.

Kage U, Madalageri WMCD, Malakannavar L and Gangashetty P, 2013. Combining ability studies for yield and its related traits in newly derived inbred lines of maize (*Zea mays* L.). Mol. Plant Breed. 4: 71-76

Katana G, Singh HB, Sharma JK and Guleria SK, 2005. Heterosis and combining ability studies for yield and its related traits in maize (*Zea mays* L.). Crop Res. 30 (2): 221-226

Khan TN, Ajmal SK, Saleem R and Nawaz MH, 2012. Yield performance of single / double crosses and F_2 progenies in maize

(*Zea mays* L.). J. Agric. Res. 50(4): 469-476.

La Rovere R, Kostandini G, Abdoulaye T, Dixon J, Mwangi W, Guo Z and Bänziger M, 2011. Potential impact of investments in drought tolerant maize in Africa. CIMMYT, Addis Ababa, Ethiopia.

Larish LLB and Brewbaker JL, 1999. Diallel analyses of temperature and tropical popcorns. Maydica 44(4): 279-284

Lee, EA, Kannenberdg LW, 2003. Effects of inbreeding methods and selection criteria on inbred and hybrid performance. Maydica 49: 191-197.

Malik HN, Malik SI, Chughtai SR and Javed HI, 2004a. Estimates of heterosis among temperate, subtropical and tropical maize germplasm. Asian J. Plant Sci. 3 (1): 6-10.

Malik SI, Malik HN, Minhas NM and Munir M, 2004b. General and specific combining studies in maize diallel crosses. Int. J. Agri. Biol. 6(5): 856-859.

Moterle LM, Braccini AL, Scapim CA, Pinto RJ, Goncalves LS, do Amaral Junior AT, and Silva TR, 2011. Combining ability of tropical maize lines for seed quality and agronomic traits. Genet. Mol. Res. 10(3):2268-2278.

Mohammad S, Khatun F, Ahmed S, Ali MA, and Begum S, 2006. Heterosis and combining ability in corn (*Zea mays* L.). Bangladesh J. Bot. 35(2): 109-116.

Oliveira GE, Von Pinho EV, Andrade T, Souza JC, Caixeta F, and Ferreira RA, 2015. Relationship among physiological quality, heterosis, and amylase gene expression in maize seeds. Genet. Mol. Res. 14(3):8623-8633.

Payne, WA, Wendt CW, Hossner LR and Gates CE, 1991. Estimating pearl millet leaf area and specific leaf area. Agron. J. 83: 937-41.

Pioneer. 2012. (online: www.pioneer.com/home/site/pakistan/products/)

Rodrigans MC, Valva FD, Brasil EM and Chaves LJ, 2001. Comparison among inbreeding systems in maize. Crop Breed. Appl. Biotechnol. 1(2): 105-114.

Shull GH, 1948.What is “Heterosis”? Genetics, 33: 439-446.

Sprague GF and Tatum LA, 1942. General vs. specific combining ability in single crosses of corn. J. Amer. Soc. Agron. 34(10): 923-932.

Uddin MS, Khatun F, Ahmed S, Ali MR and Begum SA, 2006. Heterosis and combining ability in corn (*Zea mays* L.). Bangladesh J. Bot. 35(2): 109-116

Zhang X, Lv L, Lv C, Guo B, and Xu R, 2015. Combining ability of different agronomic traits and yield components in hybrid barley. PLoS One 10(6):e0126828.

MORPHOLOGICAL CHARACTERS IN BT COTTON AS INFLUENCED BY THE SPLIT APPLICATION OF NITROGEN APPLIED AT DIFFERENT GROWTH STAGES UNDER THE AGRO-CLIMATIC CONDITION OF DERA GHAZI KHAN, PAKISTAN

Shuaib Kaleem*[1], M. Shabbir[1] and M. N. Abbas[2]

[1]*Agriculture Adaptive Research Complex, Dera Ghazi Khan, Pakistan*

[2]*Department of Agronomy, College of Agriculture, Dera Ghazi Khan, Pakistan*

ABSTRACT

Nitrogen plays a significant role to improve the growth and yield of cotton. Plant growth, assimilation, morpho-physiological expressions and photosynthate translocation are closely related to nutrients availability. A field trial to evaluate the response of different nitrogen splits on growth and yield components of cotton was conducted under the agro climatic conditions of Dera Ghazi Khan, Pakistan during 2012. Fuzzy seed (@ 20 kg ha^{-1} of cotton genotype (BT-MNH 886) was sown under Randomized Complete Block Design (RCBD) with three replications. Fixed nitrogen (N) dose (375 kg ha^{-1}) which (375 kg ha^{-1}), was applied in four treatments as a source of nitrogen; T_1=5 N splits (@75.00 kg ha^{-1} per split), T_2=6 N splits (@62.50 kg ha^{-1} per split), T_3=7 N splits (@53.57.00 kg ha^{-1} per split), and T_4=8 N splits (@46.87 kg ha^{-1} per split). The data were recorded on the basis of yield and yield attributes of cotton crop that included plant height, leaf area index (LAI), number of bolls per plant, bolls weight and seed cotton yield. On increasing N splits, vegetative characteristic of cotton i.e., plant height was improved while, features like leaf area index, bolls weight and seed cotton yield increased up to certain level of N splits, T3 (7 splits). Further, N split (8 splits in T_4) caused decrease in all these features except for plant height. So the growth and yield performance of cotton genotype (BT-886) was significantly influenced up to 7 splits of nitrogen fertilizer, improving morphological characteristics of cotton crop ultimately enhancing the seed cotton yield.

Keywords: Cotton genotype, nitrogen splits, growth and yield components

INTRODUCTION

Cotton (*Gossypium hirsutum* L.) is an important fiber crop used in textile industry (Killi et al., 2005). Cotton crop called the white gold in true sense serves as a backbone of Pakistan economy and occupies a prominent position in local textile and edible oil industry (Nadeem et al., 2010; Sawan et al., 2001). In Pakistan, cotton yield is comparatively lower than other countries which can be due to lack of advanced production practices, crop husbandry, market problems, low land holding, illiteracy, lack of capital and poor socio-economic conditions of the farming community. Poor soil fertility management resulting from imbalanced and non-judicial fertilizer application may be another reason for lower seed cotton yields (Bibi et al., 2011).

It is a big source of raw material for textile industries, spindles and oil expelling units all over the world (Ahmed et al., 2009). Nitrogen significantly enhances crop growth, chlorophyll contents and yield of cotton (Saleem et al., 2010). It has been concluded that rate of plant growth and development is mainly temperature driven, but the amount of growth and photosynthate translocation is related to nutrients availability (Munir et al., 2012).

Among plant nutrients, nitrogen plays an important role in crop productivity and is regarded as growth and yield determinant in irrigated cotton (Ahmad, 2000; Reddy et al., 1998). Cotton yield and fiber quality is associated with advanced agronomic field techniques and judicious use of soil inputs (Reddy et al., 2004). Similarly, with the increase of photosynthetic area, crop assimilation and growth is increased which ultimately increases photosynthates and crop yield (Kaleem et al., 2011).

Nitrogen application to cotton is considered to be essential to meet the basic requirements of nitrogen needed at different growth stages throughout, however excessive amount of nitrogen may reduce lint percentage (Sampathkumar et al., 2006, Amanullah et al., 2013). Nitrogen deficiency also slows down the vegetative and reproductive growth in plants which may leads to potential yield loss (Ali and

**Corresponding author:* e-mail: shuaibkaleem70@gmail.com

Hameed, 2011). Excess and wrong time application of nitrogen causes more vegetative growth that delayed crop and fruit maturity and resulted in low final crop yield (Kandil et al., 2004). Judicious increase in nitrogen to cotton applied at proper growth stage may result in more accumulation of photosynthates, higher fruit weight and nutrients uptake from soil (Anjum et al., 2007). Nitrogen has been reported to increase plant height, monopodial/ sympodial branches, boll weight and bolls per plant and ultimately seed cotton yield (Sawan et al., 2001; Nadeem et al., 2010, Saleem et al., 2010). Cotton plants require nitrogen throughout the growing season but the magnitude varies with the age of plant. Due to extensive cropping system our soils are deficient in organic matter. Nitrogen application interval must not be more than 17 days especially at critical crop growth stage otherwise crop growth will suffer badly affecting plant health, exposure and yield. As regards nitrogen application, cotton experts recommend that 28 percent of the nitrogen requirements for cotton plant are during the first 60 days of growth. The maximum absorption of nitrogen takes place during the first 60-90 days when boll formation is at its peak (Sawan et al., 2001). While, nitrogen application from 90 to 120 days of crop growth is very critical as it is stage of physiological maturity, keeping in view the development of boll size (Hussain et al., 2000; Subhan et al., 2001). Thus its application must be completed within speculated period by completing its dose through splits of normal days and applied at important crop growth stage from germination till boll formation (Sampathkumar et al., 2006).

The climatic conditions in Pakistan are severe, especially in cotton growing area of D.G. Khan where high temperature quickly volatilizes nitrogen, if applied on improper stage by improper way, similarly arid conditions badly affect applied nitrogen, so split application interval for nitrogen was kept with the view to reduce the climatic effect. In D. G. Khan district, farmers mostly apply nitrogen at improper crop stage and no proper interval for nitrogen application is kept in view, ultimately succulent crop with uncontrolled vegetative growth is achieved which attracts sucking insects and bollworms making crop unprofitable and uneconomical. This study was therefore conducted to examine the effect of split application of nitrogen at different growth stages of cotton to find out the best suit to get good crop growth, efficient nutrient use and profitable yield.

MATERIALS AND METHODS

A field experiments was conducted to study the influence of split application of nitrogen fertilizer on the growth and yield of cotton crop under the agro climatic condition of Dera Ghazi Khan, Pakistan. 20 kg ha^{-1} fuzzy seed of cotton (BT-MNH 886) was used for the experiment. Nitrogen fertilizer was applied in four different treatments (T_1, T_2, T_3 and T_4) and each treatment comprised of different N splits (T_1=5, T_2-6, T_3=7 and T_4=8 N splits). Method of application/distribution of Nitrogen (N) was as T_1=5 N splits (@75.00 kg ha^{-1} per split), T_2=6 N splits ((@62.50 kg ha^{-1} per split), T_3=7 N splits (@53.57.00 kg ha^{-1} per split), and T_4=8 N splits (@46.87 kg ha^{-1} per split). Urea was used as a source of nitrogen fertilizer and 375 kg N ha^{-1} was splitted in plots as per treatment according to schedule (Table 3a,3b) at different crop growth stages. All the P and K in the form of DAP and Sulphate of Potash (SOP) were added at the time of sowing in all the plots as per recommendations. Net plot size was maintained as 202.43 m^2. Ridges were made 75 cm apart while, 22 cm plant to plant distance was maintained. The soil analysis report was obtained before sowing (Table 1). The experiment was laid out in Randomized Complete Block Design (RCBD) with three replications. Crop was sown manually on 15th May, 2012 by using ridge technology. Pendimethyline was used as pre-emergence weedicide in wattar condition @ 2.5L ha^{-1}. Thinning was done after three weeks of dibbling the seed to maintain the required distance between the plants. All other agronomic practices were kept same for all the treatments throughout the crop growth. Seed cotton was picked in two pickings while, the second picking was done after 180 days of crop plantation. Five healthy plants (free from disease and damage) were selected from each plot of all the treatments and were tagged to calculate the morphological and yield data throughout the season. Plant samples were taken by using per unit area and different growth (plant height and leaf area index) and yield (number of bolls per plant, bolls weight and seed cotton yield) parameters were observed and recorded for data.

Data thus recorded were statistically analyzed by using Fisher's analysis of variance techniques (ANOVA) and least significant difference test (LSD) was applied to check the significance of treatment means at 5% probability level (Steel and Torrie, 1997).

Table – 1: Physio-chemical characteristics of soil at Adaptive Research Zone DG Khan, 2012.

Electrical conductivity	3.14 m cm^{-1}
Soil pH	8.10
Organic matter	0.28 %
Available phosphorous	4.00 ppm
Available potassium	215.00 ppm
Saturation	44%
Texture	Loam

Table – 2: Meterological data of the Adaptive Research Zone DG Khan, 2012.

Month	Temperature Min. (^{0}C)	Temperature Max. (^{0}C)	R.H (%)	Rain (mm)
June-2012	31.00	41.00	49.50	30 .00
July-2012	30.00	40.00	63.00	37.00
Aug-2012	26.00	35.00	68.00	40.00
Sep-2012	29.00	37.00	75.00	130.00
Oct-2012	20.00	33.20	62.00	---
Nov-2012	13.80	27.90	75.00	---
Dec-2012	09.70	22.50	70.00	---

Table – 3a: Schedule of split application of nitrogen fertilizer (Dates) to cotton crop

Total splits*	Interval days	Dates for nitrogen (N) application							
T_1=5 N splits	26	15-06-12	12-07-12	08-08-12	04-09-12	30-09-12	-	-	-
T_1=6 N splits	20.8	15-06-12	07-07-12	29-07-12	19-08-12	09-09-12	30-09-12	-	-
T_1=7 N splits	17.33	15-06-12	03-07-12	21-07-12	08-08-12	26-08-12	13-09-12	30-09-12	-
T_1=8 N splits	14.86	15-06-12	01-07-12	16-07-12	31-07-12	14-08-12	29-08-12	14-09-12	30-09-12

*It may be noted that quantity of nitrogen remained fixed (**375kg ha^{-1}**). The difference among treatments was variation in no. of splits of nitrogen fertilizer

Table – 3b: Schedule of split application of nitrogen fertilizer per split (kg $ha-^1$) to cotton crop

Total splits*	Interval days	Quantity of nitrogen application ha^{-1} per split (kg)							
T_1=5 N splits	26	75.00	75.00	75.00	75.00	75.00	-	-	-
T_1=6 N splits	20.8	62.50	62.50	62.50	62.50	62.50	62.50	-	-
T_1=7 N splits	17.33	53.57	53.57	53.57	53.57	53.57	53.57	53.57	-
T_1=8 N splits	14.86	46.87	46.87	46.87	46.87	46.87	46.87	46.87	46.87

Table 4. Effect of split nitrogen (N) application on the growth and yield of BT-cotton

S. No.	Treatment	Plant height (cm)	Leaf area index (LAI)	No. of bolls plant^{-1}	Bolls weight (g)	Seed cotton yield (Kg ha^{-1})
1	T_1= 05 'N' splits	101.13 b	2.66 c	17.45	2.41 c	2654.30 c
2	T_2= 06 'N' splits	104.77 b	2.88 b	21.47	2.47 b	3121.10 ab
3	T_3=07 'N' splits	112.00 a	3.12 a	19.38	2.55 a	3250.00 a
4	T_4= 08 'N' splits	116.22 a	2.42 d	21.31	2.44 bc	2927.00 b
Mean		**108.53**	**2.77**	**19.90 NS**	**2.47**	**2988.09**
LSD value		**5.00**	**0.14**	**6.71**	**0.04**	**285.72**

RESULTS AND DISCUSSION

Plant height (cm)

Plant height is an important vegetative growth parameter of cotton plant that is directly influenced by fertilizer especially nitrogen. The results obtained at final maturity (Table 4) indicated increasing trend of plant height with increasing number of nitrogen splits and statistically significant differences were achieved among the treatments. N fertilizer significantly ($p \leq 0.05$) and gradually improved the height of cotton plant. Maximum plant height (116.22 cm) was recorded in T_4 where (375 kg ha^{-1}) nitrogen was applied in 8 equal splits which remained statistically at par with T_3 (112.00 cm) where same nitrogen dose was applied in 7 equal splits. Minimum plant height (101.13 cm) was recorded in T_1 which was also statistically similar with T_2 (104.77 cm) where 5 and 6 splits were given respectively (Table 4). Results revealed that plant height increased with the increase in number of N splits which indicated that nitrogen even applied at latter growth stage, helped to increase plant height and vegetative growth. Our results are in close conformity with the results of Ali et al. (2010), Sawan et al. (2001) and Munir et al. (2012) those found that lesser the nitrogen application during crop life cycle, lesser is plant growth and height.

Leaf area index (LAI)

LAI, the ratio of LA (m^2) per m^2 ground, is a constitutive parameter of many yield predicting models since it is related with radiation interception (Tsialtas and Mslaris, 2008). According to the results given in Table 4, LAI of the cotton plant was also significantly ($p \leq 0.05$) influenced by application of N in splits. Maximum LAI (3.12) was recorded in T_3 and minimum (2.42) in T_4 where nitrogen was applied in equal 7 and 8 splits, respectively (Table 4). This increase in leaf area index in T_3 might be due to increase in leaf area and biomass which increased the solar intercepted area and chlorophyll contents by split application of nitrogen up to physiological maturity (7 splits) thus ultimately increasing the assimilation and photosynthetic area causing increase in plant vegetative growth and LAI which could not be further increased by increasing extra N splits in T_4. After the physiological maturity, crop shifts to maturity where already most of the assimilates and photosynthates have been translocated from vegetative to reproductive organs, affecting leaf growth, area and size. At this stage leaves start to shed being older. So last nitrogen split application (8th split of nitrogen at maturity) could not enhance LAI at this growth stage. Amanullah et al. (2013) found in an experiment that nitrogen application at latter crop growth stages badly affected leaf characteristics like number of leaves plant^{-1} (NLPP), leaf length (LL), mean single leaf area (MSLA), leaf area plant-1 (LAPP), leaf elongation rate (LER), leaf area index (LAI), specific leaf area (SLA), and leaf area ratio (LAR)] due the toxic effects of excessive Nitrogen at this particular crop growth stage. Similar findings have also been noticed by Sampathkumar et al. (2006), Sawan et al. (2001) and Kaleem et al. (2011) who found that addition of nitrogen up to certain crop growth stages enhanced net assimilation area, chlorophyll contents and leaf area which failed to increase these parameters if applied at latter crop growth stages.

Number of bolls plant^{-1}

The results obtained at final maturity (Table 4) indicated statistically non-significant differences among the treatment means. Production and fruit retention of a crop is dependent on balanced supply of the available nutrients at proper crop growth stage. Influence of nitrogen splits on number of bolls was although non-significant ($p \leq 0.05$) but results in Table 4 depict that application of N up to certain stage remained fruitful. Similar findings were reported by Sawan et al. (2001), Nadeem et al. (2010), Bednarz et al. (2005) and Nichol et al. (2004). Bolls per plant observed in present study were also in agreement with those obtained by Ali and El-Sayed (2001) who achieved maximum fruit count where N was applied @ 95 to 190 kg ha^{-1} applied up to certain crop growth stags.

Boll's weight (g)

According to the results in Table 4, statistical differences were achieved regarding boll's weight. Split application of nitrogen significantly ($p \leq 0.05$) influenced bolls weight of the cotton. Results showed that maximum boll weight (2.55 g) was achieved in T_3 (7 splits) while, minimum (2.41 g) was obtained in T_1 (5 splits) that was statistically at par with T_4 (8 splits) and T_2 (6 splits). Increase in bolls size by the addition of N up to certain crop growth stage indicated that the influence of particular nutrient with crop through enhancing net assimilation area, chlorophyll content, photosynthetic efficiency and leaf area which ultimately increased assimilate translocatin towards bolls size attainment. These results are in close conformity with the findings of Arain et al. (2001), Soomro et al. (2001), Kaleem et al. (2011) and Munir et al. (2012) who recorded maximum size of cotton bolls when nutrients especially nitrogen was applied in 6-7 equal splits up to certain crop growth stage, facilitating and improving plant growth and photosynthetic activity ultimately improving yield components.

Seed cotton yield (kg ha^{-1})

Results shown in Table 4 depicted that the split application of nitrogenous fertilizer significantly ($p \leq 0.05$) influenced the seed cotton yield. Maximum seed cotton yield (3250.00 kg ha^{-1}) was recorded in T_3 where nitrogen was applied in 7 splits which were statistically at par with T_2 (3121.10 kg ha^{-1}) (6 splits) and T_4 (2927.00 kg ha^{-1}) (8 splits). Similarly, minimum seed cotton yield (2654.30 kg ha^{-1}) was recorded in T_1 (5 splits) as depicted in Table 4. These results are in accordance with the findings of Reddy et al. (2004), Sampathkumar et al. (2006), Sawan et al. (2001), Nadeem et al. (2010) and Munir et al. (2012) those reported that cotton yield and fiber quality were associated with the timely and judicious use of fertilizer inputs. Amanullah et al. (2013) also reported in an experiment that nitrogen application at reproductive stage of the crop helps a lot in photosynthate accumulation and enhancing yield than applied at both crop stages, too earlier and latter crop growth stages. Similar results were reported by Ali et al. (2009) who found that yield was directly proportional to the timely application of the nutrients to the crops.

CONCLUSION

It may be concluded that the optimum dose of nitrogen (53.57 kg ha^{-1}) per split, applied in 7 splits per hectare at suitable crop growth stages (applied from 15.6.2012 to 30.9.2012 with 17.33 days interval of nitrogen application, applied in 7 equal splits) is pre requisite to achieve the maximum growth, vigor and biomass which improved the physiological functions of cotton plant, necessary for vigorous plant growth and higher seed cotton yield. Thus nitrogen fertilizer dose and timing of application is of prime importance. Furthermore, this dose of nitrogen, applied in 7 splits for the cultivar BT-886 may be recommended among farming community to promote for cultivation in DG Khan zone, Pakistan, to improve fertilizer use efficiency and to achieve the better seed cotton yield.

REFERENCES

Ahmad N, 2000. Fertilizer scenario in Pakistan policies and development. Proc. Conf. Agric. Fert. Use. 2010 NFDC, P&D Div. Govt. of Pakistan, February 15-16, 1999.

Ahmed AUH, Ali R, Zamir SI and Mehmood N, 2009. Growth, yield and quality performance of cotton cultivar BH-160

(*Gossypium hirsutum* L.). J. Ani. Plant Sci. 19(4):189-192.

Ali SA and El-Sayed AE, 2001. Effect of sowing dates and nitrogen levels on growth, earliness and yield of Egyptian cotton cultivar Giza 88. Egypt. J. Agric. Res. 79: 221-232.

Ali H and Hameed R, 2011. Growth, yield and yield components of American cotton (*Gossypium hirsutum* L.) as affected by cultivars and Nitrogen fertilizer. Int. J. Sci. & Eng. Res. 2(7): 1-13.

Ali H, Afzal MN Ahmad S and Muhammad D, 2009. Effect of cultivars and sowing dates on yield and quality of *Gossypium hirsutum* L. crop. J. Food, Agric. Environ. 7(3&4): 244-247.

Ali H, Afzal MN, Ahmad S, Muhammad D, Husnain Z, Perveen R and Kazmi MH, 2010. Quantitative and qualitative traits of *Gossypium hirsutum* L. as affected by agronomic practices. J Food, Agric. Environ. 8 (3&4): 945-948.

Amanullah, Hidatullah, Amanullah J and Stewart BA, 2013. Growth dynamics and leaf characteristics in oats differ at excessive nitrogen and phosphorus application. Pak. J. Bot. 45(3): 853-863.

Anjum FH, Tanveer A, Ahmad R, Ali A, Nadeem MA and Tahir M, 2007. Respose of cotton to split application of nitrogen and weed control methods. Ind. J. Agri. Sci. 77 (4): 224-9.

Arain AS, Soomoro AW and Khan RD, 2001. Cotton response to split application of nitrogen fertilizer. Online J. Bio. Sci. 1: 3-4.

Bednarz GW, Shurley WD, Anthony WSt and Nichols RL, 2005. Yield, quality and profitability of cotton produced at varying plant densities. Agron. J. 97: 235-240.

Bibi Z, Khan NU, Mussarat M and Shahee S, 2011. Response of *Gossypium hirsutum* L. genotypes to various nitrogen levels. Pak. J. Bot. 43(5): 2403-2409.

Hussain SZ, Faird S, Anwar M, Gill MI and Baugh MD, 2000. Effect of plant density and nitrogen on the yield of seed cotton-variety CIM- 443. Sarhad J. Agric. 16: 143-147.

Kaleem S, Hassan FU, Mahmood I, Ahmad M, Ullah R and Ahmad M, 2011. Response sunflower to environmental disparity. Nature and Sci. 9(2): 73-81.

Kandil AA, Bawai MA, EL-Moursy SA and Abdou UMA, 2004. Effect of planting dates, nitrogen levels and bio-fertilization treatments on 1: Growth attributes of Sugar Beet (*Beta vulgaris* L.). Scientific J. King Faisal Univ. 5(2): 1425.

Killi F, Efe L and Mustafayev S, 2005. Genetic and environmental variability in yield, yield components and lint quality traits of cotton. Int. J. Agric. Biol. 6: 1007-1010.

Munir A, Kaleem S, Qayyum A, Ahmad M and Abbas MN, 2012. Assimilate utilization I wheat crop as influenced by varying nitrogen levels in rainfed area. Life Sci. Int. J. 6 (4): 2659-2662.

Nadeem MA, Ali A, Tahir M, Naeem M, Chadhar AR and Ahmad S, 2010. Effect of nitrogen levels and plant spacing on growth and yield of cotton. Pak. J. life Soc. Sci. 8(2): 121-124.

Nichols SP, Snipes CE and Jones MA, 2004. Cotton growth, lint yield and fiber quality as affected by row spacing and cultivar. J. Cotton Sci., 8:1-12.

Reddy KR, Koti S, Davidson GH and Reddy VR, 2004. Interactive effects of CO_2 and nitrogen nutrition on cotton growth, development, yield and fiber quality. Agron. J., 96:1148-1157.

Reddy KR and Hodges HF, 1998. Photosynthesis and environment factors. Proc. Beltwide Cotton Conf., San Diego, CA. 5-9 Jan. 1998, Natl. cotton conf. USA: 1443-1449.

Saleem MF, Bilal MF, Awais M, Shahid MQ and Anjum S A, 2010. Effect of nitrogen on seed cotton yield and fiber qualities of cotton (*Gossypium hirsutum* L.) cultivars. J. Ani. Plant Sci. 20(1): 23-27.

Sampathkumar T, Krishnasamy S, Ramesh S, Prabukumar G and Gobi R, 2006. Growth, nutrient uptake and seed cotton yield of summer cotton as influenced by drip, surface irrigation methods and mulching practices. Research J. Agri. Biol. Sci., 2 (6): 420-422.

Sawan ZM, Hafeez SA and Basyony AE, 2001. Effect of nitrogen fertilization and foliar application of plant growth retardants and zinc on cottonseed, protein and oil yields and oil properties of cotton. J. Agron. Crop Sci.186: 183 - 191.

Soomro WA, Arain AS, Soomro AR,Tunio GH and Magsi MR, 2001. Evaluation of proper fertilizer application for higher cotton production in Sindh. The Pak. Cotton. 41: 6-10.

Steel RGD and Torrie JH, 1997. Principles and procedure of statistics. A biometric approach, 3rd edition, McGraw Hill, Inc. Tokyo.

Subhan M, Khan HU and Ahmed RO, 2001. Population analysis of some agronomic and technological characteristics of upland cotton (*Gossypium hirsutum* L.). Pakistan J. Biol. Sci. 1: 120-123.

Tsialtas JT and Mslaris N, 2008. Nitrogen fertilization effect on leaf morphology and evaluation of LA and LAI prediction models in sugar beet. Photosynthetica, 46 (3):346-350.

EFFICACY OF DIFFERENT WEEDS AND AGRICULTURAL SUBSTRATES FOR CULTIVATION OF *PLEUROTUS OSTREATUS* (JACQ. FR.)

Waqar Islam*[1], Asif Riaz[2], Madiha Zaynab[3] and Wu Zujian[1]

[1]*College of Plant Protection, Fujian Agriculture & Forestry University, Fuzhou, Fujian 350002, China*

[2]*University of Agriculture, Faisalabad, Pakistan*

[3]*College of Life Sciences, Fujian Agriculture & Forestry University, Fuzhou, Fujian 350002, China*

ABSTRACT

Oyster mushrooms are fleshy edible fungi which acquire huge importance due to their nutritional and medicinal properties. These mushrooms are cultivated upon various substrates which directly affect the yield and time taken to reach the maturity. So the present experiment was designed to study different weeds, agricultural wastes and their mixtures for efficacy regarding time taken to achieve 100% mycelial growth and yield attributes. Our results illustrated that cotton waste showed better performance as substrate as it took minimum days i.e. 20.5 followed by *Chenopodium album* and mixture of cotton waste + *Chenopodium album* which relatively taken 24.5 days to reach maximum growth level. These substrates also provided better yield as compared to all other substrates by recording 650.36, 583.41 and 530.43 grams yield respectively. The performance of all other substrates was below power. Hence we conclude that *Chenopodium album* anonymously or mixture of particular weed and cotton waste can be utilized for efficient cultivation of mushrooms.

Keywords: *Oyster mushroom, Yield, Maturity, Effectiveness, P1 white strain*

INTRODUCTION

Mushrooms are edible and fleshy fungi which are cultivated upon organic substrates (Ashraf et al., 2013). Due to their medicinal and nutritive properties, they acquire a very important place in human food chain (Etich et al., 2013). Their importance is continuously increasing day by day as they acquire valuable vitamins, proteins and minerals. All the edible mushrooms exhibit vitamin A, B1, B5, B6, C and D (Manzi et al., 2001; Syed et al., 2009). Stanley (2011) has reported that *Pleurotus oestreatus* (Oyster mushroom) incorporates 2-5% fats, 7-38%, 8-12% minerals, mycocellulose, 17-47% sugars and 25-50% proteins.

Various agricultural substrates such as paddy straw, vegetable residues, maize stalks and cotton waste are utilized for cultivation of oyster mushroom (Hassan et al., 2011). The growth of the mushroom is considered to be dependent upon the performance of the substrates (Iqbal et al., 2005; Kimenju et al., 2009). But the substrates which have high levels of nitrogen and carbohydrate contents are categorized as ideal for mushroom growth (Khare et al., 2010). In developing countries recycling and management of the organic wastes has become a challenge but these can be efficiently utilized in cultivation of oyster mushroom which will ultimately reduce the malnutrition problems in these countries and will also reduce the pollution (Eswaran and Ramabadran, 2000). Oyster mushroom has many industrial and medical uses as they possess anti oxidant, anti microbial, anti hypersensitive, anti inflammatory, anti tumor and food additive properties (Chang, 2007).

Pakistan being agricultural country posses 70% of its population in villages and all this population are directly or indirectly involved in cultivation of different agricultural crops (Noman et al., 2015), which lead to the production of huge agricultural waste (Anonymous, 2001). Sarwar et al (2002) estimated that Pakistan produces around 11.3 and 3.2 million tons of wheat and paddy straw per year. All this can be efficiently utilized in mushroom cultivation. China and Zimbabwe are considered as the largest producer of oyster mushroom and they use bag and tray method for its cultivation (Oei, 2003). Even after the availability of huge amount of agricultural waste, Pakistan has not accelerated the production of mushrooms which may be due to lack of knowledge and awareness (Helrich, 1990). So there is an urgent need to aware the people regarding mushroom cultivation and evaluation of different substrates for the better and efficient production of mushrooms.

**Corresponding author:* e-mail: ddoapsial@yahoo.com

Keeping in view the importance of substrates in cultivation of oyster mushrooms (P1-white strain), we tried to evaluate different weeds, agricultural substrates and their relative mixtures to identify their efficacy.

MATERIALS AND METHODS

1. Collection and execution of experiment

Weeds (*Amaranthus viridis, Chenopodium album, Trianthem portulacastrum*) and agricultural bi-products (Wheat straw, rice straw and cotton waste) were collected from industrial area of Faisalabad and agronomic research area of university of agriculture, Faisalabad. The experiment was laid out by following completely randomized design with three replications in mushroom experimental room at university of agriculture Faisalabad, Pakistan.

2. Preparation of substrates

Substrates were prepared in the following concentrations and treatments. Where T1= *Amaranthus virdis* (100%), T2= *Chenopodium album* (100%), T3= *Trianthem portulacastrum* (100%), T4= *Amaranthus virdis* (50%) + wheat straw (50%), T5= *Trianthem portulacastrum* + rice straw (50%), T6= *Chenopodium album* (50%) + cotton waste (50%) and T7= cotton waste (100%). For this process, all the weeds plants were dried under sunlight, chopped into small (1-2cm) pieces, weighed upon electronic scale and soaked for overnight. Drying out the excessive water, all the weeds were spread upon the clean floor surface. 65% moisture level was achieved through spraying of water and $CaCO_3$ @ 2% was added for enhancing the fusion of substrates for 24 hours upon the floor surface. After removing any gluts, the substrates were placed into the polypropylene bags (7 x 11 cm) according to above mentioned treatments. All the bags were closed via tying the rubber bands and were autoclaved for an hour for sterilization.

3. Spawning of bags and controlled conditions

For all the replications and treatments, spawning of polypropylene bags were done specifically at 7% dry weight and each bag was incorporated with 56 grams of spawn and incubation of these bags was done at 25^{0}C in complete darkness. Required temperature was achieved though heating via electrical heater and data was recorded till 100% growth was achieved in all bags. Similarly required humidity i.e. 80-90% was maintained through sprinkling of water upon floor surface. Furthermore, required moisture for bags was achieved by sprinkling the water on the bags thrice a day. Ventilation was maintained by operating the exhaust fan 3-4 times a day for air flushing and fulfilling the oxygen requirements for fructification of mushrooms.

4. Data recording

The data regarding growth of mycelium upon substrates in polypropylene bags was recorded upon completion of 25%, 50%, 75% and 100% mycelial growth. The data for harvesting of mushrooms were recorded till 3rd flush where first harvesting was done at maturity level while data was recorded for all the harvesting levels. All the data was subjected to analysis of variance where P= 0.05 (Steel et al., 1997).

RESULTS

(A) Days taken to achieve targeted growth.

(i) Days taken to achieve 25% mycelial growth

Achievement of 25% mycelial growth was significantly affected by all the treatments (Fig, 1). Great variation was observed regarding performance of different substrates and all the substrates exhibited different mean no. of days to attain the attributed target. But among all cotton waste (T7) proven relatively better and it taken only 6.5 days to reach 25% mycelial growth which was minimum time in comparison to all other substrates. Secondly, the performance of *Chenopodium album* (T2) was a little better as required mycelial growth was achieved in 9.5 days. Afterwards, comparative response was observed regarding T1 (*Amaranths viridis*) and T6 (*Chenopodium album* 50% + cotton waste 50%) which recorded mean days as 12 and 11.5 respectively. Response of all other treatments (T3, T4, T5) was poor as they exhibited more than 14 days to complete the 25% growth of *Pleurotus ostreatus*.

(ii) Days taken to achieve 50% mycelial growth

Comparison of means exhibited that T7 (cotton waste) which initially proven as a better substrate was again recorded better as 50% of mycelial growth was achieved in 12.5 days. But interestingly, a similar performance was recorded by T2 (*Chenopodium album*) and T6 (*Chenopodium album* 50% + cotton waste 50%) as both taken 16.5 days to reach the recorded growth (Fig, 2). The statistical data further revealed that T1 (*Amaranthus virdis*) relatively taken 18.5 days while rest of the treatments (T3, T4, T5) were again categorized poor as they recorded more than 21 days to reach the same level.

(iii) Days taken to achieve 75% mycelial growth

Fig 3 revealed that once again T7 continued to take the lead and 75% mycelial growth was achieved in just 17 days while interestingly the T2, T1 and T6 recorded results with were closed to each other by achieving the targeted growth in 20.5, 21.5 and 22 days respectively. Similarly T3, T4 and T5 continued their poor performance and these substrates taken more than 26 days to achieve the same target.

(iv) Days taken to achieve 100% mycelial growth

Comparison of means for the final attribute again exhibited that cotton waste (T7) which initially recorded better results achieved 100% mycelial growth in only 20.5 days followed by T2, T6 and T1 which were comparative to each other and recorded 24.5, 24.5 and 25.5 days simultaneously. Here again it was proved that T3, T4 and T5 were poor substrates and they relatively taken too long time i.e. > 30 days to achieve 100% growth of *Pleurotus ostreatus* (Fig 4).

(B) Yield performance

Fig, 5 extensively reveals that all the substrates exhibited significant results regarding yield of *Pleurotus ostreatus.* Statistical data and comparison of means clarify that after calculating the yield of all the three flushes, cotton waste (T7) performed extensively well by recording the yield as 650.36 grams followed by T2 (*Chenopodium album*) and T6 (*Chenopodium album* 50% + cotton waste 50%) which given 583.41 grams and 530.43 grams respectively. Rest of the treatments (T1, T4, T3 and T5) relatively showed least yield i.e. 500.37, 464.42, 429.42 and 319.44 grams.

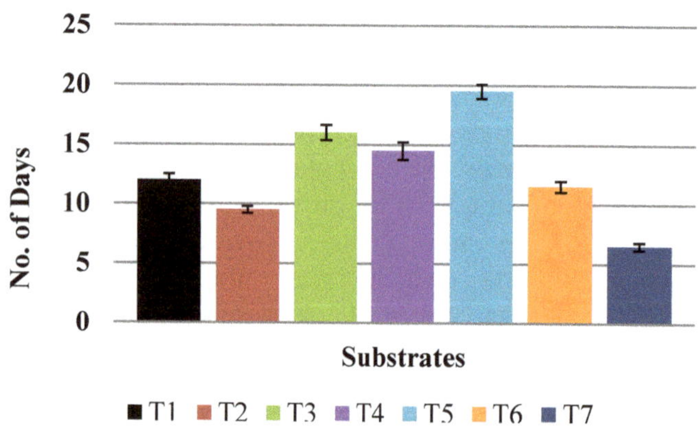

Fig, 1) Mean No. of days taken to complete 25% mycelial growth on different weed plants along with combination of agricultural waste materials

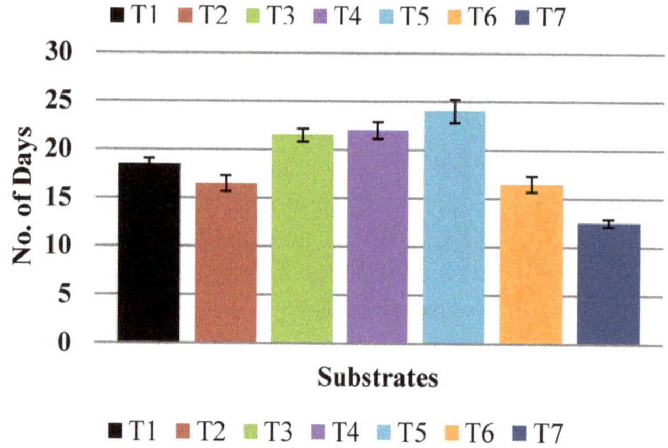

Fig, 2) Mean No. of days taken to complete 50% mycelial growth on different weed plants along with combination of agricultural waste materials.

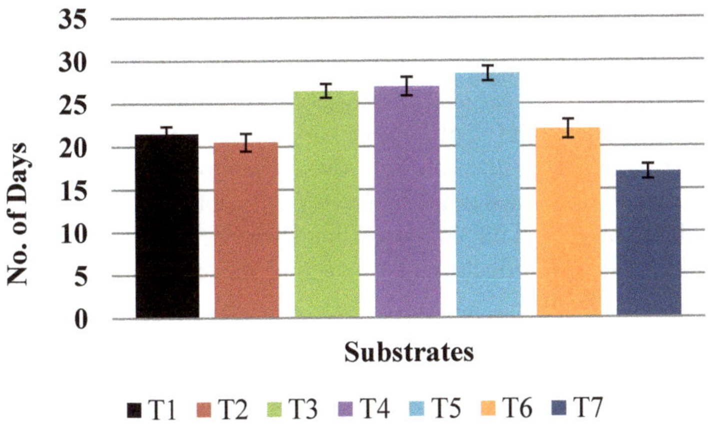

Fig, 3) Mean No. of days taken to complete 75% mycelial growth on different weed plants along with combination of agricultural waste materials.

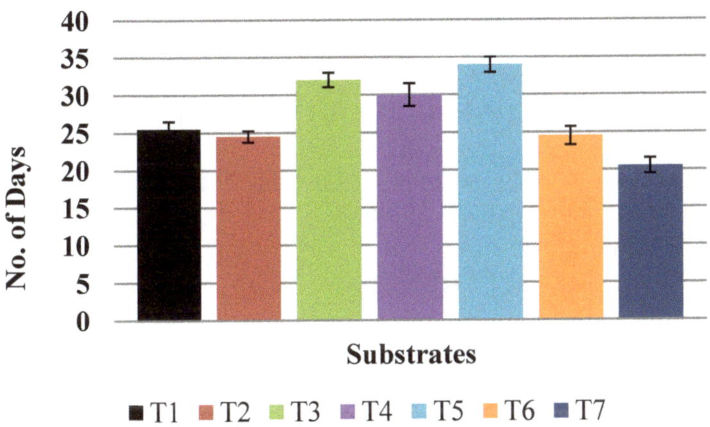

Fig, 4) Mean No. of days taken to complete 100% mycelial growth on different weed plants along with combination of agricultural waste materials.

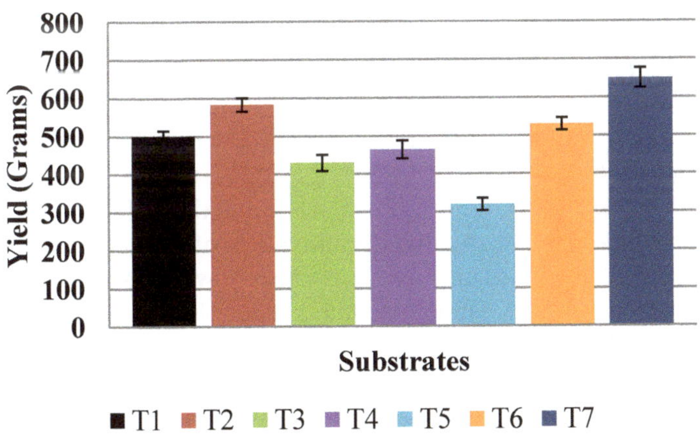

Fig, 5) Total yield of three flushes (grams) on different weed plants along with combination of agricultural waste materials

DISCUSSION

Different substrates and their mixtures can be utilized for the better cultivation of oyster mushrooms which may include various crop residues, weeds, agricultural wastes and supplements. These substrates directly affect the time frame to attain the maximum mycelial growth and also take part in the Yield attribute for oyster mushrooms. Previous reports of better yield of oyster mushrooms via utilization of different agricultural substrates attracted our mind towards it (Mendez et al., 2005; Kalmis et al., 2008; Onyango et al., 2011). Various products, bi-products, weeds and agricultural wastes which are full of nutrients can act as bio stimulants for micro organism growth such as mushrooms (Zafar et al., 2016). The organic products may include agricultural substrates, weeds and their mixtures for their effect upon time taken to achieve 100% mycelial growth and yield of the oyster mushrooms. Our results illustrated that cotton substrates relatively proven better for both the attributes as they recorded the highest yield (650.36grams) in three flushes and only taken 20.5 days to achieve the maximum mycelial growth. These results are in-line with the findings of Khan et al. (2010) who achieved a yield of 844 grams in four flushes of *Pleurotus ostreatus florida spp.* Our experiment also included the mixture of cotton waste and *Chenopodium album* substrate which also recorded significant results by giving 583.41 grams yield and completed the targeted growth in 24.5 days. Once again the findings were collaborated with Khan et al. (2010) who documented 795 grams yield of local oyster specie via using mixture of cotton waste with other substrates. Similarly improved performance of cotton waste regarding better yield of oyster mushrooms have been proved coordinating with previous reports (Fan et al., 2006; Kimenju et al., 2009). We also utilized mixture of wheat straw and *Amaranthus virdis* and observed 464.42 gram yield in three flushes and taken 30 days to reach 100% growth level. Our results are nearing to Dundar (2008) who cultivated different local and exotic species of *Pleurotus* on wheat stalk substrate and explained that the substrate took around two month to achieve maximum growth regarding all the species. Therefore our findings show that using the mixture of wheat straw along with weed (*Amaranthus virdis*) literally improved the performance of substrate. In our studies, we also incorporated the mixture of rice straw and another weed (*Trianthem portulacastrum*) and found that it took 34 days for maximum maturity and given 319.44 grams yield. Our findings approximately match with Baysal (2003) who investigated paper waste supplemented with rice husk as substrate and exhibited 350.2 grams yield of *Pleurotus ostreatus*. Recent studies of Mintesnot et al (2013) include the cultivation of different species of oyster mushrooms via utilization of various weed substrates. He recorded a total yield of 840 grams upon weed (*Parthenium hysterophorus*) substrate. Similarly, Kholoud et al., 2014 utilized date palm leaves and several other agricultural wastes efficiently to cultivate oyster mushrooms which also strengthen our results.

CONCLUSION

The results explain that the selection of better substrate not only increase the yield of the *Pleurotus ostreatus* (P1-white strain) but also reduces the time of maturity and gives early harvesting. Our results explain that cotton waste or the mixture of cotton waste and *Chenopodium album* are better substrates for the cultivation of oyster mushroom. The utilization of these weeds and agriculture wastes as substrates will not only enhance the yield of oyster mushrooms but also help in reduction of weeds and wastes.

ACKNOWLEDGMENTS

We are thankful to Professor Dr. Nasir Ali Khan from University of Agriculture Faisalabad for his value able suggestions during execution of experiment.

CONFLICTS OF INTERESTS

The authors declare that they don't have any conflicts of interest.

REFERENCES

Anonymous, 2001. Agriculture in Pakistan, Online available from http: // www. pakistaneconomist.com/ issue 2001 / issue37/ i&e 4.

Ashraf J, Ali MA, Ahmad W, Ayyub CM and Shafi J, 2013. Effect of Different Substrate Supplements on Oyster Mushroom (Pleurotus spp.) Production. Food Sci Tech. 1(3):44-51.

Baysal E, Peker H, Yalinkilic MK and Temiz A, 2003. Cultivation of Oyster Mushroom on waste paper with some added supplementary materials. Biores Tech. 89: 95-97.

Chang ST, 2007. Mushroom cultivation using the "ZERI" principle: potential for application in Brazil. Micol Aplic Int.19 (2): 33-34.

Dundar A, Acay H and Yildiz A, 2008. Yield performances and nutritional contents of three oyster mushroom species cultivated on wheat stalk, African Journal of Biotechnology. 19(7): 3497-3501.

Eswaran A and Ramabadran R, 2000. Studies on some physiological, cultural and post harvest aspects of oyster mushroom, *Pleurotus ostreatus*. Trop Agri Res J. 12: 360-374.

Etich OK, Nyamangyoku OI, Rono OI, Niyokuri JJ and Izamuhaye AN, 2013. Relative performance of Oyster Mushroom

(*Pleurotus florida*) on agroindustrial and agricultural substrate. Int J Agron Pl Prod. 4(1): 109-116.

Fan LS, Pandey A, Vandenberghe A and Soccol CR. 2006. Effect of caffeine and tannins on cultivation and fructification of *Pleurotus* on coffee husks. Braz J Microbiol. 37 (4): 420-424.

Hassan S, Mohammad AY and Kiramat K, 2011. Cultivation of the oyster mushroom (*Pleurotus ostreatus* (Jacq.) P. Kumm.) in two different agroecological zones of Pakistan. Afr J Biotech. 10: 183-188.

Helrich KC,1990. Official methods of Analysis of the AOAC. Association of Official Analytical Chemists Inc., 2: 15.

Iqbal SM, Rauf CA and Sheikh MI, 2005. Yield performance of oyster mushroom on different substrate. Int J Agri Biol. 7 (6): 900-903.

Kalmis E, Azhar N, Yildiz H and Kalyonus F, 2008. Feasibility of using olive mill effluent (OME) as a wetting agent during the cultivation of oyster mushroom, *Pleurotus ostreatus*, on wheat straw. Biores Tech. 99: 164-169.

Kimenju JW, Odero GOM, Mutitu EW, Wachira PM, Narla RD and Muiru WM, 2009. Suitability of locally available substrates for oyster mushroom cultivated in Kenya. Asian J Pl Sci. 6 (2): 648-652.

Khan NA, Rehman HU, Rehman A and Arif J, 2010. Screening of different exotic strains of oyster mushroom for yield production using cotton waste and combined with rice husk. Pak J Phytopathol. 22(1): 40-44.

Khare KB, Mutuku JM, Achwania OS and Otaye DO, 2010. Production of two oyster mushrooms, *Pleurotus sajor-caju* and *P. florida* on supplemented and un-supplemented substrates. Int J Agri Appl Scie. 6: 4-11.

Kholoud MA, Nahla AB and Nadia SA, 2014. Cultivation of oyster mushroom *Pleurotus ostreatus* on date-palm leaves mixed with other agro-wastes in Saudi Arabia. Sau J Biol Sci. 21(6): 616–625.

Manzi P, Aguzzi A and Pizzoferrato L, 2001. Nutritional value of mushrooms widely consumed in Italy. Food Chem. 73:321.

Mendez LA, Castro CAS, Casoo RB and Leal CMC, 2005. Effect of substrate and harvest on the amino acid profile of Oyster mushroom (*Pleurotus ostreatus*), J Food Comp Ana. 18: 447-450.

Mintesnot B, Ayalew A and Kebede A, 2013. Evaluation of biomass of some invasive weed species as substrate for oyster mushroom (Pleurotus spp.) cultivation. Pak J Biol Sci. 17(2): 213 9.

Noman A, Ali S, Naheed F, Ali Q, Fareed M, Rizwan M and Irshad MK, 2015. Foliar application of ascorbate enhances the physiological and biochemical attributes of maize (Zea mays L.) cultivars under drought stress. Arch Agron Soil Sci. 1: 1659-1672.

Oei P, 2003. Mushroom cultivation-Appropriate Technology for Mushroom Growers 3rd Ed. Backhuys Publishers, Netherland.

Onyango BO, Palapala VA, Arama PF, Wagai SO and Gichimu BM, 2011. Suitability of selected supplemented substrates for cultivation of Kenyan native wood ear mushrooms (*Auricularia auricula*). American J Food Tech. 6: 395–403.

Sarwar M, Khan MA and Iqbal Z, 2002. Feed Resources for Livestock in Pakistan. Int J Agric Biol.4(1): 186-192.

Stanley RP. 2011. Enumerative combinatorics, Cambridge university press. pp: 49.

Steel RGD and Torrie JH, 1997. Principles and procedures of statistics: A Biometrics Approach, 2nd ed. McGraw-Hill, New York.

Syed AA, Kadam JA, Mane VP, Patil SS and Baig MMV, 2009. Biological efficiency and nutritional contents of *Pleurotus florida* (Mont.) Singer cultivated on different Agro-wastes. Nat Scie. 7(1): 44-48.

Zafar S, Ashraf MY, Anwar S, Ali Q, Noman A, 2016. Yield enhancement in wheat by soil and foliar fertilization of K and Zn under saline environment. Soil Env. 35 (1): 46-55.

GENETIC DIVERSITY OF COMMON BEANS AS DETERMINED USING MORPHOLOGICAL MARKERS

M. E. Morojele*[1], D. N. M. Mbewe[2]

[1]*Small Grain Institute, Private Bag X29. Bethlehem. 9700. South Africa.*

[2]*The University of Zambia, Lusaka. Zambia*

ABSTRACT

Different cultivars of common bean are grown in Lesotho for home consumption originating from Zambia, South Africa and America. These have created a wide genetic diversity and duplication such that it is not easy to distinguish them. The study was conducted in Lusaka to distinguish the same common bean cultivars using morphological markers and to estimate their degree of similarity. A collection of 42 cultivars were included in the study. Seventeen morphological characters were used following International Plant Genetic Resource Unit descriptor (1982). Data collected were analysed using principal component and cluster analysis. Principal component analysis was used to identify the characters which caused major variation among cultivars. Out of 10 principal components generated from 17 characters, only the first three components which constituted 54.57% of the total variation were considered for analysis. The first, second and third components accounted for 23.23%, 16.80% and 14.54%, respectively. The characters responsible for separation along the first principal component and loadings (parenthesis) were plant height (0.57), growth habit (0.55) and seed pattern (-0.27).The characters influencing separation along the second principal component include number of flowers per node (0.50), number of locules per pod (0.44), seed colour (0.44) and leaflet length (0.30). Along the third principal component, cultivars were separated according to the pod colour (0.64) and flower colour (0.14).Thirty-five individual cultivars and two groups of cultivars were distinguished by cluster analysis. One group consisted of three cultivars from Zambia, while the other group consisted of two cultivars from Zambia and two from Lesotho.

Keywords: Common bean, morphological markers, principal component analysis, cluster analysis

INTRODUCTION

Common bean (*Phaseolus vulgaris* L.) is a leguminous food crop grown world-wide and consumed as grain, green pods or fresh vegetable. It is a rich source of protein affordable by many people and grown in almost every part of the world (Romero-Arenas et al., 2013). Moreover, the nutritional value and variable uses of dry beans in various forms appeal to both processors and consumers alike (Blair et al., 2010). According to botanical, archaeological and molecular markers, common beans originated from two gene-pools, namely; Mesoamerican and Andeans (Raggi, 2013). The third gene-pool was later discovered in the northern Andes (Debouk, 1999; Tohme et al., 1996). The route of dissemination from their origin and domestication spread to Europe, and then Africa where these were grown under varying environmental and agronomic situations (Chacon et al., 2005; Wortmann et al., 2004). These domesticated common beans further underwent evolution resulting in seven sub-groups, three in the Andean and four in the Mesoamerican (Perseguini et al., 2011; Singh et al., 1991b). In some regions, farmers have maintained common bean landraces. However, in most cases traditional cultivars have progressively been replaced with hybrids that give a high yield, resistance to diseases and pests, and most importantly income to meet farmer's needs. As a result, farmers use landraces, traditional cultivars and hybrid seeds. Similarly, a large number of different landraces, cultivars and hybrids are grown in Lesotho at both house-hold and national levels resulting in a wide genetic diversity which is not easily distinguished due to close genetic relationship. New varieties have been imported from South Africa, Southern African Development Community countries and United States of America to meet country requirements such as high yield, short maturity and high protein

**Corresponding author:* e-mail: morojelee@arc.agric.za

content. Some of these imported cultivars may be duplication; therefore it was imperative to identify all cultivars used in the country. Morphological markers have been used for a long time by farmers, agronomists and plant breeders as an effective method of identifying cultivars (Stoilova et al., 2013). Plant parts such as leaves, seeds, flowers and stems are good indications of variations.

In Lesotho, Malawi, Mozambique, Tanzania and Zambia, no attempt has been made to characterize common beans and estimate the degree of relationship among the cultivars (Wortmann et al., 2004). Nonetheless, in Bulgaria (Stoilova et al., 2013), Greece (Mavromatis, 2010), Italy (Bonnetti et al., 1995), Latin America (Singh et al., (1991) and Slovenia (Sustar-Vozlic and Maras, 2006), common bean cultivars have been characterized using morphological characters. The study attempts to provide a model that could be used to distinguish common bean cultivars and other crop species grown in Lesotho. The objective of the study was therefore to distinguish common bean cultivars using morphological markers and to estimate their degree of similarity.

MATERIALS AND METHODS

Site description

The study was conducted at the School of Agricultural Sciences Field Station, University of Zambia, Lusaka. The Field Station is located at 1140m above sea-level, at latitude 28^0 20'E and longitude 15^0 22' S. The soil is fine, loamy mixed iso-hyperthermic typic paleustalf. Soil analysis was carried out and the following results were obtained; pH ($CaCl_2$) 6.6, nitrogen 0.10%, phosphorus 26.18mg/kg, potassium 9.8mg/kg, 8.88mg/kg, calcium 14.20me/100mg and iron 9.28mg/kg.

Seed bed preparation

The land was prepared with rotavator which pulverized the soil to a fine tilt to facilitate germination, and the seed bed was levelled using hand-hoes. A compound fertilizer containing 10% nitrogen, 20% phosphorus, 10% potassium and 10% sulphur was applied as a basal dressing on the seed –bed at the rate of 300kg ha^{-1} (Welling, 1988). No top-dressing was applied.

Experimental design

Randomized Complete Block Design was used to lay-out the plots. The land was divided into forty-two plots, each measuring 2.4m x 1.2 m. Within each plot, there were four rows with inter-row and intra-row spacing of 60cm x 10cm, respectively resulting in a plant population of 166 000 plants ha^{-1}. Weeding was done twice to eradicate nutsedge (*Cyperus esculentum* L.) which was a serious problem at the site. Blister bettles (*Mylabris cinchorii*) were observed feeding on the flower at flowering stage. These were controlled with endosulfan mixed at a ratio of 15ml to 20 l of water.

Data collection

The descriptor of *Phaseolus vulgaris* L. documented by International Board of Plant Genetic Resource (1982) was used to characterize the cultivars. In the plots with four rows, the two middle rows were chosen for data collection. Five plants were randomly selected and tagged. All data were collected from the tagged plants. The following characters were recorded; height of stem from the soil surface to the tip of the apical meristem, length of the terminal leaflet on the third trifoliate leaf, growth habit, colour of the stem, colour of the flower, days from emergence to 50% flowering, number of flower buds per node, length of pod, shape of pods, locules per pod, cross-section, colour of the pod, curvature of the pod, pod suture strings, seed marginal colour, seed coat pattern, dark colour of the seed coat and seed shape.

Data analysis

A Genstat software package was used to generate multivariate analysis. Two analyses were performed, namely; (i) principal component analysis to reveal the pattern of morphological variation within the common bean cultivars, and (ii) cluster analysis to group phenotypically similar cultivars and separate dissimilar ones.

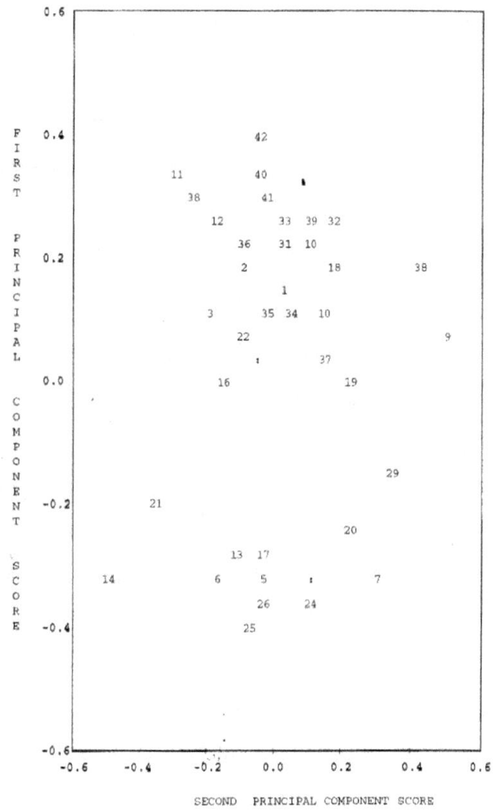

Fig. 1. Positions of Phaseolus vulgaris. L. genotypes on the first and second principal component scores based on morphological markers.

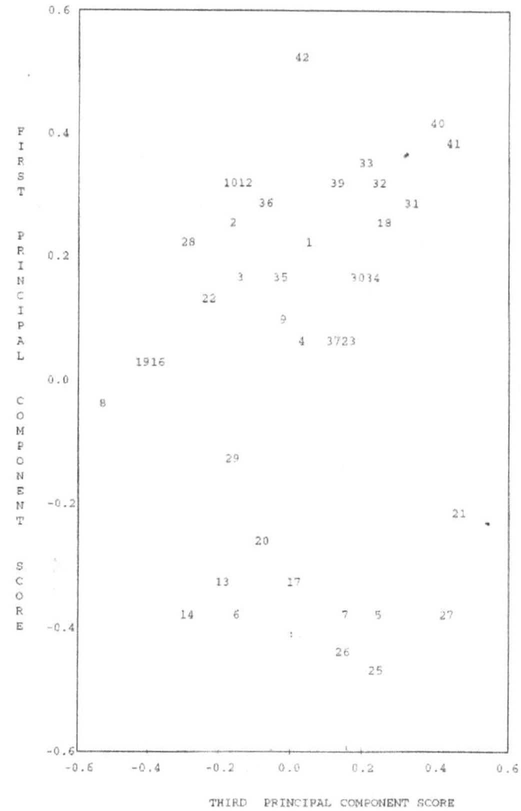

Fig. 2. Positions of Phaseolus vulgaris. L. genotypes on the first and third principal component scores based on morphological markers.

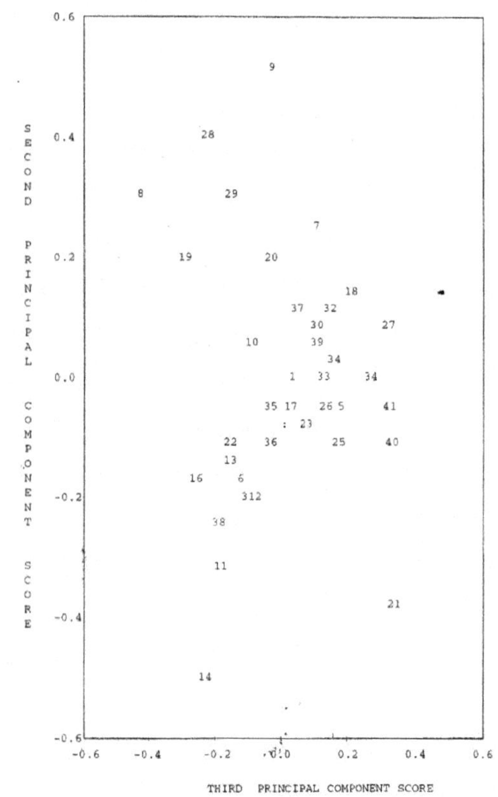

Fig. 3. Positions of Phaseolus vulgaris. L. genotypes on the second and third principal component scores based on morphological markers.

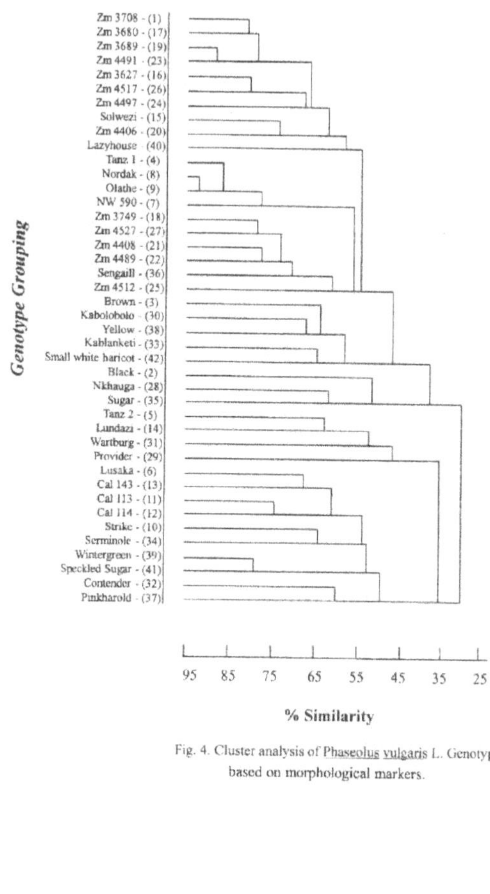

Fig. 4. Cluster analysis of Phaseolus vulgaris L. Genotypes based on morphological markers.

RESULTS

Principal component analysis

Principal component analysis was used to identify the characters which caused major variation among cultivars. Out of 10 principal components generated from 17 characters, only the first three components which constituted 54.57% of the total variation were considered for analysis. The first, second and third components accounted for 23.23%, 16.80% and 14.54%, respectively. The characters responsible for separation along the first principal component and loadings (parenthesis) were plant height (0.57), growth habit (0.55) and seed pattern (-0.270). The characters influencing separation along the second principal component include number of flowers per node (0.50), number of locules per pod (0.44), seed colour (0.44) and leaflet length (0.30). Along the third principal component, cultivars were separated according to the pod colour (0.64) and flower (0.14).

By means of the first and second principal component score, two major groups of cultivars according to growth habit and plant height emerged, namely; semi-climber and bushy types. However, there were some outliers such as Provider, Nordak, Olathe, Nkhaunya, Zm 4408, Lundazi and an extreme such as Small White Haricots. Semi-climbers consisted of 23 cultivars, most (79%) of which were obtained in Zambia (Fig.1).

The first and third principal component scores exhibited cultivars which were clustered into two groups, except outliers such as Small White Haricot, Nordak, Provider and Zm 4408. The clustering of cultivars indicated that there was a very low variability in pod colour, plant height and growth habit (Fig.2).

The second and third principal component scores produced one cluster of cultivars indicating that most of the cultivars shared similar number of flowers per node, pod colours and number of locules per pod, and variation in the characters is low. However, there were some outliers such as Lundazi, Zm 4408, Cal 113, Olathe, Nkhaunya, Nordak, Provider and NW 590 (Fig.3).

Cluster analysis

Cluster analysis was performed following principal component analysis to further examine dissimilarities and similarities among 42 cultivars collected from the three countries (Fig.4). When a cluster tree was cut at 85% level of similarity, 35 individual cultivars and 2 groups were segregated. The first group consisted of Zm 4491 and Zm 3689 which were both from Zambia. This group shared common characters such as green stem, semi-climbers, 29 days to reach flowering, purple flowers, slightly curved pods, 3 buds per inflorescence, pear-shaped cross-section, shiny green pods, moderate suture strings, medium brilliance, cuboid seed shape and no seed coat pattern. The second group consisted of Tanz1, Olathe and Nordak which shared common characters such as green stem, semi-climbing growth habit, 24 days to reach flowering, white flowers, 4 buds per inflorescence, slightly curved pods, pear-shaped cross-section, red stripes on green colour pods, moderate pod suture strings, striped seed coat pattern with dark and light brown colour, medium seed brilliance and cuboid seed shape. Tanz1 was obtained from Zambia, while Olathe and Nordak were from Lesotho.

DISCUSSION

Seventeen characters used as morphological characters were adequate to discriminate cultivars. Different combinations of these 17 characters enabled the cultivars to be differentiated, while no single character distinguished one cultivars from the other. A combination of four or more characters, for example growth habit, seed coat pattern, light and dark colour of the seed, resulted in some cultivars being distinguished. The results were consistent with the findings of Figliuolo and Spagnoletti (2000) who distinguished 57 common bean cultivars and discovered that no one character can discriminate a cultivar. Similarly, Awan et al. (2014) characterized thirteen cultivars of common bean grown in Pakistan and revealed distinguishing morphological characters that led to separation of cultivars. The importance of morphological markers in identifying cultivars is well-documented (Stoilova et al., 2013; Marzooghian et al., 2013; Berova and Stoilova, 2009). One or two characters were able to group and sub-group cultivars but these were dependent on their discriminatory power. All characters applied in this study were found to have a perceptible influence on the segregation of cultivars, although their discriminatory power differed. However, certain combinations of characters did not contribute to further sub-division of groups. In this regard, it must be emphasized that large number of cultivars share similar character such as green stems, climbing growth habit, white flowers and cross-section of the pods. Dividing the main groups into sub-groups on different characters, led to the segregation of many individual cultivars, for example, when the cultivars were grouped according to their growth habits and plant height, followed by further sub-division of groups by means of seed coat pattern and colour, cultivars were distinguished. Previous researchers also followed this procedure and found it effective (Okhii et al., 2014); Stoilova et al., 2013; Singh et al., 1991). Seed characteristics, particularly seed coat pattern and colour contributed to the clear-cut distinction between cultivars obtained from different countries occurred and this shows potential to provide a quick method of differentiating between cultivars.

The segregation of cultivars by cluster analysis was based on the seed characteristics. The three cultivars, Olathe, Nordak and Tanz1 were grouped together, while the rest were segregated. These three cultivars shared similar seed characteristics and were classified as pinto beans according to the American and Canadian classification method (Mavromatis et al., 2010). This group of bean probably consisted of duplicates or progenies from the same ancestors which did not undergo much evolution and as a result, were morphologically similar. Hornakova et al., (2003) used cluster analysis to determine degree of similarity among 82 accessions of common bean and found two main groups with sub-groups and 14 smallest sub-groupings. Awan (2014) obtained three groups from 13 common bean genotypes, each group having differing number of cultivars when analysing using dendrogram.

The second group, Zm 4491 and Zm 3689, was also formed on the basis of similarities in seed coat characteristics. The similarities in this group may be attributed to the fact that Zambia accessions were composed of different seed coat colours which were grown together over a long period of time. Consequently, cross-pollination had occurred, resulting in cultivars which were phenotypically similar. Bornnetti et al. (1995) and Roy (2001) reported that the cultivars which were morphologically similar had a close genetic relationship. Contrarily, Singh et al. (1991) argued that the morpho-agronomic characters were phenotypic traits and accessions may be similar morphologically, yet be distant genetically.

CONCLUSION

Morphological characters were able to distinguish thirty-five individual cultivars and two groups of cultivars. Plant height, growth habit, seed pattern, number of flowers per node, number of locules per pod, seed colour, leaflet length, pod colour and flower colour had a perceptible influence on segregation of cultivars.

ACKNOWLEDGEMENTS

The authors would like to thank British Council for sponsoring the research project and ensuring to become a success. The University of Zambia provided facilities and manpower, therefore, gratitude are extended to it.

REFERENCES

Awan FK, Khurshid YM and Afzal O, 2014. Agro-morphological evaluation of some exotic common bean genotypes under rainfed conditions of Islamabad, Pakistan. Pak. J. Bot. 46(1): 259 – 264.

Blair MW, Gonzalez LF, Kimani PM and Butare L, 2010. Genetic diversity, inter-gene pool introgression and nutritional quality of common beans from Central Africa. Theor. Appl. Genet. 121(2): 237 – 248.

Bonnetti A, Miggiano A and Lovato A, 1995.Identification of beans (*Phaseolus vulgaris* L.) Cultivars grown in Italy by field and electrophoresis tests: a comparative study. Seed Sci. Technol. 23:123-146.

Chacon MI, Pickersgill SB and Debouck DG, 2005. Domestication pattern in common bean and the origin of the Mesoamerican and Andean cultivated races. Theor. Appl. Genet. 110:432-444.

Debouck DG, 1999. Diversity in *Phaseolus* species in relation to the common bean, In: S.P. Singh (ED). Common bean improvement in the twenty-first century. Kluwer Academic Publication. Dordrecht. The Netherlands.

Figliuolo G and Spagnoletti PL, 2000. A nested analysis to detect relationships between genetic markers and germplasm classes of durum wheat. Plant Genet. Res. Newslett. 22: 24 – 50.

Hornakova O, Zavodna M, Zakova M, Zakova MA, Kraic J and Debre F, 2003. Diversity of common bean landraces collected in the Western and Eastern Carpatien. Czech. J. Genet. Plant Breed. 39(3): 73-83.

International Plant Genetic Resource Unit, 1982. The descriptors of common beans. IPGRU, Rome, Italy.

Mavromatis AG, Arvanitoyannis I S, Korkovelas AE, Giakountis A, Chatzi-theodorou VA. and Goulas CK, 2010. Genetic diversity of common bean, Greek landraces and commercial cultivars. Span. J. Agric. Res. 8(4) 986-994.

Okii D, Tukamuhabwa P, Odong T, Namayanja J, Paparu P and Gepts P, 2014. Morphological diversity of tropical common bean germplasm. Afr. Crop. Sci. J. 22(1): 59-67.

Perseguini CKJ, Chiorado FA and Zucchi MI, 2011. Genetic diversity in cultivated common Beans based on molecular analysis. Genet. Mol. Biol. 34(1): 88-102.

Poehlman JM, 1987. Breeding field crops. Van Nostrand Reinhold, New York, USA.

Raggi L, 2013. Italian common bean landraces: diversity and population structure. Genet. Res. Crop. Evol. 60: 1515 – 1530.

Romero-Arenas O, Damian Huato MA, Rivera Tapia JA, Baez Simon A, Huerta-Lara M and Cabrera-Huerta E, 2013. The nutritional value of Beans and its importance for feeding of rural communities in Puebla-Mexico. Int. Res. J. Biol. Sci. 21((8): 59–65.

Roy D, 2001. Plant Breeding: Analysis and exploitation of variation. Alpha Science International, Pangbourne, India.

Singh PS, Gutierrez JA, Molina A, Urrea C and Gepts P, 1991b. Genetic diversity in cultivated Common bean: Marker-based analysis of morphological and agronomic traits. Crop Sci. 31: 23-29.

Singh PS, 2001. Broadening the genetic base of common bean cultivars. Crop Sci. 41: 1659-1667.

Soura E and Sorrels S, 1991. Relationships among 70 North American Oat Germplasm 1. Cluster analysis using quantitative characters. Crop Sci. 31: 599-605.

Stoilova T, Pereira G, Tavares-de-Sousa M, 2013. Morphological characterization of a small common bean (*Phaseolus vulgaris* L.) collection under different environments. J. Cent. Eur. Agri. 14(3): 854-864.

Tohme J, Gonzalez DO, Beebe S and Duque MC, 1996. AFLP analysis of gene-pools of a wild bean core collection. Crop Sci. 36:1375-1384.

Wortmann SC, Kirkby RA, Ledu CA and Allen D, 2004. Atlas of Common Bean production in Africa. CIAT, Colombia, Latin America. pp129.

INFLUENCE OF ORGANIC MANURING ON GROWTH AND YIELD OF STRAWBERRY Cv. CHANDLER

Muhammad Younas [1] and Rifat Ali [2]*

[1]Department of Horticulture, [2]Department of Plant Pathology,
The University of Agriculture, Peshawar-Pakistan

ABSTRACT

The research work was conducted at Newly Developmental Farm (NDF), Horticulture Section, The University of Agriculture Peshawar, Pakistan during 2012. The objective was to find out the optimum dose of Farm yard manure and poultry manure for the growth and yield of Strawberry plants and to find out the survival rate of cultivar Chandler of Strawberry during the frost of the fall season. In this study, organic matter, farmyard manure and poultry manure were used for percent plant survival, plant height (cm), number of branches plant^{-1} and number of leaves plant^{-1} of Strawberry Cv. Chandler where the greater percentage of plant survival, plant height, number of branches and leaves were obtained in plants treated with farmyard manure compared to poultry manure.

Keywords: *Frageria ananassa* Duch, Chandler, Farmyard Manure (FYM), Poultry manure

INTRODUCTION

Strawberry (*Fragaria* x *ananassa* Duch.) is one of the most important fruits of the world which has attained a premier position in the world fruit market not only as fresh fruit but also in the processing industries (Sharma and Sharma, 2003). There are more than 20 described species and many hybrids and cultivars of strawberry. The most common strawberries grown commercially are cultivars of the garden strawberry. Strawberries are not true berries. The fleshy and edible part of the fruit is receptacle and the parts that are sometimes mistakenly called "seeds" are achens (E-Flora, 2008).

The fruit is appreciated for its characteristic aroma, bright red color, juicy texture and sweetness. The flowers are white and grow in clusters on thin stalks. In addition to the seeds produced by these flowering plants, strawberries are spread by stolons as they get older. Nutritionally strawberries are considered to be good low caloric and are a potential source of vitamin C and fibers. It has maximum vitamin C content. It is also considered as a good source of minerals, iron and sodium. A mature fruit contains 89.9% water (Anonymous, 2011)

The modern Garden strawberry is believed to have been first cultivated in the late 13th century. The strawberry culture is mostly suitable for the farmers having small land holdings, as it fetch maximum economic returns. Nowadays strawberries are grown in many parts of the world including Japan, USA, Mexico, Italy and Germany. In Pakistan the work on the strawberry culture was started in 1986 at NARC, Islamabad. Where it produced Only 274 tons strawberry on 78 hectares land during the year 2008 to 2009 (GOP, 2009). Also Korona and Tuft have been identified as the two best varieties for cultivation in the Malakand division (Ayaz *et al*., 1998).

Strawberry is a cool season crop. It grows in hilly areas especially in Swat and Murree with little tendency to strawberry cultivation in plain areas of Peshawar, where it is cultivated in November. Flowering occurs in April and early May. It continues till late May-June. Fruits should be consumed just after picking for home consumption because it is highly perishable. The average temperature in May-June is about 30- 35°C, during which it is difficult for the farmers to shift strawberries into the market because the chances of deterioration are more during transportation (Amin, 1996).

Among the various factors which contribute towards the growth and yield of strawberry, nutrition is the important term of crop production (Umar *et al.*, 2008).

Integrated nutrient management includes the use of inorganic, organic and microbial sources of nutrients which ensure balanced nutrient proportion by enhancing nutrient response efficiency and maximizing crop productivity of desired quality.

**Corresponding author:* e-mail: rifatali644@yahoo.com

At present, synthetic chemicals contribute a lot in fulfilling the nutrient requirement but their regular, excessive and unbalanced use may lead to health, polluting our environment i.e. water, land and air resources, depletion of physico-chemical properties of the soil and ultimately poor yields. Hence application of organic manures like FYM and poultry manure to soil not only improve soil physical properties, pH, water holding capacity but also add important nutrients to the soil, thus increase the nutrient availability and its ultimate absorption by plant. Keeping this in view, a study was conducted to analyze the effect of FYM and Poultry manure on growth, yield of strawberry cv. Chandler.

MATERIALS AND METHODS

An experiment was conducted in Newly Developmental Farm (NDF), Horticulture Section, The University of Agriculture Peshawar, Pakistan. Three plots were prepared in the field. Runners of strawberry plants were taken from Swat and uniform size were transplanted on ridges at a spacing of 15 x 30 cm in first week of October during the year of experimentation. Organic fertilizer was applied to the two plots while one plot was remained without fertilizer (Controlled).The plots were treated with the following organic fertilizers:

Plot no.1 was treated with Farmyard manure (FYM).
Plot no.2 was treated with Poultry manure.
Plot no.3 was remained without fertilizer (control).

Parameters Studied

Data was recorded on the following growth parameters:

Percent plant survival: The percent plant survival was calculated at the end of the experiment by using the formula:

$$\% \text{ Survival} = \frac{\text{No. of plants survived}}{\text{Total No. of plants}} \times 100$$

Plant height: The average height of the plants from the soil to the tip of the plant was measured with a scale and average was calculated.

Number of branches plant^{-1}: The total number of branches plant^{-1} was calculated in randomly selected plants and average was calculated.

Number of leaves plant^{-1}: The total number of leaves plant^{-1} was calculated in randomly selected plants and average was calculated.

Table 1. Organic fertilizers affects the Percent plant survival, Plant height (cm), Number of branches plant^{-1} and Number of leaves plant^{-1} of Strawberry Cv. Chandler:

Treatments	Percent Survival (%)	Plant height (cm)	No. of branches plant^{-1}	No. of leaves plant^{-1}
FYM	92%	10.44 cm	6.4	18.8
Poultry Manure	86.5%	9.9 cm	5.6	16.8
Control	75%	8.5 cm	4.5	13.5

RESULTS AND DISCUSSIONS

Percent plants survival

The data recording Percent plant survival (%) is presented in the Table 1. The data indicated maximum percentage of plants survived (92%), in FYM treated plants, followed by (86.5%) in the plants treated by poultry manure. The minimum percentage of plants survival was recorded in the 3rd plot i.e. (75%) in control treatment.

The maximum percentage of plants survival in FYM treated plots might be due to sufficient nutrients supply which helped in good photosynthesis, while the minimum percentage of plants survival in controlled plots might be due to deficiency of nutrients and low photosynthesis rate.

Plant height (cm)

The data recorded on average plant height is presented in the Table. The data indicated the maximum plant height (10.44cm) in FYM treated plants, followed by the plants which were treated by poultry manure (9.9cm). The minimum plant height (8.5cm) was recorded in the control treatment. The maximum plant height in FYM treated plants might be due to sufficient nutrients supply which helped in good photosynthesis, while the minimum plant height in controlled plot might be due to deficiency of nutrients with low photosynthesis rate.

Number of branches plant^{-1}

The data recorded on number of branches plant^{-1} is presented in the Table. The data indicated the maximum number of branches plant^{-1} (6.4) in FYM treated plants, followed by the plants which were treated by poultry manure (5.6). The minimum numbers of branches plant-1 (4.5) were recorded in the control treatment. The maximum number of branches in FYM treated plants might be due to good availability of nutrients which help the plants in vegetative growth. While the minimum number of branches in the plants which were in the control treatment was due to the unavailability of nutrients, leading to poor growth of plants.

Number of leaves plant^{-1}

The data recorded on number of leaves plant^{-1} is presented in the Table. The data indicated the maximum number of leaves plant^{-1} (18.8) in FYM treated plants, followed by the plants which were treated by poultry manure (16.8). The minimum numbers of leaves plant-1 (13.5) were recorded in the control plot. The maximum number of leaves in FYM treated plants might be due to good availability of nutrients which helped the plants in vegetative growth, while the minimum number of leaves in the plants which were in control plot was due to the unavailability of nutrients, due to which the growth of plants was retarded.

CONCLUSION AND RECOMMENDATIONS

Greater percentage of plant survival, plant height, number of branches and leaves were obtained in plants treated with FYM compared with poultry manure and can resist the extreme cold of winter season. FYM as the suitable organic fertilizer is recommended for the best production and high yield of strawberry runners under the agro-climatic conditions of Peshawar.

REFERENCES

Amin, NU. 1996. Evaluation of different strawberry cultivars for off-season production under plastic tunnels. P.3 M.Sc (Hons) Thesis, Department of Horticulture KP Agricultural University Peshawar.

Anonymous. 2011. United States Department of Agriculture (USDA). National nutrient database for standard reference.

Ayaz, M., Ahad SF and Rehman K. 1998. Evaluation of strawberry for Malakand division. Sarhad J. of Agric. 14(4): 317.

E-Flora BC. 2008. Electronic atlas of the plants of British Columbia. Copy right 2008 E-Flora, BC.

GOP. 2009. Fruits, vegetables and condiments statistics of Pakistan. Ministry Food Agric. Econ. Div. pp. 1-2

Nestby, R. 1998. Effect of N fertilization on fruit yield, leaf N and sugar content in fruits of strawberry cultivars. J. Hort. Sci. and Biotechol. 73(4): 563-568.

Sharma VP and Sharma RR. 2003. The Strawberry. Indian Council of Agricultural Research, New Delhi, pp. 166.

Umar I, Wali VK., Kher R and Sharma A. 2008. Impact of Integrated nutrient management on strawberry yield and soil nutrient status. Appl. Biol. Res. 10: 22-25.

TILLAGE PRACTICES AND SOWING METHODS AFFECT YIELD AND RELATED ATTRIBUTES OF MAIZE

Ehsanullah[1], UmairAshraf[2], Shakeel Ahmad Anjum[1*]
Farrukh Ehsan[3], Imran Khan[1] and Abdul Ghaffar[1]
[1]*Deaprtment of Agronomy, University of Agriculture, Faisalabad*
[2]*Department of Crop Science and Technology, College of Agriculture, South China Agricultural University, Guangzhou, 510642 P.R. China*
[3]*Deaprtment of Farm Machinery and Power, University of Agriculture, Faisalabad*

ABSTRACT

Adequate tillage practices and sowing of crops at suitable time with proper sowing methods are among the main factors that regulate the crop growth and ultimate yield. This experiment was conducted to assess the effects of different tillage practices and sowing methods on maize hybrid '32T16' that was grown under field conditions where treatments were assigned in split plot arrangement by keeping tillage practices (conventional tillage, minimum tillage and deep tillage) in main plots while sowing methods (flat sowing, ridge sowing and bed sowing) in sub-plots and repeated thrice. Conclusively, tillage practices and sowing methods significantly improved maize growth, yield and yield contributing traits. Ridge sowing under deep tillage resulted in maximum grains and grain weight per cob, 1000-grain weight, grain and biological yield as well as harvest index. Regarding economics of the crop, maximum marginal rate of return (2232.79%) was recorded in ridge sown maize under conventional tillage.

Keywords: Maize, Tillage, Sowing methods, Growth, Yield

INTRODUCTION

Field crops depend largely on soil conditions to show their full potential regarding growth and yield. For improving soil condition, tillage plays a crucial role in improving maize production. A compacted soil, due to its high strength and low porosity, limits the crop roots in the upper layer and decreases the soil volume that can be searched by the plants for water and nutrients (Lipiec et al., 2003). Among different crop production factors, tillage contribution is up to 20% in crop production (Ahmad *et al*., 1996). Tillage is very effective farm activity for improving the soil physical condition and soil tilth, which enhances nutrient uptake and ultimately good final yield of crops (Bahadar *et al*., 2007). For a double cropping system, reducing tillage is very important due to short time for seedbed preparation and also to keep the low cost of production (Limon-Ortega *et al*., 2002). Most economical, feasible and time saving cultural practices are required for raising the maize crop. Moreover, Tillage practices affect evaporation and also dry out soil to tillage depth. No till soils reduces surface residue incorporation and ultimately soil aeration in to soil. Soil disturbance and tillage activity usually can cause an increase in the residue decomposition, soil aeration, organic nitrogen mineralization and accessibility for plant to use nitrogen (Dinnes *et al*., 2002). For the remediation of subsoil compaction, deep tillage and crop rotations with the help of deep-rooted crops are best management options (Motavalli *et al*., 2003). Due to low soil organic matter percentage and poor crop establishment, maize grain yield is very low (Amanullah *et al*., 2010). Tillage breaks the hard subsoil layer that is formed due to repetitive tillage practices on the same depth for year after year. The presence of hard subsoil layer had bad impact on soil bulk density, penetration resistance, nutrient status and soil porosity that directly or indirectly disturbs the crop yield by lowering soil porosity and increasing bulk density of soil (Ahmad *et al*., 2009).

Sowing method is another key factor which has important role in maize crop production. The sowing methods which are mostly being used in Pakistan are ridge sowing, bed sowing and flat sowing. There are so many disadvantages of old broadcast method i.e. irregular seed distribution, uneven seed depth and seed picking up by the birds. Ashique *et al*. (1997)

**Corresponding author:* e-mail: shakeelanjum1034@gmil.com

found several reasons for low production of maize in Pakistan among them high weed infestation, poor weed management practices and improper planting methods are common problems. Low yield potential of maize is due to many constraints like cultivar selection, late sowing, use of traditional sowing methods, improper plant protection measures and poor crop establishment. Proper sowing method has several benefits like improved inter-culturing, uniform irrigation, weeding, pest management and mechanical harvesting (Memon *et al*., 2007). Among the various agronomic practices, planting technique is of considerable importance because of optimum plant population and also proper use of the land and input resources (Ali *et al*., 1998). Improper sowing methods result in unproductive plants while improved sowing methods are important to get the maximum potential of maize hybrids (Alias *et al*., 2003). So, the aim of the study was to assess the variation in yield and yield contributing factors due to various tillage practices and sowing methods.

MATERIALS AND METHODS

The proposed study was conducted at Agronomic Research Area, University of Agriculture, Faisalabad, Pakistan (31.25°N latitude, 73.09°E longitude, altitude 184 m). Experimental soil belongs to Lyallpur soil series (Haplargid, hyperthermic Ustalfic, Aridisol-fine-silty, mixed according to USDA classification). The prevailing environmental conditions during crop growing period are presented in the fig. 1. Randomized complete block design (RCBD) with split plot arrangement was used to layout the experiment in three replications with a net plot size of 9.0 m × 4.8 m. Three different tillage practices i.e., TP_1: Conventional tillage (Disc harrow twice + cultivator once + planking), TP_2: Minimum tillage (Cultivator twice + planking) and TP_3: Deep tillage (Chisel plough twice + cultivator once + planking) were kept in main plots while three sowing methods i.e., SM_1: Flat sowing (R × R = 60 cm), SM_2: Ridge sowing (R × R = 60 cm) and SM_3: Bed sowing (Bed = 120 cm) were kept in sub-plots. The experiment was hand planted using 30 Kg ha^{-1} seed rate of maize hybrid 'Pioneer-32T16', intending to achieve good stand establishment. Two seeds were dropped per hill and the space between two adjusted hills within each row was 20 cm. Thinning was done at four leaves stage, to adjust the population. NPK was applied @ 250:125:125 kg ha^{-1}. Crop was irrigated when required and necessary plant protection measures were practiced to avoid pest attack. The crop was harvested manually on 16th June, 2012 and cobs (sun dried) were shelled with the help of a maize sheller to detach grains from pith. Data were recorded by using standard procedures and statistically analyzed by using Fisher's analysis of variance technique and least significant difference (LSD) test at 5% probability level was used to compare treatment means (Steel *et al.,* 1997) whereas marginal analysis was done to calculate the net benefits and marginal rate of returns according to procedures devised by CIMMYT (1988).

RESULTS AND DISCUSSION

Plant population at maturity

Tillage practices and sowing methods did not affect plant m^{-2} at maturity (Table 2). Moreover, Interaction between tillage practices and sowing methods for plant population was also found non-significant. This might be due to maintaining plant population after germination of crop through thinning and using uniform seed rate to maintain plant to plant distance in all plots. These results are supported by Amin *et al.* (2006) who concluded that plant population was not significant under different sowing techniques.

Cob length (cm)

Cob length affected appreciably by various sowing methods while there were non-significant differences in cob length when maize hybrid was sown by different tillage practices (Table 2). Maximum cob length was observed under deep tillage while minimum cob length was recorded in minimum tillage but they were not varying up to the level of significance. Furthermore, maximum cob length was observed in bed sowing that was similar with ridge sown maize while minimum was recorded in flat sowing. This might be due to more moisture conservation and better crop stand in ridge sowing method because crop can get more porous soil on ridges than soil surface. The interactive effect of various tillage practices and sowing methods on maize cob length was also found to be non-significant (Table 2). Khan *et al.* (2010) by supporting these results concluded that maize ridge sowing influenced the cob length more as compared to other sowing methods.

Cobs per plant

The individual and interactive effect of various tillage practices and sowing methods on cobs per plant was found non-significant (Table 2). It might be a reason that the number of cobs per plant is a genetic character of maize hybrid which is not affected by cultural practices. Planting methods did not affect the number of cobs per plant considerably (Bakht *et al.*, 2011).

Cob length (cm)

Cob length as affected appreciably by various tillage practices and sowing methods (Table 2) which further illustrated that there were non-significant differences in cob length when maize hybrid was sown under different tillage practices. Maximum cob length was observed deep tillage while minimum cob length was recorded in minimum tillage but they were not varying up to the level of significance. Furthermore, maximum cob length was observed in bed sowing that was similar with ridge sown maize while minimum was recorded in flat sowing (17.76 cm). This might be due to more moisture conservation and better crop stand in ridge sowing method because crop can get more porous soil on ridges than soil surface. The interactive effect of various tillage practices and sowing methods on maize cob length was also found to be non-significant. Khan *et al.* (2010) by supporting these results concluded that maize ridge sowing influenced the cob length more as compared to other sowing methods.

Grain rows per cob

Non-significant difference regarding number of grain rows per cob was recorded when maize hybrid was sown under different tillage practices. Maximum number of grain rows per cob was observed when deep tillage was practiced while minimum grain rows per cob were recorded in minimum tillage but they were not varying up to the level of significance. Moreover, highest number of grain rows per cob was observed in ridge sowing that was statistically similar with bed sown maize while minimum grain rows were recorded in flat sowing. However, the interactive effect of various tillage practices and sowing methods on number of grain rows per cob was found to be non-significant (Table 2). These results were supported by Shahzad and Khan (2003) who reported that ridge sown maize significantly increased number of grain rows per cob as compared to other sowing methods.

Grains per cob

A significant interaction between tillage practices and sowing methods regarding grains per cob was observed (Table 2). In conventional tillage, maize sown on ridges and beds produced maximum number of grains per cob against flat sown maize. Similar trend was recorded in minimum tillage. Under deep tillage, maize ridge sowing performed best and produced maximum number of grains per cob that was statistically similar with bed sowing while flat sown maize produced minimum number of grains per cob which was statistically at par with bed sowing. Bakht *et al.* (2007) concluded that sowing methods had a significant effect on the number of grains per cob and they obtained maximum grains per cob in ridge sowing method as compared to other sowing methods.

Grain weight per cob (g)

Combined effect of various tillage practices and sowing methods on grain weight per cob was found to be significant (Table 2). Grain weight per cob was highest in ridge sown maize under deep tillage while lowest was recorded in flat sowing under minimum tillage. At conventional tillage, bed sown maize produced maximum grain weight per cob that was statistically alike with the ridge sown maize while flat sown maize gave minimum grain weight per cob. Under deep tillage, maize ridge sowing performed best and produced maximum grain weight per cob that was statistically alike with bed sowing while flat sown maize produced minimum grain weight per cob. Our results are corresponding to Shahzad and Khan (2003) who reported that maize ridge sowing increased grain weight per cob significantly as compared to other sowing methods.

1000-grain weight (g)

The significant interaction (tillage practices × sowing methods) revealed that 1000-grain weight was highest in ridge sown maize under deep tillage while the lowest was recorded in flat sowing under minimum tillage while conventional tillage, bed sown maize produced maximum1000-grain weight that was statistically at par with the ridge sown maize. In minimum tillage, maize on beds produced higher 1000-grain weight while flat sown maize produced lower 1000-grain weight. Similarly under deep tillage, maize on ridges produced higher 1000-grain weight however flat sown maize produced lower 1000-grain weight (Table 2). In sum, all tillage practices were found statistically similar regarding 1000-grain weight in case of ridge and bed sowing followed by flat sown maize. Anjum *et al.* (2014) and Bakht *et al.* (2007) supported these results and found maximum 1000-grain weight in ridge sowing method as compared to other sowing methods.

Grain and biological yield (t ha^{-1})

All tillage practices were found statistically similar regarding grain and biological yield of maize in case of ridge followed by bed sowing but differs in flat sown maize. Flat sown maize yielded less than ridge and bed sown maize. Over all the effectiveness of all three tillage practices were: deep tillage > minimum tillage > conventional tillage, while for sowing methods, it may be stated as: Ridge sowing > bed sowing > flat sowing. The higher grain yield of maize on ridges might be ascribed to more number of grains per cob with higher grain weight than the crop sown with other planting methods. These results were supported by Musambasi *et al.* (2003) who stated that maize sowing on ridges provided the maximum grain yield. Anjum *et al.* (2014) also reported that ridge sowing method significantly enhanced the biological yield than other sowing methods.

Harvest index (%)

The interaction of tillage practices and sowing methods explicated that harvest index was higher in ridge sown maize under deep tillage while lower was recorded in flat sowing under minimum tillage. At conventional tillage, bed sown maize furnished maximum harvest index found statistically allied to ridge sowing while flat sown maize gave minimum harvest index. All the sowing methods including flat, ridge and bed sowing showed statistically non-significant difference regarding harvest index at minimum tillage. However under deep tillage, all the sowing methods showed substantial differences regarding harvest index. At deep tillage, ridge sown maize produced maximum harvest index followed by bed sowing while flat sown maize produced minimum harvest index. Balasubramaniyan *et al.* (2001) and Graybill *et al.* (1991) also reported that harvest index was affected significantly by sowing methods.

Dominance and marginal analysis

Final recommendation for the production technology cannot be specified only on the basis of net field benefit (NFB) as NFB does not indicate the rate of return in relation to investment. Domination is the mechanism for identification of good alternatives. Data presented in table 3 revealed that NFB of some treatments were less with lowering cost and as a result these treatments were dominated (D). The remaining (un-dominated) treatments were further considered for the marginal analysis. It is clear from the table 3 that the treatments in which flat sowing was practiced under conventional and deep tillage were dominant due to their lower net field benefits as compared to other treatments.

As real differences were found in the yield among different treatments, therefore a marginal analysis was done. Table 3 presents the marginal analysis of un-dominated treatments. Maximum marginal rate of return of 2232.79% was obtained in treatment where maize sown on ridges under conventional tillage. It is clear from the results that farmers with poor resources can accomplish maximum benefits by sowing the maize hybrid on ridges under conventional tillage system.

CONCLUSION

Conclusively, ridge sowing after deep tillage proved better regarding growth and yield of maize and provided higher net returns as compared to other tillage practices and sowing methods. Hence, maize could be planted on ridges after deep ploughing to get maximum benefits.

AKNOWLEDGEMENT

The authors are highly thankful to the co-workers of Agro-Biology Lab, Department of Agronomy, University of Agriculture, Faisalabad, Pakistan for their help in completion of this research.

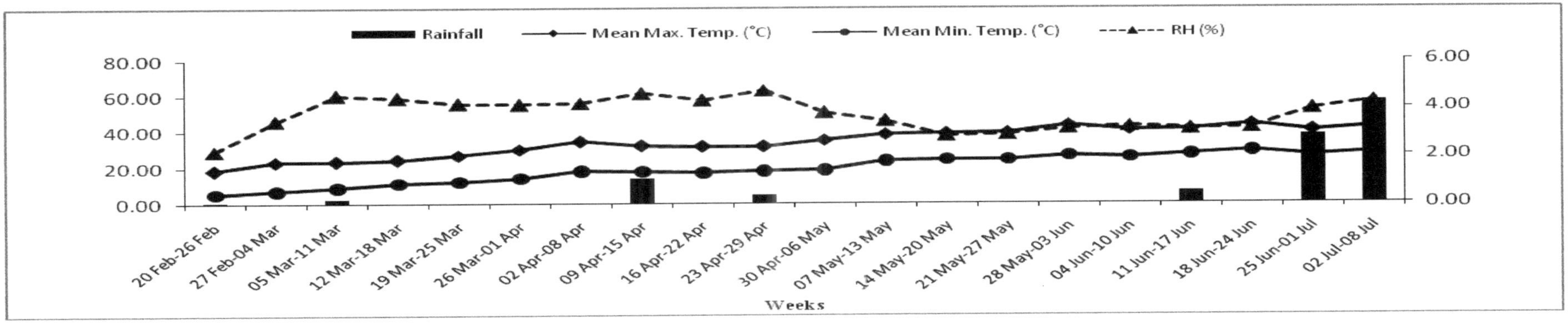

Figure 1. Mean maximum temperature (°C), Mean minimum temperature (°C), Relative humidity (%) and Rainfall (mm) during the growth period of crop.

Table 1 Affect of different tillage practices and sowing methods on yield and yield related attributes of maize

Treatments	Plant population at maturity	Cob length (cm)	Number of cobs per plant	Number of grain rows per cob	Number of grains per cob	Grain weight cob (g)	1000-grain weight (g)	Biological yield (t ha^{-1})	Grain yield (t ha^{-1})	Harvest index (%)
TP_1	6.9	18.83	1.0	15.53	534.3 B	129.8 B	242.4 B	16.30 B	5.82 B	35.68 A
TP_2	7.0	18.45	1.0	15.39	485.5 C	124.2 C	231.8 C	15.87 B	5.44 C	34.28 B
TP_3	7.0	18.85	1.1	15.80	590.9 A	134.0 A	256.5 A	17.18 A	6.08 A	35.39 A
LSD($p \leq 0.05$)	**NS**	**NS**	**NS**	**NS**	**11.12**	**3.47**	**5.08**	**0.44**	**0.13**	**0.92**
SM_1	6.4	17.76 B	1.0	15.19 B	505.3 B	123.9 B	234.8 B	15.80 C	5.43 C	34.33 B
SM_2	7.6	19.14 A	1.1	15.84 A	555.0 A	132.2 A	247.3 A	16.92 A	6.05 A	35.74 A
SM_3	6.9	19.23 A	1.0	15.69 A	550.5 A	131.9 A	248.6 A	16.63 B	5.87 B	35.28 A
LSD($p \leq 0.05$)	**NS**	**0.96**	**NS**	**0.36**	**11.29**	**2.18**	**3.88**	**0.24**	**0.06**	**0.49**
$TP_1 \times SM_1$	6.3	17.73	1.0	15.13	507.6 b	125.1 b	236.8 b	15.83 b	5.52 b	34.87 b
$TP_2 \times SM_1$	6.3	17.90	1.0	15.06	433.0 b	115.7 b	225.9 b	14.83 c	5.04 c	33.96 a
$TP_3 \times SM_1$	6.7	17.66	1.1	15.40	575.4 b	130.8 b	241.8 b	16.73 b	5.72 c	34.17 c
$TP_1 \times SM_2$	7.7	19.20	1.1	15.86	548.7 a	132.0 a	242.6 ab	16.67 a	6.01 a	36.04 a
$TP_2 \times SM_2$	7.3	18.66	1.1	15.46	513.6 a	127.9 a	233.4 a	16.62 a	5.74 a	34.55 a
$TP_3 \times SM_2$	7.7	19.56	1.0	16.20	602.6 a	136.8 a	265.9 a	17.47 a	6.40 a	36.62 a
$TP_1 \times SM_3$	6.7	19.56	1.0	15.60	546.6 a	132.3 a	247.9 a	16.40 a	5.92 a	36.13 a
$TP_2 \times SM_3$	7.3	18.80	1.0	15.66	510.1 a	128.9 a	236.1 a	16.17 b	5.55 b	34.34 a
$TP_3 \times SM_3$	6.7	19.32	1.1	15.80	594.7 ab	134.4 ab	262.0 a	17.33 a	6.13 b	35.37 b

LSD($p \leq 0.05$)	NS	NS	NS	NS	19.54	3.78	6.72	0.42	0.11	0.86

Means not sharing a letter in a common within a column differ significantly ($p \leq 0.05$); NS: Non-Significant. TP_1: Conventional tillage (Disc harrow twice + cultivator once + planking), TP_2: Minimum tillage (Cultivator twice + planking) and TP_3: Deep tillage (Chisel plough twice + cultivator once + planking); SM_1: Flat sowing ($R \times R = 60$ cm), SM_2: Ridge sowing ($R \times R = 60$ cm) and SM_3: Bed sowing (Bed = 120 cm)

Table 2 Economic and dominance analysis of tillage practices and sowing methods of maize

Treatments		Variable cost (Rs. ha^{-1})	Marginal cost (Rs. ha^{-1})	Net field benefits (Rs. ha^{-1})	Marginal net benefits (Rs. ha^{-1})	Marginal rate of return (%)
Minimum Tillage	Flat Sowing	19110	--	153823	--	--
	Bed Sowing	21112	2002	164344	10521	525.52
	Ridge sowing	21635	523	167030	2686	513.57
Conventional Tillage	Flat Sowing	25180	3545	145610	**D***	**D**
	Bed Sowing	26880	1700	159193	13583	799
	Ridge sowing	27127	247	164709	5515	2232.79
Deep Tillage	Flat Sowing	29730	2602	156183	**D**	**D**
	Bed sowing	31457	1727	167002	10819	626.46
	Ridge Sowing	32200	742	173788	6785	914.42

*D = Dominance

REFERENCES

Ahmad N, Rashid M and Vaes AG, 1996. Fertilizer and their use in Pakistan, p. 274. NFDC. Pub. No. 4/96, 2nd ed. Islamabad.

Ahmad N, Hassan FU and Belford RK, 2009. Effect of soil compaction in the sub-humid cropping environment in Pakistan on uptake of NPK and grain yield in wheat. Field Crops Res. 110: 54-60.

Ali M, Khalil SK, Ayaz S and Marwat MI, 1998. Phenological stages, flag leaf area, plant height, and leaves per plant of corn influenced by phosphorus levels and plant spacing. Sarhad J. Agri. 14: 515-522.

Alias A, Usman M, Ehsanllah, and Warraich EA, 2003. Effect of different phosphorus levels on growth and yield of two cultivars of maize. Int. J. Agri. Biol. 5:142-147.

Amin M,RazzaqA, Rehmatullah and RamzanM, 2006. Effect of planting methods, seed density and nitrogen phosphorus (NP) fertilizer levels on sweet corn (*Zea mays* L.). J. Res. Sci. 17:83-89.

Amanullah, Zakirullah and Khalil SK, 2010. Timing and rate of phosphorus application influence maize phenology, yield and profitability in north-west Pakistan. Int. J. Plant Prod.4: 281-292.

Anjum SA, Ehsanullah, Ashraf U, Tanveer M, Qamar R and Khan I, 2014. Morphological and phenological attributes of maize affected by different tillage practices and varied sowing methods. Am. J. Plant Sci. 5: 1657-1664.

Ashique M, ShahML and ShafiM, 1997. Weeds of maize and their eradication. Zarat Nama, pp. 35-89.

Balasubramaniyan P and Palaniappan SP, 2001. Nutrient management, p: 185-188. In P. Balasubramaniyan, P.P. Balasubramaniyam and S. Palaniappan (eds.), Principles and Practices of Agronomy. Agrobios, India.

Bahadar KM, Arif M and Khan MA, 2007. Effect of tillage and zinc application methods on weeds and yield of maize. Pak J. Bot. 39: 1583-1591.

Bakht J, Siddique MF, Shafi M, Akbar H, Tariq M, Khan N, Zubair M and Yousef M, 2007. Effect of planting methods and nitrogen levels on the yield and yield components of maize. Sarhad J. Agri. 23: 553-559.

Bakht J, Shafi M, Rehman H, Raziuddin and Anwar S, 2011. Effect of planting methods on growth, phenology and yield of maize varieties. Pak. J. Bot.43: 1629-1633.

CIMMYT. 1988. From agronomic data to farmer recommendations: An Economic Training Manual. Completely revised edition, Mexico. Available online at: http://apps.cimmyt.org/, Accessed on: 28-02-2011.

Dinnes DL, Karlen DL, Jaynes DB, Kasper TC, Hatfield JL, Colvin TS and Cambardella CA, 2002. Nitrogen management strategies to reduce nitrate leaching in tile-drained mid western soils. Agron. J. 94: 153-171.

Graybill JS, Cox WJ and Otis DJ, 1991. Yield and quality of forage maize as influenced by hybrid, planting date and plant density. Agron. J. 83: 559-564.

Khan MAH and Parvej MR, 2010. Impact of conservation tillage under organic mulches on the reproductive efficacy and yield of quality protein maize. J. Agric. Sci. 5:52-63.

Lipiec J, Medvedev VV, Birkas M, Dumitru E, Lyndina TE, Rousseva S and Fulajtar E, 2003. Effect of soil compaction on root growth and crop yield. Int. Agrophy. 17: 61-69.

Limon-Ortega A, Sayer KD, Drijber RA and Francis CA, 2002. Soil attributes in a furrow-irrigated bed planting system in North-West Mexico. Soil and Tillage Res.63: 123-132.

Memon SQ, Baig MB and Mari GR, 2007. Tillage practices and effect of sowing methods on growth and yield of maize crop. Agri. Tropica et Subtrop. 40: 89-99.

Motavalli PP, Stevens WE and Hartwig G, 2003. Remediation of subsoil compaction effects on corn N availability by deep tillage and application of poultry manure in a sandy-textured soil. Soil and Tillage Res.71: 121-131.

Musambasi D, Chivinge OA and Mariga IK, 2003. Effect of ridging treatment and two early maturing maize cultivars on witch weed density and maize yield under dry land maize based cropping system in Zambabwe. Crop Res. 25: 37-45.

Shahzad Z and Khan S, 2003. Comparative evaluation of planting and fertilizer application methods on maize (*Zea mays* L.). University of Agriculture, Peshawar, Pakistan.

Steel RGD, Torrie JH and Dicky DA, 1997. Principles and procedures of statistics, pp. 352-358. A biological approach. 3rd Ed. McGraw Hill, Inc. Book Co. N.Y. USA.

INFLUENCE OF BIOENGINEERED ZINC NANOPARTICLES AND ZINC METAL ON *CICER ARIETINUM* SEEDLINGS GROWTH

K. V. Pavani*, V. Divya, I. Veena, M. Aditya, G. V. S Devakinandan
Department of Biotechnology, Gokaraju Rangaraju Institute of Engineering and Technology, Bachupally, Hyderabad 500090, Telangana, India

ABSTRACT

Metal nanoparticles are newly emerging alternate source of carriers to transmit and maintain the desired plant food sources available for plant growth and development. Their accumulation in agriculture lands is accompanied by both positive and negative effects on plants. Considering their impact on plants, it is necessary to evaluate them on seed germination and seedling growth. The main objective of this study was to investigate whether the treatment of *Cicer arietinum* seeds with relatively low concentrations of bioengineered ZnNPs and zinc metal would affect the seed germination and seedlings growth. Zinc nanoparticles were characterized by transmission electron microscopy (TEM). There were differences in germination (2%), root length (0.18cm), shoot length (0.17cm), fresh weight (0.52cm) and dry weight (0.03cm) between seeds grown in control and Zn NPs. There also existed significant differences in germination (2.67%), root length (4.94cm), shoot length (1.29cm), fresh weight (0.50cm) and dry weight (0.09cm) between seeds grown in control and $ZnSO_4$. *Cicer arietinum* seeds grown in zinc nanoparticles showed increase in germination, shoot length, root length, fresh weight and dry weight, whereas seeds grown in zinc metal decreased growth. Our findings suggest that bioengineered zinc nanoparticles showed positive effect on seedlings growth.

Keywords: Zinc nanoparticles, zinc *metal, Cicerarietinum,* positive effect, seedlings growth

INTRODUCTION

Nanotechnology has become a dynamically developing industry with increasing numbers of commercial products such as materials manufacturing, computer chips, medical diagnosis, energy and cosmetics (Ahsan et al., 2007). Due to rapid industrialization and intensive agriculture contamination of heavy metals in soil has been increased. Excessive level of heavy metals in the soil environment adversely affect the germination of seeds, plant growth and interfere with the activities of enzymes related to normal metabolic and developmental processes (Ali and Madhavan, 2012; Arora et al., 2012) . Engineered nanoparticles (ENPs) synthesized by Microorganisms in contaminated soils play a critical role in plant growth.

Manufactured nanomaterials can be accidentally or incidentally release into waste water streams and agriculture fields (Auffan et al., 2010). The impact of MNMs on different plant species can vary greatly, and there are reports of both positive and negative effects (Blaser et al 2013). Positive effects on germination of aged spinach seeds and on the growth of seedlings were obtained if the seeds were soaked in high-strength TiO2-nanoparticles-solution (Elena and Katarina, 2013). Increase of the electron transfer, oxygen evolution, and photophosphorylation was observed in the chloroplasts of spinach treated with anoanatase-TiO2 (John et al.., 2009). Carbon nanotubes (CNTs) enhanced root growth of onion (*Allium cepa*) and cucumber (*Cucumis sativa*) (Kuriakosa and Prasad, 2008). However, majority of the reports available in the literature indicate phytotoxicity of NPs (Laure et al., 2011). Application of MNPs such are silica, palladium, gold and copper nanoparticles significantly influenced the growth of lettuce plants (Lee et al. 2010). *Lecanicillium lecanii* based silver nanoparticles showed less effect on seedling emergence, shoot length, leaf surface area, chlorophyll content and phyllosphere microflora of economic important pulses such as cowpea (*Vigna unguiculata*), black gram (*Vigna mungo*), green gram (*Vigna radiate)* , sorghum (*Sorghum vulgare*) and horse gram *Macrotyloma uniflorum*) (Ling and watts, 2005). Scientific investigation on plant uptake and accumulation of ENPs in the environment

**Corresponding author:* e-mail: pavani_20042003@yahoo.co.in is still in its rudimentary stage, few publications have been reported in the past few years in the area of ENP toxicology and uptake by plants. The use of engineered nanoparticles is one of the recent advances in the field of agriculture biotechnology.The objective of our study is to assess the effect of bioengineered Zinc nanoparticles synthesized by fungi and Zinc metal on the growth of *Cicer arietinum* seeds.

MATERIALS AND METHODS

1. Bioengineered nanoparticles

Aspergillus species (BankIt ID: 1526725) isolated from the soil samples collected from Hyderabad Metal Plating Industry, I.D.A, Balanagar, Hyderabad, A.P. were preserved at 4°C in a refrigerator. A loop full of fungal culture was inoculated in a 250 ml of Erenlymer flask containing 100 ml of fungal media. The flask was incubated at 28^0 C, for 2 days on a rotary shaker at 160 rpm, and the cells were filtered using filter paper. The filtered solution was autoclaved and the size of the nanoparticles was determined by Transmission Electron Microscopy (TEM) (JEOL JEM 2010, Japan, operated at 120 kV).

2. Seed germination

Cicer arietinum seeds were immersed in a 0.1% mercuric chloride solution for 5 min for surface sterilization. After rinsing three times in sterile distilled water, they were soaked in the media containing zinc nanoparticles, deionised water (control), $ZnSO_4$ metal solution (5mM/L) for 48hours. A piece of filter paper (Whatman No. 42) was put into each Petri dish (90 mm × 15 mm), 4 ml of deionized water or nanoparticle media and ZnSo4 solution were added, and 25 seeds were then transferred onto each dish. Petri dishes were sealed with parafilm and placed in an incubator. The seeds were kept under a photoperiod of 12 hr, and 25/18 °C day/night temperature. The seedlings were harvested after 14 days and the germination percentage, root and shoot length, fresh weight and dry weight was measured for all replicates. The number of seeds germinated was counted and the germination percentage was calculated by using the following formula

$$\text{germination percentage} = \frac{\text{number of seeds germinated}}{(\text{total number of seeds sown})} \text{ X } 100$$

Fresh weight (g/plant) was taken with the help of an electrical single pan balance. The collected plant materials were kept in hot air oven at 50°C for 48 hours and their dry weight (g/plant) was recorded.

STATISTICAL ANALYSIS

Each treatment was conducted in at least triplicate, and the results were presented as mean values with respective standard errors mean (SEM). Statistical significance of differences between treatments was determined by using paired T test at 95% confidence level.

Table.1: The impact of bioengineered zinc nanoparticles and zinc metal on growth of *Cicer arietinum* seedlings growth. Each value represents the mean of 3 replicates ± standard error in mean.

	% of germination	Root length	Fresh weight	Dry Weight	Shoot length
Control	94.67 ± 3.528	6.308 ± 0.312	1.378 ± 0.238	0.216 ± 0.021	5.38 ± 0.580
Zn NPS	96.00 ± 2.309	6.48 ± 0.652	1.896 ± 0.0145	0.24 ± 0.016	5.550 ± 0.399
$ZnSO_4$	92.00 ± 4.619	1.36 ± 0.356	0.772 ± 0.055	0.125 ± 0.0098	4.09 ± 0.420

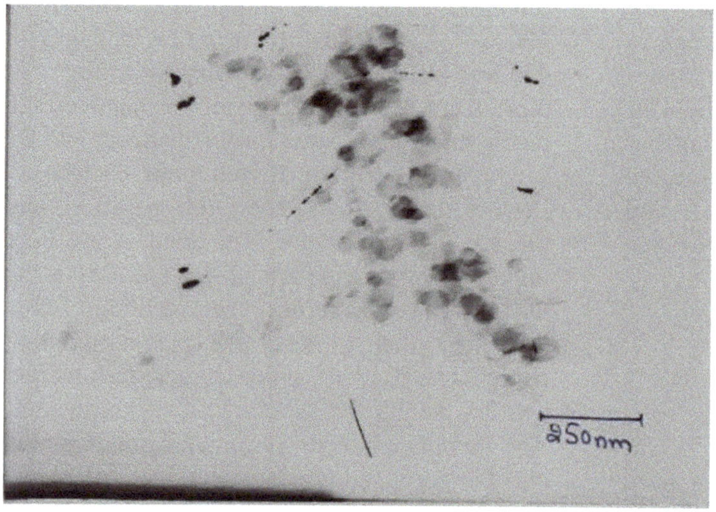

Fig.1: Transmission Electron Microscope showing the size of Zinc nanoparticles. Magnification 77200 X

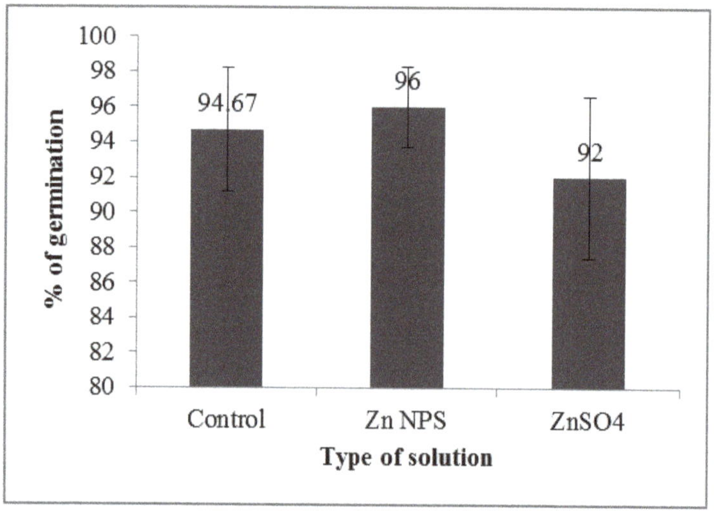

Fig.2: Changes in the germination % of *Cicer arietinum* grown in Zn and Zn NPS

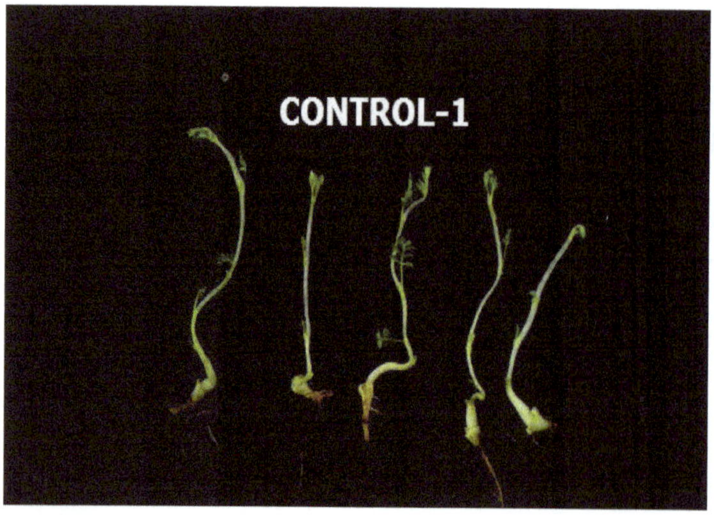

Fig.3: Comparative morphology of *Cicer arietinum* grown in water

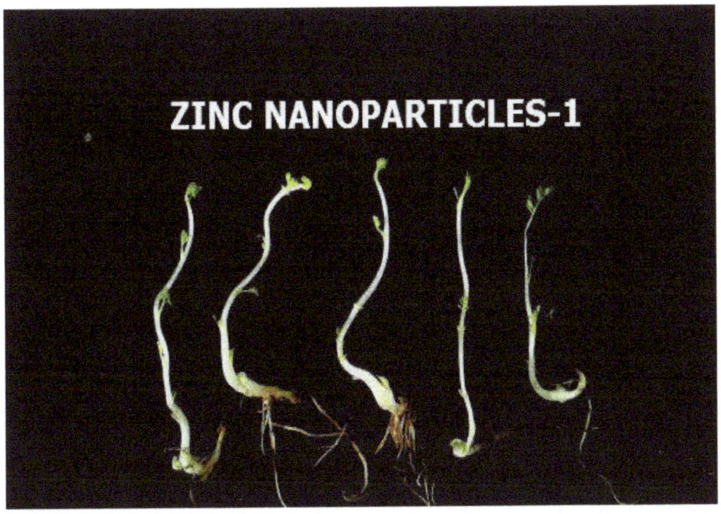

Fig.4: Comparative morphology of *Cicer arietinum* grown in zinc nanoparticles

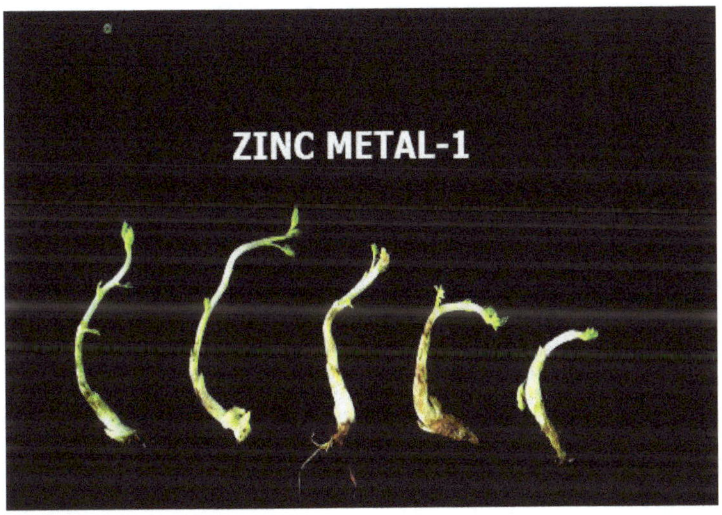

Fig.5: Comparative morphology of *Cicer arietinum* grown in zinc sulphate

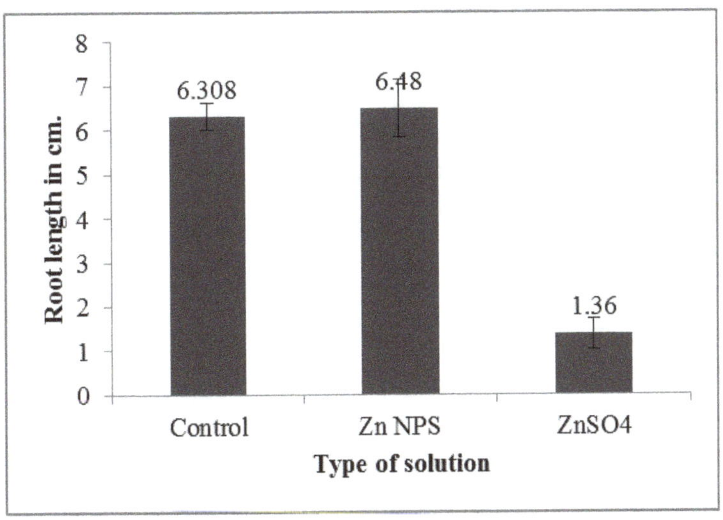

Fig.6: Changes in the root length of *Cicer arietinum* grown in Zn and Zn NPS

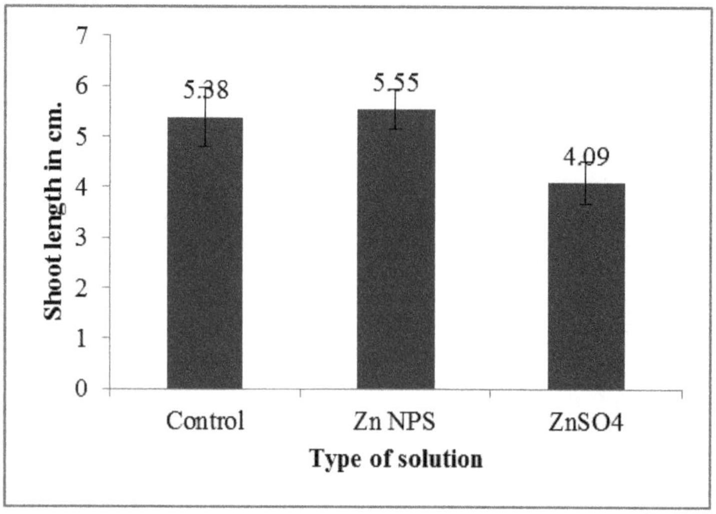

Fig.7: Changes in the shoot length of *Cicer arietinum* grown in Zn and Zn NPS

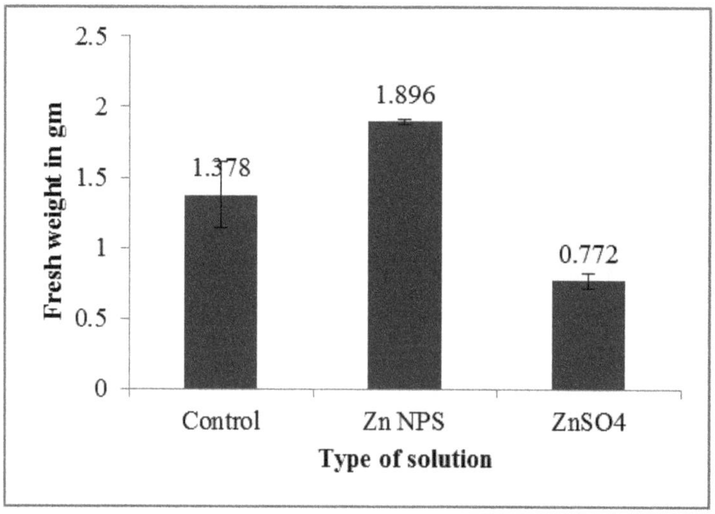

Fig.8. Changes in the fresh weight of *Cicer arietinum* grown in Zn and Zn NPS

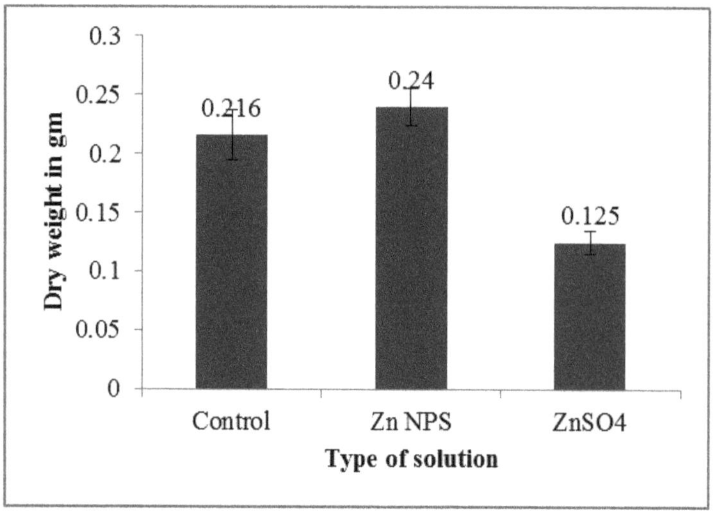

Fig.9: Changes in the dry weight of *Cicer arietinum* grown in Zn and Zn NPS

RESULTS AND DISCUSSION

Nanoparticles are introduced into the soil as a result of human activities, and by conversion of metals into nanoparticles by microorganisms present in the soil. So in the present study we tried to analyse the effect of zinc nanoparticles synthesized by *Aspergillus species* on bengal gram, as it is important to analyse the connections among nanoparticles, microbes and plants.

TEM studies

TEM (Transmission Electron Micrograph) revealed the presence of mono and polydispersive nanoparticles. The zinc nanoparticles synthesized are spherical and the size ranges from 13.2nm – 19.2nm. The average particle size of 15.332 nm was observed. It was also observed that bioengineered nanoparticles are aggregated (Fig.1). The surface properties of engineered NPs are of essential importance for their aggregation behavior, and thus for their mobility in aquatic and terrestrial systems and interactions with microorganisms and plants (Mihoub et al. 2005; Muhammad et al. 2008).

Germination percentage

Germination percentage was calculated after 14 days of seed germination. Number of seeds germinated was counted from each triplicate and the percentage was calculated by the formula mentioned in the method. Exposure to zinc sulphate significantly affected seed germination rates of Bengal gram, compared to the seeds exposed to zinc nanoparticles and control. (Fig. 2-5). Germination response was more in seeds soaked in zinc nanoparticles than the control. The decrease in seed germination of bengal gram under metal treatment can be attributed to the metal stress (Arora et al. 2012), osmotic effect (Navarro, 2008) or alterations in permeability properties of cell membrane (Punz and seighardt, 1993) hydrolytic enzymes activity, which led to impaired hydrolysis of complex food material (starch) or obstruction in mobilization of sugars (Radha et al. 2011; Rahoui, 2008; Remedios et al. 2012).

The reason for reduced seedling growth under metal exposure may be due to low water potential and nutrient uptake (Karthick et al. 2011). Hydrolytic enzymes which break down the food reserves during seedling growth might be affected by metal stress which may lead to the reduction in seedling growth. Similar observations have been made by several authors under various metal toxicity (Ali and Madhva, 2012; Arora et al. 2012; Shah and Belozerova, 2009).

Root length

Length of the root was measured for all triplicates in centimetres (cm) using a measurement scale. Highest root length is taken among many roots. Root length was increased in seeds exposed to Zn NPS. The root length is significantly affected in the seeds exposed to Zn metal (Fig.6). This may be due to the roots are the first target tissue to the toxic pollutants (Shaukat et al. 1999). The inhibition of root growth can be attributed to the inhibition of mitosis, reduced cell wall synthesis and changes in metabolism (Smiri et al. 2013).

Al_2O_3 NPs at concentrations up to 4000 mg/L had no significant toxic effects on seed germination, root elongation, and number of leaves of *Arabidopsis thaliana* (Thul et al. 2013). Larue and collaborators (Yang et al. 2006) studied the effects of TiO_2-NPs on *Triticum aestivum*, *Brassica napus*, and *Arabidopsis thaliana*. They showed that the TiO_2-NPs did not affect the germination and root elongation.

Shoot length

Length of the shoot was measured for all triplicates in centimetres (cm) using a measurement scale. As shown in (Fig. 7) shoot growth was affected by NPs. Shoot lengths in the presence of Zn NPs were significantly larger than in control. The same results were obtained when seeds of *Triticum aestivum* were grown in the presence of 50 mg/L alumina nanoparticles (Zheng et al. 2007). Field experiments with foliar sprays of gold-nanoparticles on Brassica *juncea* plant shows positively affect the growth profile; including plant height, stem diameter, number of branches, number of pods and seed yield (Arora et al. 2012).

Fresh weight & Dry weight

Fresh weight (g/plant) was taken with the help of an electrical single pan balance (Fig.8). The collected plant materials was kept in hot air oven at 50°C for 48 hours and their dry weight (g/plant) was taken by using an electrical single pan balance(Fig.9). Fresh and dry weights were found to be in accordance With shoot and root lengths for corresponding treatments. The fresh and dry weights were both significantly higher than those of the untreated plants (controls) and Zn treated plants.

CONCLUSION

The present study clearly reveals nontoxic effect of bioengineered zinc nanoparticles on seedlings emergence, and toxic effect of zinc metal on seedlings emergence. Nanoparticles size plays important roles in the behaviour, in the reactivity and in the toxicity of NPs on plants. Considering these aspects it is not strange to find both positive and negative effects of nanoparticles on higher plants. Our findings suggest important new avenues of research for understanding the fate and transport of NPs in natural media, the interactions between NPs and plant tissues, and indirect and direct effects of NPs in mixed communities. Further studies will be helpful to assess toxicity of released nanoparticles on plant growth and other components of ecosystem.

REFERENCES

Ahsan N, Lee DG, Lee SH, Kang KY, Lee JJ. and Kim PJ, 2007. Excess copper induced physiological and proteomic changes in germinating rice seeds. Chemosphere. 67: 1182-1193.

Ali Riahi-Madvar, FatemehRezaee and VahidrezaJalali, 2012. Effects of alumina nanoparticles on morphological properties and antioxidant system of *Triticumaestivum*. Appl. Biochem Biotechnol.167:2225-2233.

Arora S, Sharma P, Kumar S, Nayan R, Khanna PK, Zaidi MGH, 2012. Gold-nanoparticle induced enhancement in growth and seed yield of *Brassica juncea*. Plant Growth Regul. 66: 303-310.

Auffan M, Bottero JY, Chaneac C and Rose J, 2010. Inorganic manufactured nanoparticles: how their physicochemical properties influence their biological effects in aqueous environments. Nanomedicine. 5: 999–1007.

Blaser SA, Scheringer M, Macleod M, Hungerbuhler K, 2008. Estimation of cumulative aquatic exposure and risk due to silver: Contribution of nano-functionalized plastics and textiles. Sci. Total. Environ. 390:396–409.

Elena Masarovicova and KatarínaKralova, 2013. Metal nanoparticles and plants. Ecol. Chem. Engs. 20:9-22.

John RP, Ahmad K, Gadgil and Sharma S, 2009. Heavy metal toxicity: effect on plant growth, biochemical parameters and metal accumulation by *Brassica juncea* L. Int. J. Plant. Prod. 3: 65-76.

Karthick Raja Namasivayam S and Chitrakala K, 2011. Ecotoxicological effect of Lecanicilliumlecanii (Ascomycota: Hypocreales) based silver nanoparticles on growth parameters of economically important plants. J. Biopest. 4: 97 - 101.

Kuriakosa SV, Prasad MNV, 2008. Cadmium stress affects seed germinationand seedling growth in *Sorghum bicolor* (L.) Moench by changingthe activities of hydrolyzing enzymes. Plant Growth. Regul. 54:143-156.

Larue C, Khodja H, Herlin-Boime, N, Brisset F, Flank AM, Fayard B, Chaillou S, Carriere M, 2011. Investigation of titanium dioxide nanoparticles toxicity and uptake by plants. J. Phys. 304: 1-4.

Lee CW, Mahendra S, Zodrow K, Dong Li, Yu-chang T, Janet B, Pedro JJ, 2010. Developmental phytotoxicity of metal oxide nanoparticles to Arabidopsis thaliana. Environ. Toxic. Chem. 29: 669–675.

Ling Yang, Watts DJ, 2005. Particle surface characteristics may play an important role in phytotoxicity of alumina nanoparticles, Toxic. Lett. 158:122–132.

Mihoub A, Chaoui A, and Ferjani E, 2005. Biochemical change associated with cadmium and copper stress in germinating pea seeds (*Pisum sativum* L.). C.R. Biol. 328: 33-41.

Muhammad SI, Zafar M and Mohammad A, 2008. Effect of lead and cadmium on germination and seedling growth of *Leucaena leucocephala*. J. Appl. Sci. Environ. Manag. 12: 61-66.

Navarro E Baun, A. Behra R, Hartmann NB, Filser J, Miao AJ, Quiqq A, Santschi PH, 2008. Environmental behavior and ecotoxicity of engineered nanoparticles to algae, plants, and fungi. Ecotoxicology. 17: 372–386.

Punz WF and Sieghardt H, 1993. The response of roots of herbaceous plant species to heavy metals. Environ. Exp. Bot. 33: 85-98.

Radha Solanki, AnjuPoonam, Dhankhar R. 2011. Zinc and copper induced changes in physiological characteristics of *Vigna mungo* (L.). J. Environ. Biol. 32: 747-751.

Rahoui S, Chaoui A, Ferjani E, 2008. Differential sensitivity to cadmium in germinating seeds of three cultivars of faba bean (*Vicia faba* L.). Acta. Physiol. Plant. 30(4): 451-456.

Remedios C, Rosario F and Bastos V, 2012. Environmental Nanoparticles Interactions with Plants: Morphological, Physiological, and Genotoxic Aspects. J. Botany. Article ID 751686, 8 pages.

Shah V and Belozerova I, 2009. Influence of metal nanoparticles on the soil microbial community and germination of lettuce seeds. Water Air Soil Pollut. 197:143-148.

Shaukat SS, Mushtaq M and Siddiqui ZS, 1999. Effect of cadmium, chromium and lead on seed germination, early seedling growth and phenolic contents of *Parkinsonia aculeate* L. and *Pennisetum americanum* (L.) Schumann. Pak. J. Biol. Sci. 2: 1307-1313.

Smiri M, Chaoui A and Ferjani E, 2009. Respiratory metabolism in the embryonic axis of germinating pea seed exposed to cadmium. J. Plant Physiol. 166: 259-269.

Sresty TVS and Rao KVM, 1999. Ultrastructural alterations in response to zinc and nickel stress in the root cells of pigeonpea. Environ. Exp. Bot. 41: 3–13.

Thul ST, Sarangi BK. and Pandey RA, 2013. Nanotechnology in Agroecosystem: Implications on plant productivity and its soil environment. Expert. Opin. Environ. Biol. 2: 1-7.

Yang F, Hong F, You W, Liu C, Gao F, Wu C and Yang P, 2006. Influences of nano-anatase TiO2 on the nitrogen metabolism of growing spinach. Biol. Trace Elem. Res.110:179-190.

Zheng L, Su MG, Liu C, Chen L, Huang H, Wu X, Liu XQ, Yang F, Gao FQ and Hong FH, 2007. Effects of nanoanatase TiO2 on the photosynthesis of spinach chloroplasts under different light illumination. Biol. Trace. Elem. Res. 119:68-76.

MICROBIOLOGICAL EVALUATION OF AN EDIBLE ANTIMICROBIAL COATINGS ON MANGOES FRUIT STORED UNDER EV APORATIVE COOLANT SYSTEM (ECS)

Adetunji Charles Oluwaseun[1*], Williams Jospeh[1], Olayemi Folorunsho[1], Omojowo Funsho Samuel[3]

[1]Nigerian Stored Product Research Institute, Km 3 Asa dam road, P.M.B. 1489, Ilorin, Nigeria
[2]National Institute of Freshwater Fisheries Research (NIFFR), P.M.B. 6006, New-Bussa, Nigeria

ABSTRACT

Two different coatings were developed from the mucilage of Cactus and their effects were investigated on the microbial qualities of mango fruits. The two experimental coatings were: Pure mucilage extracts (ME) and Mucilage extract mixed with 5ml glycerol (MEG) which served as plasticizer. Samples of the mangoes were immersed into these coatings and they were stored for seven weeks at an average temperature of 27±2°C and relative humidity55-60% under Evaporative Coolant System(ECS) . The *Cactus* mucilage treatment of mangoes significantly attenuated the microbial growth on fruits compared to control. Concentration of yeast and mold was found to be decreased from 10.23log CFU g^{-1} to 0.34 log CFU g^{-1} (MEG) and 2.11log CFU g^{-1} (ME),also the concentration of aerobic Psychrotrophic bacteria was decreased from 23.56log CFU g^{-1} in control to 1.23 log CFU g^{-1} (MEG) and 5.30 log CFU g^{-1} (ME) mucilage treated fruits while the concentration of aerobic mesophilic bacteria was decreased from 20.84log CFU g^{-1} in control to 1.07 log CFU g^{-1} (MEG) and 3.12 log CFU g^{-1} (ME) mucilage treated fruits. The overall result showed that Cactus mucilage coating hindered the growth of microorganisms significantly ($P<0.05$) and extended the shelf-life of mango fruits when compared to untreated in the order, MEG>ME>Control.

Keywords: Antimicrobial coatings, Evaporative Coolant System, *Cactus* mucilage, microbial qualities.

INTRODUCTION

Fruits and vegetables are an extraordinary dietary source of nutrients, micronutrients, vitamins and fiber for humans and are thus vital for health and well being. Well balanced diets, rich in fruits and vegetables, are especially valuable for their ability to prevent vitamin C and vitamin A deficiencies and are also reported to reduce the risk of several diseases (Kalia and Gupta, 2006).

Fruits and vegetables are widely exposed to microbial contamination through contact with soil, dust and water and by handling at harvest or during postharvest processing. They therefore harbor a diverse range of microorganisms including plant and human pathogens (Nguyen-the and Carlin, 1994; Dunn et al., 1995; Carmo et al., 2004). Differences in microbial profiles of various fruits and vegetables result largely from unrelated factors such as resident microflora in the soil, application of nonresident microflora via animal manures, sewage or irrigation water, transportation and handling by individual retailers (Ray and Bhunia, 2007; Ofor et al., 2009). Thus despite their nutritional and health benefits, outbreaks of human infections associated with the consumption of fresh or minimally processed fruits and vegetables have increased in recent years (Hedberg, 1994; Altekruse and Swerdlow, 1996; Beuchat, 1996, Beuchat, 2002). Enteric pathogens such as *Escherichia coli* and *Salmonella* are among the greatest concerns during food-related outbreaks (Buck et al., 2003).

Even though pharmacological industries have produced a number of new antibiotics in the last three decades, resistance to these drugs by microorganisms has increased. In general, bacteria have the genetic ability to transmit and acquire resistance to drugs, which are utilized as therapeutic agents (Gislene et al., 2000). For a long period of time, plants have been a valuable source of natural products for maintaining human health. The use of plant extracts and phytochemicals, both with known antimicrobial properties, can be of great significance in therapeutic treatments (Seenivasan et al., 2006). Many plants have been used because of their antimicrobial traits, which are due to compounds synthesized in the

**Corresponding author:* e-mail: charliguitar@yahoo.com

secondary metabolism of the plant. These products are known by their active substances e.g. the phenolic compounds which are part of the essential oils, as well as tannin (Tyagi and Malik, 2010).

Numerous studies have been published on the antimicrobial activities of plant extracts and many of these studies have included plants used in ethno medicine. Some of the compounds responsible for the antimicrobial activity observed from the plants have been isolated by activity-guided fractionation and characterized using various spectroscopic methods. The antimicrobial agents from plants can, therefore, be classified into groups such as essential oils and terpenoids, tannins, phenols and alkaloids to name a few (Cowan, 1999).

The greatest losses in food are due to microbiological alterations. Many chemical and physical processes have been developed to preserve food quality. Among such processes, adequate packaging is a fundamental factor in the conservation and marketing phases. Thus, packaging plays a prominent role in maintaining food quality (Debeaufort et al ., 1998). Antimicrobial films and coatings have innovated the concept of active packaging and have been developed to reduce, inhibit or delay the growth of microorganisms on the surface of foods in contact with the packaged product (Appendini & Hotchkiss, 2002). In most fresh or processed foods, microbial contamination occurs at a higher intensity on the food surface, thus requiring an effective microbial growth control (Padgett et al., 1998). Traditionally, antimicrobial agents are added directly to the foods, but their activity may be inhibited by many substances in the food itself, diminishing their efficiency. In such cases, the use of antimicrobial films or coatings can be more efficient than adding antimicrobial agents directly to the food since these may selectively and gradually migrate from the package onto the surface of the food, thereby high concentrations being maintained when most necessary(Ouattara et al., 2000).

Edible films and coatings can be used to help in the preservation of fruit and vegetables because they provide a partial barrier to moisture, O_2 and CO_2, also improving mechanical handling properties, carrying additives, avoiding volatiles loss and even contributing to the production of aroma volatiles (Olivas & Barbosa-Ca´ novas, 2005).

Edible coatings may be composed of polysaccharides, proteins, lipids or a blend of these compounds (Mahmoud and Savello, 1992; Park et al., 1994a, b; Guilbert et al., 1996; Li and Barth, 1998; Arvanitoyannis and Gorris, 1999). Their presence and abundance determine the barrier properties of material with regard to water vapor, oxygen, carbon dioxide and lipid transfer in food systems (*Guilbert et al.*, 1996). However, none of the three constituents can provide the needed protection by themselves and so are usually used in a combination for best results (McHugh and Krochta, 1994a, b; Guilbert et al., 1996).

Cactus mucilage is an example of edible coating containing polysaccharides that is commonly used in food, cosmetics, pharmaceutical and other industries. The complex polysaccharide is part of dietary fibre and has the capacity to absorb large amounts of water, dissolving and dispersing itself and forming viscous or gelatinous colloids (Dominguez-Lo´pez, 1995).

Mango (*Mangifera indica* L) is a very delicious tropical fruit which belongs to family *Anacardiaceae.* It is an abundant source of vitamins, minerals and is famous for its excellent flavor, attractive fragrance and nutritional value. It is as an emerging tropical export crop and is produced in about 90 countries in the world with a production of over 820,877MT. The magnitude of post harvest losses in fresh fruits and vegetables is an estimated 5-25% in developed countries and 20-40% in developing countries, depending upon the commodity (FAO, 2001).

The present study is aimed to evaluate the effect of prickly pear cactus (*O. ficus indica*) mucilage as an edible coating on the microbiological properties of mango fruits during storage in ECS.

MATERIALS AND METHODS

Source of fruits and Coating materials

Ogbomoso mangoes, common consumer varieties, were purchased from a local market on the day after harvest and were immediately placed in ambient storage (27°C ± 2).Uniform sized, defect-free fruits were selected. Cactus stems were obtained from a local farmer and were stored at 23°C ± 1 prior to formation of the coating solution. Glycerol (99.5%) was purchased from Sigma Chemical Co.

Preparation of Coating solution

Cactus stems were peeled and cubed (1 cm^3). Samples were homogenized (20% w/v) in distilled water. The slurry was centrifuged for 10 min at 4500 × g and the supernatant obtained was used to prepare the edible coating (Sa´enz, Va´squez, Trumper, & Fluxa´, 1992). It was then pasteurized to form a pure mucilage extract(ME) .Mangoes were dipped in coating solution for 30secs, the excess coating was drained and the coated mangoes were dried in a forced-air dryer (20 °C) for 30 min. Mangoes dipped in distilled water were used as a blank.

Treatments

T_o (control):-Mangoes that wasn't coated with cactus mucilage ; $T_{1:-}$ Mangoes coated with Pure mucilage extract (ME); T2:-Mangoes coated with mucilage extract(ME) as described above was mixed with 5ml glycerol (MEG) serving as plasticizer for the antimicrobial compounds present on the pure mucilage extract to be directly attached to mango fruits. The treated and untreated fruits were packed in small plastic baskets and each basket contained 20 mango fruits. The baskets were stored at ECS temperature and relative humidity (27±2°C and 55-60%).

The evaporative cooling system

The evaporative cooler used for the study consisted of a double-walled rectangular brick construction with the interspaces filled with riverbed sand saturated with water. The clay brick used in the wall construction was factory baked (at 600 °C) and was of dimensions 25.5 cm x 12.0 cm x 6 cm thick. The sand was previous to water. The water used was clean and free of foreign matter so as to maintain previousness and avoid clogging of the sand.
The interior surfaces of the cooling chamber walls were given a smooth finish with 1.2 cm thick cement plaster, while a heat insulating cover of 1.9 cm thick particleboard closed the top. The walls were built on a short plinth of concrete to prevent water seepage into the soil. Two framed doors of sawn wood were fixed one to each of the adjacent walls on a side to provide access to the 1.38 m^3 capacity chamber. These two doors were considered adequate for the needed thermal insulation against heat flux but an additional polystyrene foam board could be installed for more insulation. The permanent structure was erected in open space exposed to free air but shaded from direct solar radiation with an open sided shed of thatch

Microbial analysis

Thirty grams of mango fruit pulps were removed aseptically from each treatment. The sample was then homogenized in peptone saline solution (8.5 g l^{-1} NaCl+1 g l^{-1} peptone (Oxoid, L34)) for 1 min in a stomacher (S400, Shanghai Scientific Instrument Co., Ltd., Shanghai, China). After making serial dilutions in peptone water, the samples were plated on different media as follows: (1) plate count agar (PCA, OxoidCM325) for isolating total aerobic psychrotrophic micro- organisms was incubated at 30°C for 72 h; (2) Sabouraud media (Oxoid CM41) for isolating yeasts and molds was incubated at 25 °C for 120 h. (3) Mesophilic aerobic counts were enumerated using Plate Count Agar (Oxoid) and incubation at 30°C for 3 days. Colonies were counted and the results expressed as $CFUg^{-1}$ of mangoes. Analyses were carried out periodically in randomly sampled pairs of trays within 7 weeks. Two replicate counts were performed for each tray.

Statistical analysis

Statistical analysis of the data was carried out employing analysis of variance (ANOVA) package included in StatSoft Statistica 8.0 (Hill and Lewicki, 2007).

RESULTS AND DISCUSSION

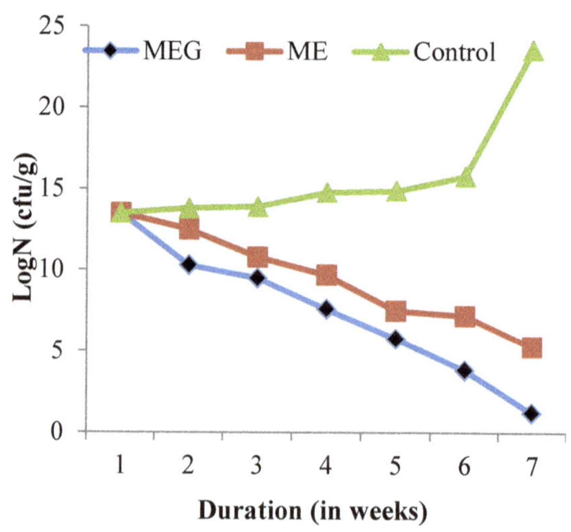

Fig 1: Effect of mucilage coatings on Total aerobic psychrotrophic count of mango fruit during storage in ECS

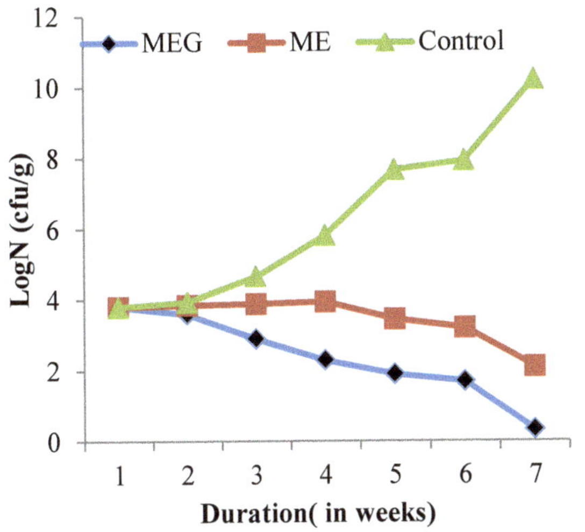

Fig 2: Effect of mucilage coatings on Moulds and yeast count of mango fruit during storage in ECS

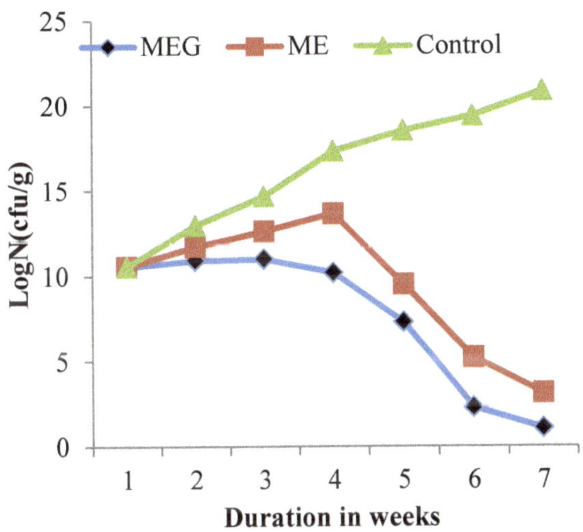

Fig 3: Effect of mucilage coatings on Total aerobic psychrotrophic count of mango fruit during storage in ECS

The predominant microflora which influences the shelf life of fruits and vegetables are psychrotrophic bacteria (Garcia- Gimeno & Zurera-Cosano, 1997; Hotchkiss & Banco, 1992). Changes in the total aerobic Psychrotrophic count, total number of yeasts and moulds in mangoes stored for seven weeks at an average temperature of 27±2°C and relative humidity55-60% are shown in Figs. 1 and 2. During the period of storage coating significantly hindered the increase in total aerobic psychrotrophic count compared with the control samples (Fig.1). Similar effect of coating was observed in reducing the growth of yeasts and moulds during the storage (Fig. 2). At the end of 7 weeks of storage, virtually apparent differences were observed between coated MEG and ME and the control samples ($p<0.05$) at both temperatures.

At harvest, mangoes had 13.5 and 3.8 log CFU g^{-1} for total aerobic psychrotrophic and yeast and mold counts, respectively. Following 7 weeks of ECS storage at temperature of 27±2°C and relative humidity55-60%, the microbial populations of *Cactus* mucilage-treated mangoes were significantly reduced, the reduction being slightly more effective for yeast and mold counts from coated mangoes MEG and ME which were (0.34 log CFU g-1) and (2.11log CFU g^{-1}) respectively than for aerobic Psychrotrophic counts from coated mangoes MEG and ME which were (1.23 log CFU g^{-1}) and (5.30 log CFU g^{-1}) respectively. Moreover, the yeast and mold from uncoated mangoes fruit significantly increases to (10.23log CFU g^{-1}) while the aerobic Psychrotrophic in the uncoated mangoes fruit significantly increases to (23.56log CFU g^{-1}), but which were significantly reduced in *Cactus* mucilage-coated mangoes.

Fig 3. shows that the Cactus mucilage reduces the microbial load of mesophilic population on the coated mangoes MEG and ME which were (1.07 log CFU g^{-1}) and (3.12 log CFU g^{-1}) compared to the uncoated mango which had (20.84 log CFU g^{-1}) at the end of storage.

The microorganisms present in fruits and vegetables are a direct reflection of the sanitary quality of the cultivation water, harvesting, transportation, storage, and processing of the produce (Beuchat, 1996; Ray and Bhunia, 2007).

The number of mesophilic aerobic microflora present was very similar to that of psychrotrophic microflora and therefore most of the microorganisms were able to grow at storage temperatures. Garg et al. (1990) pointed out that mesophilic and psychrotrophic counts may be of similar magnitude at the time of processing. Storage at refrigeration temperatures generally selects in favour of the growth of psychrotrophic microorganisms, including the pectinolitic pseudomonads (Nguyen-the and Carlin, 1994). Nevertheless, mesophilic microorganisms may continue to grow at low temperature, albeit at reduced growth rates (Vescovo et al., 1996; Carlin et al., 1990).

Observed numbers of yeasts and moulds were lower than bacteria, during storage in this study. Tournas (2005) obtained similar results with samples of fresh and minimally-processed

vegetables, and sprouts. The role of yeasts in the spoilage of fresh vegetables is not well studied, although they have been implicated in the spoilage of fermented vegetable products and the development of soft rot (Fleet, 1992).
However, some authors (Tournas, 2005; Tournas and Katsoudas, 2005) have pointed out the possible health problems associated with the presence of moulds in fruit and vegetables, as some may produce mycotoxins and others are known to cause allergies when they are able to produce large numbers of conidia.
The Cactus mucilage treatment of mangoes significantly attenuated the microbial growth on fruits compared to control. Concentration of yeast and mold was found to be decreased from 10.23log CFU g-1 to 0.34 log CFU g^{-1} (MEG) and 2.11log CFU g^{-1} (ME),also the concentration of aerobic Psychrotrophic bacteria was decreased from 23.56log CFU g^{-1} in control to 1.23 log CFU g^{-1} (MEG) and 5.30 log CFU g^{-1} (ME) mucilage treated fruits while the concentration of aerobic mesophilic bacteria was decreased from 20.84log CFU g^{-1} in control to 1.07 log CFU g^{-1} (MEG) and 3.12 log CFU g^{-1} (ME) mucilage treated fruits.
Rojas-Gra¨u et al. (2007) found that lemongrass, oregano, and vanillin essential oils in alginate coatings reduced the growth of psychrophilic aerobes, yeast, and molds on apple cuts by more than 2 log cfu/g. Natamycin in a bilayer coating of chitosan significantly decreased fresh melon decay caused by two strains of spoilage fungi (Cong et al. 2007). Clove, cinnamon, and oregano essential oils totally inhibited the growth of *Candida albicans*, *Aspergillus flavus*, and *Eurotium repens* in vitro when they were used in paraffin coating of paper packaging materials and completely protected strawberry from visible fungal growth during storage for 7 days at 4°C (Rodriguez et al. 2007). Chitosan is a natural antimicrobial compound and is capable of forming stable coatings on fresh-cut papaya that suppress microbial growth (Gonz′alez-Aguilar et al. 2009).
Addition of glycerol to the cactus mucilage extract greatly improves the increase the shelf life of mangoes during storage compared to the mangoes without plastizicer.
Edible films and coatings need to have good elasticity and flexibility, a low brittleness, a high toughness and to prevent cracking during handling and storage (Barreto et al. 2003). Therefore, plasticizers of low molecular weight (non volatile) are typically added to hydrocolloid film forming solutions to modify the flexibility of edible films. Plasticizers with characteristics such as small size, high polarity, more polar groups per molecule, and greater distance between polar groups within a molecule generally impart greater plasticizing effects on a polymeric system. Indeed, they act by increasing the free volume or in other word by decreasing intermolecular attractions between adjacent polymeric chains by reducing hydrogen bonding between polymers chains (Myllarinen et al. 2002).
Several studies on plasticization of chitosan films revealed that poly (ethylene glycol) (PEG) could improve the elastic properties of chitosan. Caner et al. (1998) observed that chitosan plasticization using PEG was stable until 9 weeks of storage. On the contrary, Butler *et al.* (1996) found the water barrier and mechanical properties of plasticized chitosan films with glycerol changed during storage. Other authors used plasticizers in chitosan blends. Hosokawa et al. (1990) used glycerol to plasticize chitosan/cellulose composites, whereas Arvanitoyannis et al. (1998) used sorbitol and sucrose to plasticize chitosan/poly (vinyl alcohol) blends. They stated that the elongation of blended films increased with increasing plasticizer contents, but at high plasticizer contents there were decreases in both tensile strength and modulus.

CONCLUSIONS

The results of this experiment showed that the use of an antimicrobial coating consisting Pure mucilage extract (ME) and mucilage extract mixed with 5ml glycerol (MEG) , is a viable alternative in controlling the microbiota present in minimally processed mango, since the growth of psychrotroph sand yeasts and molds, was substantially inhibited by the application of MEG and ME .
Based on the concept of protection barrier technology, the use of such coating may contribute to improve safety in mango fruit thereby prolonging its shelf life. Coating may be applied on fruits and vegetables, combined to other types of controls, such as quality raw material, hygienic processing conditions and storage temperatures. The combination of these treatments as barrier offers a greater potential for shelf-life extension of fruits and vegetables.

REFERENCES

Appending P and Hotchkiss JH, 2002. "Review of antimicrobial food packaging". Innovative Food Science and Emerging Technologies. 3:113–126.

Arvanitoyannis I and Gorris LGM, 1999. "Edible and biodegradable polymeric materials for food packaging or coating in processing foods: Quality optimization and process assessment". CRC Press, Boca Raton, FL., 357-371.

Arvanitoyannis I, Nakayama A and Aiba S, 1998. Edible films made from hydroxypropyl starch and gelatin and plasticized by polyols and water. Carbohydrate Polymers. 36:105-119.

Altekruse SF and Swerdlow DL, 1996. The changing epidemiology of food borne diseases. American Journal of Medical Science. 311:23-29.

Barreto PLM, Pires ATN and Soldi V, 2003. Polymer Degradation and Stability. 79:147–152.

Beuchat LR, 1996. Pathogenic microorganisms associated with fresh produce. Journal of food Protection. 59:204-216.

Butler BL, Vergaro PJ, Testin RF, Bunn JM, and Wiles JL, 1996. Mechanical and barrier properties of edible chitosan films as affected by composition and storage. Journal of Food Science. 61: 953-961.

Beuchat LR, 2002. Ecological factors influencing survival and growth of human pathogens on raw fruits and vegetables. Microbes Infection. 4:413-423.

Buck JW, Walcott RR and Beuchat LR, 2003. Recent trends in microbiological safety of fruits and vegetables. Plant Health Progress. 10:1094.

Cowan MM, 1999. Plant products as antimicrobial agents. Clinical Microbiology Review. 12:564-582.

Carmo LS, Cummings C, Linardi VR, Dias RS, Souza JM, Sena MJ, Santos DA, Shupp JW, Pereira RK and Jett M, 2004. A case study of a massive *Staphylococcal* food poisoning incident. Foodborne Pathogen Diseases. 1:241-246.

Caner C, Vergano PJ and Wiles J.L, 1998. Chitosan film mechanical and permeation properties as affected by acid, plasticizer and storage. Journal of Food Science. 63(6):1049-1053.

Carlin F, Nguyen-the C, Chambroy Y, Reich M, 1990. Effects of controlled atmospheres on microbial spoilage, electrolyte leakage and sugar content of fresh 'ready to use' grated carrots. International Journal of Food Science and Technology. 25:110–119.

Cong F, Zhang Y, and Dong W, 2007 .Use of surface coatings with natamycin to improve the storability of Hami melon at ambient temperature. Postharvest Biology and Technology. 46:71–75.

Dunn RA, Hall WN, Altamirano JV, Dietrich SE, Robinson-Dunn B and Johnson DR, 1995. Outbreak of *Shiegella flexneri* linked to salad prepared at a central commissary in Michigan. Public Health Reports. 110(5):580-586.

Debeaufort F, Quezada-Gallo JA, and Voilley A, 1998. "Edible films and coatings: tomorrow packaging: a review". Critical Reviews in Food Science. 38(4):299–313.

Dominguez-Lo´pez, A, 1995. "Review: use of the fruit and stems of the prickly pear cactus (*Opuntia spp.*) into human food". Food Science and Technology International. 1: 65–74.

FAO, 2001. Food Balance Sheets: A Hand Book. Economic and Social Development Department, Food and Agriculture Organization of United Nations, Viale delle Terme di Caracalla, 00153 Rome, Italy, pp: 14.

Fleet GH, 1992. Spoilage yeasts. Critical Reviews in Biotechnology. 12: 1–44.

Guilbert S. Gontard N. and Gorris LGM, 1996. Prolongation of the shelf-life of perishable food products using biodegradable films and coatings. Lebensmittel-Wissenschaft und-Technologie., 29:10-17.

Gonz´alez-Aguilar GA, 2009. Effect of chitosan coating in preventing deterioration and preserving the quality of fresh-cut papaya "Maradol". Journal of the Science Food Agriculture. 89:15–23.

Garg N, Churey JJ, Splittsotesser DF. 1990 Effect of processing conditions on the microflora of fresh-cut vegetables. Journal of Food Protection. 53:701–703.

Garcia-Gimeno RM, and Zurera-Cosano G, 1997. "Determination of ready-to-eat vegetable salad shelf-life". International Journal of Food Microbiology. 36:31–38.

Gislene GF, Locatelli N.J, Paulo CF and Giuliana LS, 2000. Antibacterial activity of plant extracts and phytochemicals on antibiotic resistant bacteria. Brazilian Journal of Microbiology.,31:247-256.

Herdberg CW, McDonald KL and Osterholm MT,1994. Changing epidemiology of foodborne disease: a Minneasota perspective. Clinical Infectious Diseases., 8:671-682.

Hill T. and Lewicki P. (2007). Statistics Methods and Applications. Tulsa, OK. StatSoft, USA.

Hotchkiss J and Banco M,1992. "Influence of packaging technologies on the growth of microorganisms in produce". Journal of Food Protection.,55:815–820.

Hosokawa J, Nishiyama M, Yoshihara K, Kubo T. 990 Biodegradable film derived from chitosan and homogenized cellulose. Industrial and Engineering Chemistry Research .,29(1):800-805.

Kalia A and Gupta RP. Fruit Microbiology, in Hui Y.H, J., Cano, M.P., Gusek, W., Sidhu, J.W., Sinha, N.K,2006. Handbook of Fruit and Fruit processing. 1st Edition, Blackwell publishing., pp 3-28.

Li P and Barth MM, 1998. "Impact of edible coatings on nutritional and physiological changes in lightly processed carrots". Postharvest Biology and Technology., 14:51-60.

McHugh TH and Krochta JM, 1994a. "Milk protein based edible films and coatings". Food Technology., 48:97-103.

McHugh TH and Krochta JM,1994b. "Sorbitol vs glycerol plasticized whey protein edible films: Integrated oxygen permeability and tensile property evaluation". Journal of Agricultural and Food Chemistry., 42: 41-45.

Mahmoud R and Savello PA ,1992. "Mechanical properties of and water vapor transferability through whey protein films". Journal of Dairy Science., 75: 942-946.

Ouattara B, Simard R, Piette G, Be´gin A, and Holley RA. 2000 "Inhibition of surface spoilage bacteria in processed meats by application of antimicrobial films prepared with chitosan". International Journal of Food Microbiology.,62:139–148.

Olivas GI and Barbosa-Canovas GV, 2005. "Edible coatings for fresh-cut fruits". Critical Reviews in Food Science and Nutrition., 45:657-670.

Ofor MO, Okorie VC, Ibeawuchi II, Ihejirika GO, Obilo OP and Dialoke SA. 2009 Microbial Contaminants in Fresh Tomato Wash Water and Food Safety Considerations in South-Eastern Nigeria. Life Science Journal., 1:80-82.

Padgett T, Han LY and Dawson, P. L, 1998. "Incorporation of food-grade antimicrobial compounds into biodegradable packaging films". Journal of Food Protection., 61(10):1330–1335.

Park HJ, Chinnan MS and Shewfelt RL, 1994a. "Edible coating effects on storage life and quality of tomatoes". Journal of Food Science., 59:568-570.

Park HJ, Chinnan MS and Shewfelt RL, 1994b. "Edible corn-zein film coatings to extend storage life of tomatoe"s. Journal of Food Processing and Preservation., 18:317-331.

Ray B and Bhunia AK. 2007 Fundamental Food Microbiology. 4th Edn., CRC Press, USA. p. 492.

Myllarinen P, Partanen R, Sepalla J and Forsell P, 2002. Carbohydrate. Polymer., 50:355–61.

Rodriguez A, Batlle R and Ner´ın C, 2007. The use of natural essential oils as antimicrobial solutions in paper packaging. Part II. Progress in Organic Coatings., 60:33–38.

Rojas-Grau, M., Grasa-Guillem, R., Martin-Belloso, O., 2007. "Quality changes in fres-cut Fuji apple as

effected by ripeness stage, antibrowning agents and storage atmosphere. Journal of Food Science., 72(1):36-42.

Sa´enz, C, Va´squez, M, Trumper S, and Fluxa´ , C, 1992. "Extractio´n y Composicio´n Quı ´mica del Mucı´lago de Tuna (Opuntia ficus indica). *In Proceedings II Congreso Internacional de Tuna y Cochinilla"*.pp 93–96.

Seenivasan P, Manickkam J and Savarimuthu I, 2006. *In vitro* antibacterial activity of some plant essential oils. Complementary alternative medicine ., 6: 39.

Tyagi AK and Malik A, 2010. Liquid and vapour-phase antifungal activities of selected essential oils against *Candida albicans*. Complementary alternative medicine.,10:65.

Tournas VH, 2005. Moulds and yeasts in fresh and minimally processed vegetables, and sprouts. International Journal of Food Microbiology., 99:71–77.

Tournas VH and Katsoudas E. 2005 Mould and yeast flora in fresh berries, grapes and citrus fruits. International Journal of Food Microbiology. 105:11–17.

Nguyen-the C and Carlin F. 1994 The microbiology of minimally processed fresh fruits and vegetables. Critical Reviews in Food Science and Nutrition. 34:371–401.

Vescovo M, Torriano S, Orsi C, Macchiarolo F, and Scolari G, 1996. Application of antimicrobial-producing lactic acid bacteria to control pathogens in ready-touse vegetables. Journal of Applied Bacteriology. 81:113–119.

APPLICATION OF DNA MARKERS FOR THE DETECTION OF AMYLOSE CONTENT IN SRI LANKAN RICE (*ORYZA SATIVA* L.) VARIETIES

N.S. Kottearachchi[1*], R.K. Peiris[1] and S. Rebeira[2]

[1]*Department of Biotechnology, Faculty of Agriculture and Plantation Management, Wayamba University of Sri Lanka, Makandura, Gonawila (NWP), Sri Lanka.* [2]*Department of Grain Quality, Rice Research and Development Institute, Bathalagoda, Sri Lanka*

ABSTRACT

Grain quality improvement of rice is considered as prioritized research area in rice breeding programs in Sri Lanka. Amylose Content (AC) is a key determinant of end quality of rice. Amylose Content is controlled by granule bound starch synthase (GBSS) enzyme which is encoded by *waxy* gene, located on chromosome 6. Similarly another character associated with grain quality, Gelatinization Temperature (GT), is controlled by the alkali degeneration locus (*alk*), which encodes for a soluble enzyme called starch synthase IIa (SSIIa). Two SSR markers RM 190 and RM 314 were selected from the previously developed QTL maps that showed association with AC and GT. This study demonstrates the feasibility of RM 190 and RM 314 for the detection of polymorphism associated with AC and GT. In total, 18 different rice varieties were analysed. The relationship between marker genotype and amylose content was detected by comparing the pooled AC under 2 different allele categories using Mann-Whitney Test. The results indicated that RM 190 was able to detect genetic variation among the rice varieties quite efficiently, on the basis of which, the rice varieties could be identified with two categories; with low AC and with intermediate and high AC. Therefore, the SSR marker, RM 190, could be efficiently employed in large-scale screening programmes to predict AC through Marker Assisted Selection (MAS).

Keywords: Amylose content, Gelatinization temperature, Microsatellite Markers, *Oryza Sativa* L.

INTRODUCTION

Rice is one of the most important staple food crops of more than one-half of population with approximately 95% of production in Asia (Bhattacharjee *et al.*, 2002). Starch is the major component of the rice endosperm, which consists of amylose and amylopectin. The amylose is made up of linear molecule composed of α (1,4)- linked glycosidic chains, whereas, the amylopectin exhibits highly branched glucan with α (1,6) glycosidic bonds that connect linear chains. Amylose content (AC) (Juliano, 1985), gel consistency (GC) (Cagampang *et al.*, 1973) and Gelatinization Temperature (GT) (Little *et al* ., 1958) are the three physiochemical characteristics of starch, involve in determination of eating, cooking and processing quality of rice.

Amylose content is considered as one of the most important indicators of rice grain quality. The AC is determined as the ratio of amylose amount present in endosperm to total starch content. The high AC leads to a dry, fluffy and separated cooked rice which becomes hard upon cooling (Juliano, 1985), whereas, the low AC gives a moist, sticky and glossy rice grain after cooking. Rice cultivars are commonly categorized as glutinous/ waxy (0-5% of amylose), low AC (<20% of amylose), intermediate AC (21-25% of amylose) and high AC (>25% amylose) (Kongseree and Juliano, 1972).

Earlier work on rice genetics, has identified major genes as well as QTLs (Quantitative trait loci), which are considered responsible for AC in rice endosperm. The QTLs are present on chromosome 5 and 6 and, the major gene, *waxy*, located in chromosome 6, accounts for 91.9% of the total variation of AC (He *et al.*, 1999). The *waxy* gene plays a key role in amylose synthesis by encoding the enzyme, granule bound starch synthase (GBSS) (Tan *et al.*, 1999 and Fan *et al.*, 2005). Amylose content polymorphism in rice has been explained by the *waxy* alleles: wx^a and wx^b (Sano, 1984). The wx^a allele is predominantly involved in the synthesis of GBSS. According to the study performed by Wang and co-workers (1995), the variation in AC accumulation in rice is primarily governed at post transcriptional levels. Gelatinization

**Corresponding author:* e-mail: kottearachchins@yahoo.com

temperature (GT) of rice starch is defined as the range of temperature at which nearly all the starch granules start to swell irreversibly in hot water with simultaneous loss of birefringence and crystallinity (Virmani, 1994). GT of rice is controlled by the alkali degeneration locus (*alk*) on chromosome 6, which encodes soluble starch synthase IIa (SSIIa) isoform (Fan *et al.*, 2005). Low GT rice needs lesser energy input during cooking than high GT rice (Bao *et al.*, 2004). Based on GT values, rice cultivars can be broadly categorized into three classes: low (<70 °C), intermediate (70-74 °C) and high (>74 °C) (Kongseree and Juliano, 1972). Development of cultivars with improved agronomic traits while maintaining the desired AC and GT is of major concerns of rice breeders. Conventional methods for measuring of these quality traits are time consuming and expensive. DNA markers provide a useful, easy and cost-effective way to identify plants with desired level of AC and GT values (DeOliveira *et al.*, 2010 and Tabkhkar *et al.*, 2012). Theretofore, this study was carried out to detect a suitable DNA marker, which is closely linked with AC and GT in order to facilitate rice breeding programs.

MATERIALS AND METHODS

Plant Materials

Both mature seeds and husked and polished grains of 18 different rice varieties were collected from Rice Research and Development Institute, Bathalagoda, Sri Lanka (see Table 2 for details).

Analysis of Amylose Content

Initially rice samples were husked and polished prior to milling. Twenty whole-milled rice kernels of 18 rice samples were ground separately in a Udy cyclone mill (sieve mesh size 60). Amylose content per 100 mg was determined through measuring blue value of rice varieties as described by Juliano (1971). A 100-mg of rice sample was put into a 100 ml volumetric flask and 1 ml of 95% (v/v) ethanol was added. Then 1 ml of 1 N (w/v) NaOH was added. The content was then boiled for 20 min to gelatinize starch. After cooling the content, the volume was made up to 100 ml. A 5-ml of starch solution was pipetted out into a 100-ml volumetric flask. 1 ml of 1 N acetic acid and 2 ml of iodine solution (0.2 g iodine and 2.0 g potassium iodide in 100 ml aqueous solution) were added. Volume was made up to 100 ml with distilled water and solution was allowed to stay for 20 min after shaking. The absorbance of the solution was measured at 620 nm with spectrophotometer (JENWAY 6305, UK).

The standard curve was prepared using 40 mg of potato amylose to calculate amylose content of rice varieties. Fourty (40) mg of potato amylose was put into a 100 ml of volumetric flask and 1 ml of 95% (v/v) ethanol and 9 ml of 1N NaOH were added and content was heated for 20 min at boiling temperature. After cooling, the content volume of the solution was made up to 100 ml using distilled water. Then 1, 2, 3, 4, and 5 ml of amylose solution were pipetted out into 100 ml flasks. Then 0.2, 0.4, 0.6, 0.8 and 1 ml of 1 N acetic acid were added to the flasks respectively. Finally, 2 ml of iodine solution was added to each flask and volumarized up to 100 ml with distilled water. Solutions were stood up for 20 min after shaking. Then absorbance values were measured at 620 nm. Measured absorbance values were plotted at 620 nm against the concentration of anhydrous amylose (mg).

Analysis of Gelatinization Temperature

The GT was indirectly measured on rice by the alkali spreading value (Little *et al* ., 1958). Husked and polished rice samples were used for the analysis. Selected duplicate sets of six milled grains without cracks of each sample were put into petridishes. 10 ml of 1.7% (w/v) KOH was added and grains were spread in the petridish to provide enough space. Constant temperature of 30 °C was maintained to ensure better reproducibility. After 24 hrs, GT was identified based on the score given from 1 to 7, by visual appearance of the gelatinization. Then the gelatinization temperature was identified based on the score value of the samples as reported by Cruz and Khush (2000).

Genomic DNA Extraction and Quantification

Genomic DNA was extracted from 2 to 3-week-old leaves. Approximately, a 3-cm tender leaf piece was ground with 300 µl of DNA extraction buffer (1 M KCl, 1 M Tris HCl, 0.5 M EDTA) in a micro centrifuge tube separately for each variety. Then homogenate was incubated at 70 °C for 15 min at room temperature and centrifuged at 15000 rpm for 15 min immediately after the incubation. 100 µl of isopropanol was added into a new eppendorf tube. Then supernatant was transferred into the tube containing isoproponol and mixed gently. Tubes was kept at 4 °C overnight and centrifuged at 15000 rpm for 15 min at 4 °C. The supernatant was removed and DNA pellet was washed with 150 µl of 70% (v/v) ice cold ethanol by centrifuging for 10 min at 15000 rpm at 4 °C. Finally, the supernatant was removed and pellet was air dried. After the removal of all the ethanol traces, the DNA pellet was dissolved in 300 µl of 1/10th TE buffer (10 mM Tris, 1 mM EDTA) and used for PCR.

.

Selection of SSR Markers

The SSR markers linked with AC and GT were selected by QTL maps developed by Fan *et al.* (2005) and linkage association studies as reported previously (Septiningsih *et al.*, 2003; Shu *et al.*, 2006; Tabkhkar *et al.*, 2012). Accordingly two SSR primers; RM 190, RM 314 (Table 1) were selected for PCR amplification

Table 1: DNA sequence information of SSR markers used in the study

Marker	Motif	Primer Sequence	AC*	GT**	Reference
RM190	(CT)11	F: 5' TTTGTCTATCTCAAGACAC 3' R: 5' TTGCAGATGTTCTTCCTGATG 3	*ac6a*	*asv6a*	Fan *et al.* (2005), Tabkhkar *et al.* (2012)
RM314	(GT)8(CG)3 (GT)5	F: 5' CTAGCAGGAACTCCTTTCAGG 3' R: 5' AACATTCCACACACACACGC 3'	*ac6b*	*alk2(t)*	Septiningsih *et al.* (2003), Fan *et al.* (2005), Shu *et al.* (2006), Tabkhkar *et al* . (2012)

*QTL reported for Amylose content, **QTL reported for Gelatinization temperature

No.	Variety	AC (%)	Allele type produced by RM 190	GT
1	At 303	28.72	A	High
2	At 306	23.93	B	High
3	At 308	28.03	A	In
4	At 402	21.54	A	High
5	At 405	13.29	B	Low
6	Bg 250	26.15	A	In
7	Bg 300	26.41	A	In
8	Bg 305	27.35	A	High
9	Bg 358	25.21	A	H/I
10	Bg 379-2	31.88	A	High
11	Bg 403	31.88	A	Low
12	Bg 406	21.41	A	High
13	Bg 407-H	26.84	A & B	Low
14	Basmati 370	16.24	B	In
15	Ld 356	25.3	A	In
16	Ld 365	21.67	A	High
17	Ld 408	21.79	A	Low
18	Pusa basmati	22.09	A	Low

Table 2. Characteristics of rice varieties used in the study

GT; In: Intermediate, H/I: High/Intermediate Allele type produced by RM 190; A allele-amplified short DNA fragment, B allele-amplified long DNA fragment

Table 3: Analysis of genotypes by Mann-Whitney test

Primer	Median AC(%) for allele types		p value
	A	B	
RM 190	16.239	25.470	0.0041

Figure 1. PCR profile generated by markers, RM 190 (upper gel) and RM314 (lower gel) for the tested rice varieties

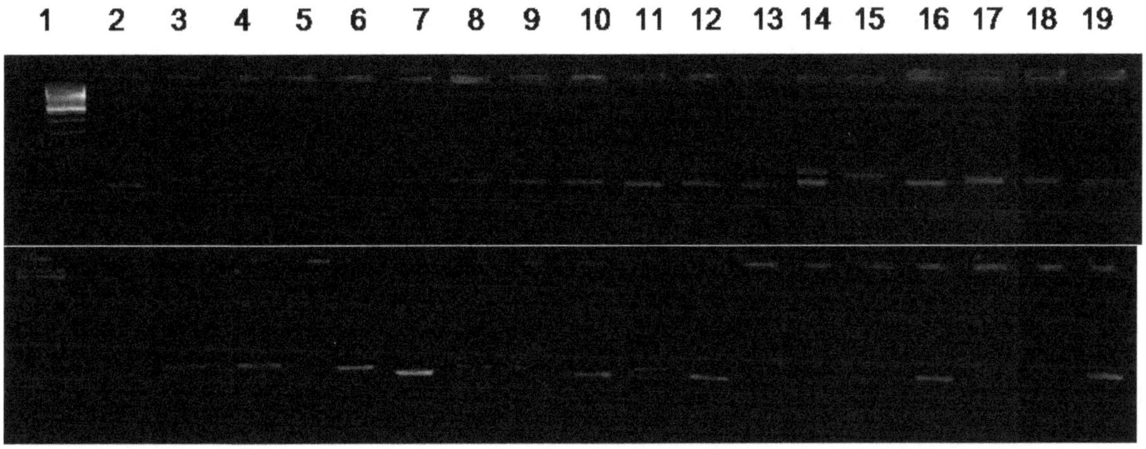

Lane 1- 100 bp ladder, 2-19: At 303, At 306, At 308, At 402, At 405, Bg 250, Bg 300, Bg 305, Bg 358, Bg 379-2, Bg 403, Bg 406, Bg 407-H, Basmati 370, Ld 356, Ld 365, Ld 408, Pusa basmati respectively

Figure 2. Dendrogram constructed based on genotypes of RM190 and groups of Amylose Content showing similarities among rice varieties

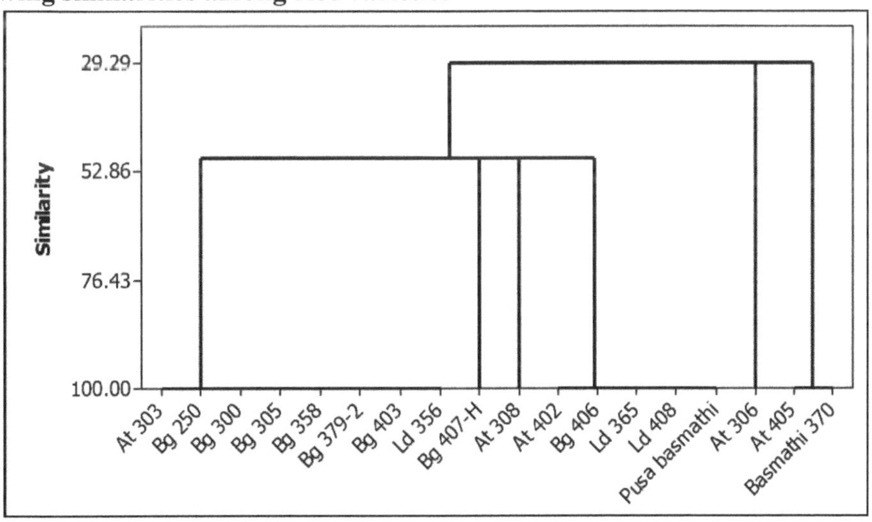

PCR Amplification

The PCR amplification was performed in a 12-μl reaction volume, which consisted of 5 μl of diluted DNA (10ng/μl), 1.5 μl of 10X PCR buffer (provided with Dream Taq polymerase, Fermentas, USA), 1.2 μl of dNTPs (2.5 mM), 0.9 μl of primers (20 pmol/μl), 0.12 μl of Taq DNA polymerase (Dream Taq, Fermentas, USA) and 3.28 μl distilled water. Amplification was carried out in a BioRad (My Cycler^{TM}) thermal cycler with following parameters: initial denaturation at 95 °C for 5 min followed by 35 cycles of 95 °C for 1 min, 52 °C (for RM 190) or 57 °C (for RM 314) for 30 sec, 72 °C for 1 min and final extension at 72 °C for 5 min. The amplified PCR products were detected under agarose gel electrophoresis using 3.8 % gel by precast staining with ethidium bromide. After the electrophoresis, the gels were visualized by using a gel documentation system.

Statistical Analysis

The banding patterns of amplified alleles from SSR markers were observed and based on the allele size, plants showing the presence of similar alleles were categorized viz. A (short allele), B (long allele) etc. The Mann-Whitney test (Fay *et al.*, 2010) was performed to detect whether RM190 marker shows strong association with AC and allele type using Minitab 15. 1. A dendrogram was constructed using the hierarchical clustering procedure based on RM190 genotype and AC values and 'euclidean' distance was calculated by single linkage method using Minitab version 15.1.

RESULTS

Analysis of Amylose Content and Gelatinization Temperature

The AC and GT were measured in 18 locally prevalent rice varieties by Blue value method described by Juliano (1971). According to the analysis At 405 and Basmati 370 were categorized as low AC rice varieties as they were in the range of 10-19%, while At 306, At 402, Bg 406, Ld 365, Ld 408 and Pusa basmati were categorized as intermediate AC rice varieties as they exhibited AC values in the range of 20-25%. Other rice varieties, which showed more than 25% of AC were categorized as high AC containing rice varieties.

Similarly, based on GT values the varieties were grouped into three categories: high, intermediate and low GT varieties (Table 2).

Genotyping by RM 190

With the RM 190 primer pair the PCR amplification reaction produced two distinguishable alleles. Both alleles were between 100-150 base pairs (bp). Except in Bg 407-H, all the other rice varieties had one allele, either A or B (A allele- amplified a shorter DNA fragment, B allele- amplified longer DNA fragment) (Figure 1.). All the low AC rice varieties tested here showed the presence of allele 'B' while At 306, whose AC values were in intermediate range also amplified allele 'B' exceptionally. All the other intermediate and high AC rice varieties showed the presence of allele 'A'.

In order to detect whether there is significant relationship between genotype produced by marker RM190 and the AC values, the pooled amylose categories were compared by Mann-Whitney test. As a result it appeared that RM 190 could significantly ($p<0.01$) distinguish between low AC rice varieties from intermediate and those with high AC (Table 3). All varieties that possessed 'A' allele of RM190 marker were separately clustered with nearly 50% similarity index giving four subclusters further. All intermediate AC varieties, At402, Bg406, Ld365, Ld408 and Pusa basmati except Bg306 were separately subclustered with 100% similarity index. All varieties that contained high amylose content were separately subclustered except Bg308 and Bg407-H. The dendrogram showed that three varieties that contained allele 'B' of RM190 marker were separately clustered and of those, At 405 and Basmati 370, which had low amylose, were separately sub-clustered with 100% similarity index (Figure 2).

Genotyping by RM 314

The marker RM314 was found polymorphic among all the varieties used in the study and produced three distinguishable alleles ranging from bp 111-128. The marker RM 314 locates close to the QTL of AC, *ac6b* (Fan *et al.*, 2005) and QTL of GT *alk2 (t)* (Shu *et al.*, 2006). However, phenotypic variation of AC and GT in rice varieties tested in this study has not linked with any of the RM 314 alleles. Bg 407-H, the hybrid rice variety, also expressed heterozygosis nature through producing two alleles by RM 314.

DISCUSSION

In this study, we have tested the power of two rice markers associated with rice grain quality. Our results show that at least one of the two markers, RM 190, could be efficiently used for the identification of rice population with a desired amylose content (AC). Hence, it could be a useful component of the future breeding programmes aimed at rice crop improvement against various biotic and abiotic stresses in a background of desired AC.

Ayres *et al.* (1997), Bergman *et al.* (2001), Lang and Buu (2004) and Chen *et al.* (2008)

suggested that RM 190 primer was suitable to use as a marker for the detection of AC polymorphism during breeding process. It flanks major QTL, *ac6a,* linked with AC (Fan *et al.*, 2005; Tabkhkar *et al.*, 2012). Therefore, RM 190 primer was mainly used in this study for the detection of AC polymorphism in Sri Lankan rice varieties. The AC values were determined using blue value method. The RM 190 was able to distinguish two types of alleles among tested varieties and showed a strong association with AC values.

Fan *et al.*, 2005 have reported that marker RM190 is one of the closer markers to the main QTL (*asv6a*) responsible for GT. However, our study could not find any relationship between RM 190 and GT. Another marker, RM 314, one of the closest markers to the another AC QTL, *ac6b* and another GT QTL, *alk2 (t)*, located in the same chromosome 6, showed a good level of polymorphism in the rice varieties used. However, once again this marker did not show any linkage with AC and GT values. According to Fan *et al.*, 2005 the effect of *ac6b* was relatively small, giving contributions of about 0.85% to the phenotypic variation of AC in their tested population. This fact indicates that the effect of such minor QTLs are not likely to be distinguished among diverse population.

CONCLUSION

The RM 190 was able to distinguish two types of alleles among tested varieties and showed a strong association with AC values of tested rice. On the basis of our results it can be concluded that RM 190 has the potential to be used in distinguishing low AC rice varieties over the intermediate and high AC. Local rice variety At 405 may have the potential to be used as parental line in developing low amylose content varieties in addition to the donor parent Basmati 370 following RM190 marker based selection. Analysis of segregation population using RM190 would be useful for further validation.

ACKNOWLEDGEMENTS

The authors express their gratitude to the research grant (RG/2011/BT/02) of National Science Foundation, Sri Lanka for donating primers for this study.

REFERENCES

Ayres NM, McClung AM, Larkin PD, Bligh HFJ, Jones CA and Park WD, 1997. Microsatellites and a single-nucleotide polymorphism differentiate apparent amylose classes in an extended pedigree of US rice germplasm. Theor. Appl. Genet. 94: 773-781.

Bao JS, Sun M, Zhu LH and Corke H, 2004. Analysis of quantitative trait loci for some starch properties in rice (*Oryza sativa* L.), thermal properties, gel texture, swelling volume. J. Cereal Sci. 39: 379-385.

Bergman CJ, Delgado JT, McClung AM and Fjellstrom RG, 2001. An improved method for using a microsatellite in the rice *waxy* gene to determine amylose class. Cereal Chem. 78: 257-260.

Bhattacharjee P, Singhal RS and Kulkarni PR, 2002. Basmati rice: A review. International J. Food Sci. Tech. 37: 1-12.

Cagampang GB, Perez CM and Juliano BO, 1973. A gel consistency test for eating quality in rice. J. Sci. Food Agric. 24: 1589-1594.

Chen MH, Bergman C, Pinson S and Fjellstrom R, 2008. *Waxy* gene haplotypes: Association with apparent amylose content and the effect by the environment in an international germplasm collection. J. Cereal Sci. 47: 536-545.

Cruz N D and Khush GS, 2000. Rice grain quality evaluation procedures, In: Aromatic rices. Eds: Singh RK. Singh US and Khush GS. New Dhelhi, Oxford & IBH Publishing Co. Pvt. Ltd, New Delhi. pp. 15-28.

DeOliveira TC, Brondani RPV, Breseghello F, Coelho ASG, Mendonca JA, Rangel PHN, Brondan IC, 2010. Association mapping for yield and grain quality traits in rice (*Oryza sativa* L.). Genet. Mo. Biol. 33(3): 515–524

Fan CC, Yu XQ, Xing YZ, Xu CG, Luo LG and Zhang QF, 2005. The main effects, epistatic effects and environmental interactions of QTLs on the cooking and eating quality of rice in a doubled-haploid line population. Theor. Appl. Genet. 110: 1445-1452.

Fay MP, and Proschan MA, 2010. Wilcoxon-Mann-Whitney or t-test? On assumptions for hypothesis tests and multiple interpretations of decision rules. Statistics Survey. 4: 1–39

He P, Li SG, Qian Q, Ma YQ, Li JZ, Wang WL, Chen Y and Zhu LH, 1999. Genetic analysis of rice grain quality. Theor. Appl. Genet. 98: 502-508.

Juliano BO, 1971. A simplified assay for milled- rice amylose. Cereal Sci. Today. 16 (4): 334-338.

Juliano BO, 1985. Criteria and test for rice grain quality. Rice chemistry and technology. American Association of Cereal Chemists, 2nd edition. Incorporated Saint Paul, Minnesota, USA. American Association of Cereal Chemists, 443-513.

Kongseree N. and Juliano BO, 1972. Physiochemical Properties of Rice Grain and Starch from Lines Differing in Amylose content and Gelatinization Temperature. J. Sci. Food Agric. 20 (3): 714-718.

Lang NT and Buu BC, 2004. Quantitative analysis of Amylose content by DNA Markers through backcross population of Rice (*Oryza sativa* L.). Omonrice. 12: 13-18.

Little RR, Hilder GB and Dawson EH, 1958. Differential effect of Dilute alkali on 25 varieties of milled white rice. Cereal Chem. 35: 11-126.

Sano Y, 1984. Differential regulation of waxy gene expression in rice endosperm. Theor. Appl. Genet. 68: 467-473.

Septiningsih EM, Trijatmiko KR, Moeljopawiro S, McCouch SR, 2003. Identification of quantitative trait loci for grain quality in an advanced backcross population derived from the *Oryza sativa* variety IR64 and the wild relative *O. rufipogon*. Theor. Appl. Genet. 107: 1441-1433

Shu XL, Shen SQ, Bao JS, Wu DX, Nakamura Y, Shu QY, 2006. Molecular and biochemical analysis of the gelatinization temperature characteristics of rice (*Oryza sativa* L.) starch granules. J. Cereal Sci. 44: 40-48.

Tabkhkar N, Rabiei B and Sabouri A, 2012. Genetic diversity of rice cultivars by microsatellite markers tightly linked to cooking and eating quality. Australian J. Crop Sci. 6 (6): 980-985.

Tan YF, Li JX, Yu SB, Xing YZ, Xu CG and Zhang Q, 1999. The three important traits for cooking and eating quality of rice grains are controlled by a single locus in an elite rice hybrid, Shanyou 63. Theor. Appl. Genet. 99: 642-648.

Virmani SS, 1994. Heterosis and hybrid rice breeding. In; Monographs on Theoretical and Applied Genetics 22, Eds: Frankel R, Grossman M and Maliga P, Springer-Verlag Berlin Heidelberg. Germany. p. 129

Wang ZY, Zheng FQ, Shen GZ, Gao JP, Snustad DP, Li MG, Zhang JL and Hong MM, 1995. The amylose content in rice endosperm is related to the post transcriptional regulation of the *waxy* gene. The Plant J. 7: 613-622.

EFFECT OF PHOSPHATE SOLUBILIZING BACTERIA ALONG WITH ACC-DEAMINASE ON WHEAT GROWTH UNDER AXENIC CONDITIONS

Shabana Ehsan[1*], Ifra Saleem[1] and Hafsa Zafar[2]

[1] *Soil Chemistry section, ISCES, Ayub Agricultural Research Institute, Faisalabad*

[2] *Biochemistry Section, PHRC, Agricultural Research Institute, Faisalabad*

ABSTRACT

Phosphorus is one of the major plant growth limiting nutrients and in Pakistan its deficiency found in 90% soils due to alkaline calcareous nature of soils. Plant growth promoting rhizobacteria are efficient in supplying nutrients to plants by solubilizing the unavailable/insoluble (P) in soil or by promoting extensive root growth through 1-aminocyclopropane-1-carboxylate (ACC-deaminase) activity. This study was conducted to assess the effects of phosphate-soluibilizing rhizobacteria and their impact on wheat growth under axenic conditions. Initially rhizobacteria (S1 to S10) were isolated through enrichment by rock phosphate [$Ca_3(PO_4)_2$] and ACC as sole P and N source, respecitively. After isolation, identification and characterization of these strains was done for root colonization, P-solubilization, in vitro ACC-deminase ctivity, auxin production and chitinase activity. The results revealed that rhizobacterial isolate S8 was best strain and showed 2.6 fold higher fresh weight, 2.7 fold higher shoot length and 4 fold higher root length of wheat over uninoculated control. Similarly, S8 was best root colonizer in wheat rhizosphere (8.0×10^{-7} cfu g^{-1}) while isolate S9 showed the highest P- solubilization i.e. 14.36 ppm, S2 showed maximum ACC-deaminase activity (1.8 μmol α-ketobutyrate) and S4 showed highest auxin production (36.77 mgL^{-1}). This study concluded that S8 strain efficiently solubilize phosphate alongwith ACC-deaminase trait which improved the growth of wheat under axenic conditions.

Keywords: Rock phosphate, Phosphate Solubilizing bacteria (PSB), ACC-deaminase.

INTRODUCTION

Phosphours (P) is one of the major nutrients limiting plant growth. In contrast to other essential elements, P is required for growth and development of plants. It is involved in photosynthesis, energy transfer, singal transduction, macromolcular biosyntheis and respiration (Fernandez et al. 2007). However many agricultural soils worldwide are P-deficient (Arcand and Schneider, 2000) and therefore require P to replenish the P-demands by crop plants (Hamdali et al. 2012). Rock phosphate is a natural, cheap and clean compound but unfortunately it is a poor fertilizer since its solubilization is to slow to satisfy plant needs (Zapata and Zahara, 2002) Rock phosphate or phosphorite is a non-detrial sedimentary rock having phosphate in fluoroapatite $Ca_5(PO_4)_3F$ (CFA) and hydroxyapatite $Ca_5(PO_4)_3OH$ (Balte and Robert, 1996).

Plant growth promoting rhizobacteria (PGPR) encompass all the bacteria that inhabit plant roots and affect the promotion by mechanisms ranging from direct influence e.g. increased solubilization and uptake of nutrients or production of growth regulators (Steenhoudt and Vanderleyden, 2000) to direct influence e.g., suppression of pathogens by producing siderophores or antibiotics (Kloepper et al. 1989; Arshad and Frankenberger , 1998). Many PGPR are known as solubilizer of insoluble rock phosphate through the production of organic acids or root exudation that results in lowering soil pH (Gyneshwar et al. 2002; Puente et al. 2004). Organic acids mainly synthesized are citric, lactic,gluconic, oxalic etc (kucey 1983; Vazques et al. 2000) These acids act on insoluble phosphates and convert the same into soluble from, thus providing phosphorus to plants (Ponmurugan and Gopi, 2006; Nautiyal et al. 2000). Other strategie that lead to better uptake or acquisition of nutrients include increased root surface area through enhanced root growth and root hair development (Lynch and Brown, 1998; Gilory and Jones, 2000). Identification and characterization of soil PSB for the effective plant growth promotion broadens the spectrum of phosphate solubilizers available for field application (Chen et al. 2006).

**Corresponding author:* e-mail: shabana.ehsan@yahoo.com

Higher concentrations of ethylene are considered inhibitory for normal plant growth particularly for root growth. Phosporus deficiency could lead in production of higher levels of ethylene in roots and inhibit root and shoot growth (Sarquis et al. 1991; Morgan et al. 1993). ACC serves as precursor of ethylene. Inoculation of seeds or roots with specific inoculants containing ACC- deaminase enzyme could suppress the endogeneous ethylene synthesis which subsequently creates physiological responses (Glick et al. 1998). The enzyme which has been found in soil/some microoganism catalyses the hydrolysis of ACC to NH_3 and a-ketobutyrate (Mayak et al. 1999; Penrose and Glick, 2003).

Keeping in view the importance of P for plants this study was planned to isolate, characterize the phosphate solubilizing rhizobacteria with and/or without ACC- deaminase trait and to assess their effets on growth of wheat under P-deficient environment using rock phosphate (as sole P source) and ACC (as sole N source).

MATERIAL AND METHOD

This experiment was conducted under controlled P conditions to compare the effectiveness of phosphate solubilizing rhizobacteia and for this purpose initially isolation and characterization process was done.

Isolation of strains

For isolation of rhizobacteria different soils were collected from wheat plant rhizosphere. Rhizobacteria were isolated by dilution plate technique using salt minimal medium (Dworkin and Foster, 1958) containing ACC as sole nitrogen source or NBRI-P medium (Nautiyal, 1999) using rock phosphate as sole phosphorus source. S1, S2 strains had ACC- deaminicase activity, S3, S4 had phosphate solubilizing activity while S5-S10 strains had both the activites. The collected rhizobacterial isolates were purified by further streaking on fresh plates. All the isolates were stored in 20% glycerol at-20 °C for further studies.

Preparation of inocula

For preparation of inocula D.F salt minimal media (Dworkin and Foster, 1958) containing ACC as sole N source was used for strains containing ACC-deaminase activity. Modified NABRI-P media was used for preparation of inocula for strains containing phosphate solubilizing activity. Selected strains of bacteria were used for inoculation of broth and incubated for 48 hrs under shaking (100 rpm) conditions. After incubation, optical density of 0.5 at λ 550nm was achieved by dilution to maintain uniform cell density (10^7-10^8 cfu mL^{-1}). The suspension of selected rhizobacteria was used as inocula for wheat seed coating.

Seed Germination Assay

Wheat seed were surface sterilized by dipping in 95% ethanol solution for 5 min, 0.2% $HgCl_2$ solution for 3 min and washed thoroughly with distilled water. After sterilization these seeds were sown on sterilized filter paper sheets placed in Petri plates. These plates were incubated in a growth chamber at 28±1 °C for uniform germination. After seed germination assay, different parameters were studied to characterize rhizobacterial strains e.g. gnotobiotic root colonization assay (Simon's *et al.*, 1996) was done to calculate colony forming unit CFU/g of root biomass, solubilization of inorganic phosphate (Mehta and Nautiyal, 2001), ACC-deaminase activity (penrose and Glick, 2003), Chitinase activity (Chernin *et al.*, 1998) and auxin production was measured as indole acetic acid (IAA) equivalent in the presence or absence of L-tryptophan (Khalid et al. 2004).

Jar Experiment

Ten PGPR strains containing phosphate solubilizing and/or ACC deaminase activity were used in sand jar and this sand was dipped in 5% HCl solution and washed thoroughly with distilled water. Each glass jars contained 500g approx. sand and 1 g rock phosphate. Pregerminated seeds were sown in this sand at the rate of 3 seeds per jar and inoculated with respective strains. Two control treatments were also maintained. There were three replications for each treatment. In uninoculated rock phosphate control (So) only rock phosphate was added and P-deficient Hoagland solution was supplied. In uninoculated KH_2PO_4 control (N) only KH_2PO_4 was added and normal Hoagland solution was used. Appropriate temperature was maintained with 16 hours light and 8 hours dark period. After 14 days data regarding root length, shoot length, fresh and dry weights and number of tillers/plant were recorded. All the data were subjected to statistical analysis (Steel and Torrie, 2002).

Charaterization of Rhizobacterial isolates

The results depicted in Table 1. revealed that ten rhizobacterial strains S1, S2, S3, S4, S5, S6, S7, S8, S9 and S10 were taken for the purpose of characterization. In case of ACC-deaminase activity, all rhizobacterial strains showed this activity except S3 and S4 but S2 has maximum ACC-deaminase activity (1.8 umol a-ketobutyrate g- biomass 1/2 h^{-1}) followed by S1 (1.5 umol a-ketobutyrate g- biomass 1/2 h^{-1}) giving an indication that these are AAC-deaminase containing bacteria only because these have negligible amount of phosphate solubilizing activity. Rhizobacterial strains like S7, S9 and S10 have maximum ability to solublize phosphorous i.e., 10.07, 14.36 and 10.57mg/kg. Further S3 and S4 have not shown ACC-deaminase activity giving an indication that these are phosphate solubilizing rhizobacteria only. All rhizobacteria strains have the ability to colonize roots as compared to uninoculated control but S8 (8×10^7cfug^{-1}) has more ability to colonize roots than other strains. Indole acetic acid (IAA) production was also maximum in case of S4 (36.15 mg L^{-1}) followed by S8 (30.92 mgL^{-1}) with L-TRP as compared to all other strains. While without L-TRP maximum IAA production was showed by S6 (1.23 mg L^{-1}) and S7 (1.0 mg L^{-1}) There was significant linear correlation between auxin produced by rhizobacteria in vitro and growths of wheat seedlings (particularly root and shoot parameters) under control condition. All the rhizobacterial isolates possessed the ability to produce auxins (IAA) in the presence of L-TRP while in the absence of L-TRP only S5, S6, S7, S8 and S9 have showed auxin activity. This IAA might have directly or indirectly affected the plant growth. High IAA stimulates more ACC which is converted into NH_3 and a-ketobutyrate, thus reducing inhibitory effect of ethylene (Glick et al. 1998). Okon and Vanderleyden (1997) suggested that the secretion of plant growth promoting substances by the bacteria could be responsible for the beneficial effects of PGPR (Neol et al. 1996). So the results showed that rhizobacterial isolate S8 was the best strain in relation to wheat seedling shoot/root growth because it has more ability to solubilizs phosphorus, colonize roots, produce more IAA and have optimum ACC-deaminase activity.

Effect of selected isolates on growth of wheat seedlings under axenic conditions

The screening of rhizobacterial isolates for wheat growth promotion under axenic condition was done under P-deficient environment. Phosphorus deficiency could lead in production of stress conditions i.e. higher levels of ethylene in roots and inhibit root and shoot growth. In this trial inoculation with phosphate solubilizing rhizobacteria containing ACC-deaminase activity S8, S7, S9 and S10 has increased wheat seedlings growth significantly than rhizobacteria containing phosphate solubilizing or ACC-deaminase activity alone i.e. S1, S2, S3 and S4. The results depicted in Fig.1 shows that S8 (20 cm) produced the more shoot length where as S0, S1 and S2 had minimum shoot lengths. These results clearly indicated that inoculated strains produced 64% more shoot length than uninoculated control. Similarly in case of root length, the roots of strain S8 proliferated 75% more area than all other strains (Fig.2). This result implies that rhizobacertia grown in rock phosphate, under P-deficient environment, utilized it as a sole phosphorus source through solubilization of insoluble compounds by decreasing pH through excretion of organic acids and by the activity of phosphate enzyme (Nautiyal et al. 2000; Puente et al. 2004; Marschner et al. 1986).

In case of fresh shoot weight (Fig.3) 49% increase in weight was found in case of S8 followed by S7 which cased 46% over uninoculated control (So). While the maximum dry weight of shoot (Fig. 4) was found in S8 i.e. 83% and minimum in S1 (20%) over uninoculated control. Similarly other scientists also relate that PSM inoculants with lettuce and maize give higher biomass increase up to 20% (Chabbot et al. 1993). Bacterial strains isolated from alkalin soil have been able to solubilize phosphate in higher salt, pH and temperature stress condition (Nautiyal et al. 2000). Both rock phosphate addition and microbial inoculantion improved biomass production and P accumulation in test plants (Barea et al. 2005) Phosphate solubilizing microorganisms can promote plant growth directly by solubilizing this unavailable/insoluble P in soil through plant microbe interaction and indirectly enhance P acquisition by plant through promoting extensive root growth because of their ACC-deaminase activity. The results of fresh weight of root (Fig.5) showed that rhizobacterial isolate S8, S7 and S10 increased weights up to 48%, 43% and 42% over uninoulated control. While the loose in weight in case of root (Fig.6) was also found maximum in S8 i.e. 94% followed by S9 and S10 which caused 93% decrease in weight of root than fresh weight over uninoculated control. This implies that rhizobacteria grown on ACC utilized it as a nitrogen source via deaminase trait i.e., ACC is coverted into ammonia and a-ketobutyrate instead of ethylene. Thus it is highly like that the ability of ACC enriched rhizobacteria isolated to deaminate ACC was the responsible mechanism of action for promoted root shoot growth because lowering of ACC levels results in decreased endogenous ethylene. This contention is strongly supported by the work reported by several other researchers (Glick et al. 1998; Hall et al. 1996; Mayak et al. 1999; Penrose and Glick, 2003). Further, maximum number of tillers (Fig.7) was also found in S8 (4.2) followed by S9, S7 and S10 which produced 4.0, 3.4 tillers, respectively.

Table 1. Characterization of rhizobacterial isolates

PGPR strains	ACC – deaminase activity (μmol α-ketobutyrate)	IAA production (mgL^{-1})		Phosphorus solubility (mg/kg)	Root colonization (cfu g^{-1} root)	Chitinase activity
		Without L-TRP	With L-TRP			
Uninoculted control (S_0)	0	0	0	0	0	0
S1	1.50	0	1.62	0	2×10^{-7}	+ve
S2	1.80	0	8.15	0	4×10^{-7}	+ve
S3	0	0	19.54	8.98	1×10^{-7}	+ve
S4	0	0	36.77	8.13	3×10^{-7}	+ve
S5	1.40	0.15	22.00	7.72	4×10^{-7}	-ve
S6	1.40	1.23	18.92	7.45	3×10^{-7}	-ve
S7	0.83	1.00	6.92	10.07	6×10^{-7}	+ve
S8	0.77	0.92	30.92	9.98	8×10^{-7}	+ve
S9	0.25	0.77	5.54	14.36	7×10^{-7}	+ve
S10	0.24	0	18.08	10.57	5×10^{-7}	+ve

Effect of phosphate solubilizing rhizobacteria on wheat growth under axenic conditions

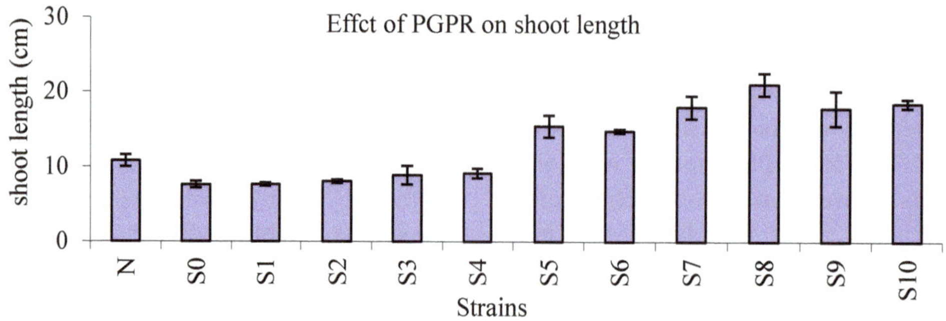

Fig.1: Effect of phosphate solubilizing rhizobacteria on shoot length of wheat seedlings

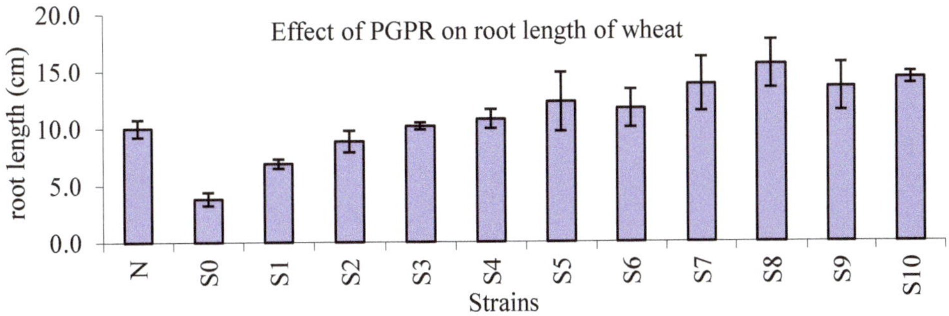

Fig.2. Effect of phosphate solubilizing rhizobateria on root length of wheat seedlings

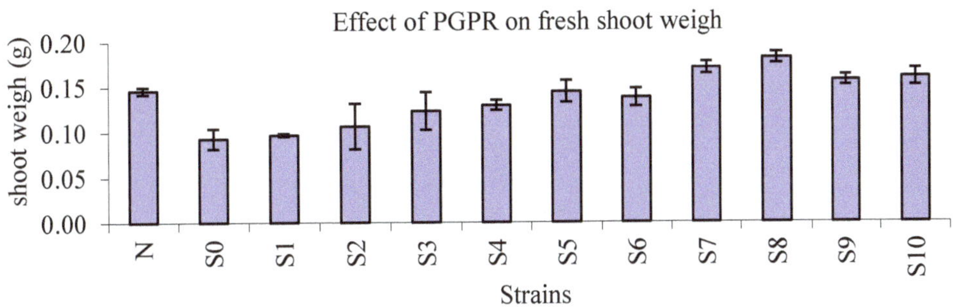

Fig.3. Effect of phoaphate solubilizing rhizobacteria on fresh weigh of shoot

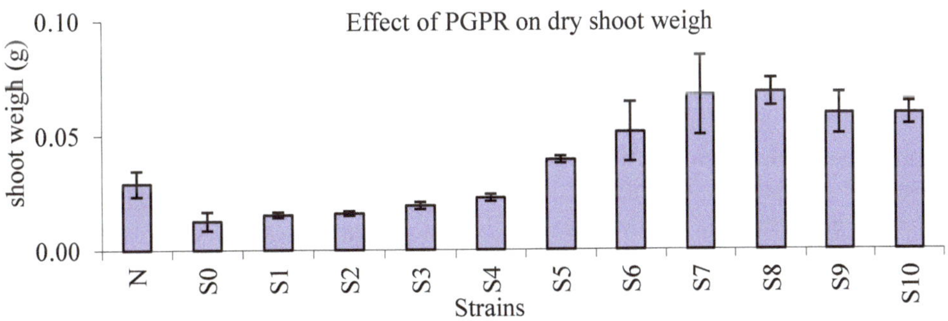

Fig.4. Effect of phosphate solubilizing rhizobacteria on dry weigh of shoot of wheat seedlings

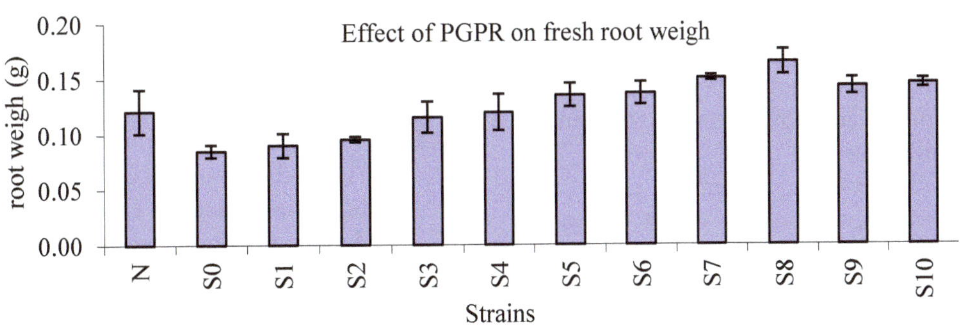

Fig.5. Effect of phosphate solubilizing rhizobacteria on fresh weigh of root of wheat seedlings

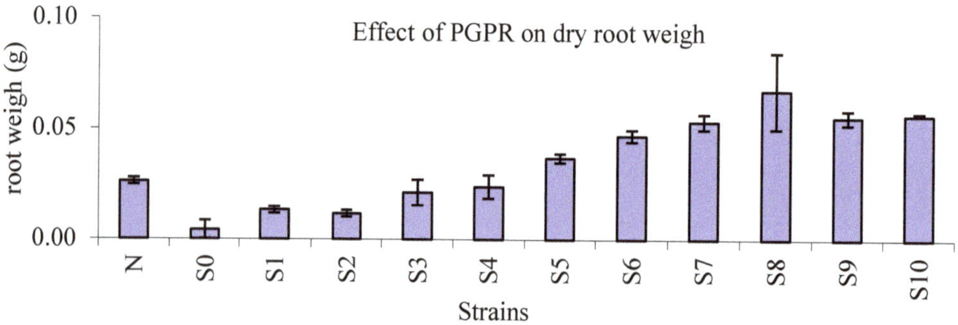

Fig.6. Effect of phosphate solubilizing rhizobacteria on dry root weigh of wheat seedlings

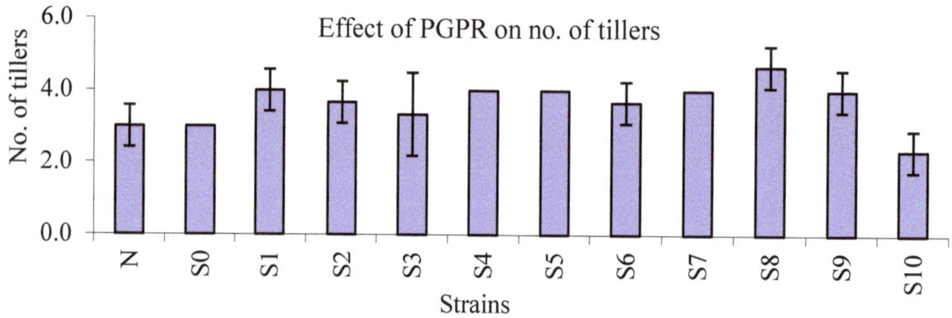

Fig.7. Effect of phosphate solubilizing rhizobacteria on number of tillers of wheat seedlings

CONCLUSION

This is concluded that inoculation with PGPR containing phosphate solubilizing activity alone could increase P availability to plants as it caused decreased endogeneous ethylene levels. Also inoculation with rhizobacteria containing ACC-deaminase activity alone enhances plant growth and could decrease endogenous ethylene levels. As this study showed that S8 strain increased 64%, 75%, 49%, 83%, 48% and 94% shoot length, root length, shoot fresh and dry weigh, fresh root and dry weighs respectively. Any factor/stimulus, with causes a change in the endogenous ethylene in a plant results in modified growth and development. While utilization of ACC-deaminase and phosphate solubilizing (organic acids and phosphatases) biotechnology could be useful tool for sustainable agriculture.

REFERENCES

Arcand MM, and Schneider KD. 2006. Plant and microbial based mechanisms to improve the agronomic effectioness of phosphate rock: a review: Ann. Acad. Bras. Cienc. 78: 791-807.

Arshad M, and Frankenberger WT. 1998. Plant growth regulating substances in the biosphere. Microbial production and funtions. Adv. Agron. 62: 46-151.

Balte H, and Robert JT. 1996. Defination of phosphorite, synonymous, antonymous and derivatives of phosphorite. Petrology. 2nd ed.: 345-349

Chabbot R, Antoun H, and Cecas P. 1993. Stimulation de la croissancedu mais et de la laitue romaine per des microorganisms dissolvent ie Phosphore inorganiqu. Can. J. Microbiol. 39: 941-947.

Chen YP, Rekha PD, Arun AB, Shen FT, Lai WA, and Young CC. 2006. Phosphate solubilizing bacteria from subtropical soil and their tricalcium phosphate solubilizing abilities. App. Soil Eco. 34: 33-41.

Chernin LS, Winson MK, Thompson JM, Haran S, Bycroft BW, Chet I, Williams P, and Stewart GSAB. 1998. Chitionlytic activity in *Chromobacterium violaceum*: substrate analysis and regulation by

Quorum sensing. J. Bacteriol. 180: 4435-4441.

Dworkin M, and Foster J. 1958. Experiments with some microorganisms which utilize ethane and hydrogen. J. Bacteriol. 75: 592-60.

Fernandez LA, Zalpa P, Gomez MA, and Sagardoy MA. 2007. Phosphate solubilization activity of bacterial strains in soil and their effect on soybean growth under greenhouse conditions. Biol. Fert. Soils. 43: 805-809.

Gilroy S, and Jones DL. 2000. Thorough form to function Brook hair development and nutrient uptake. Trends Plant Sci. 3: 56-60.

Glick BR, Penrose DM, and Li J. 1998. A model for the lowering of plant ethylene concentrations by plant growth_promoting bacteria. J. Theor. Biol.190: 63-68

Gyaneshwar P, Kumar GN, Parekh LJ, and Poole PS. 2002. Role of soil microorganisms in improving P nutrition of plants. Plant & Soil.245: 83-93.

Hall JA, Peirson D, Ghosh S, and Glick BR. 1996. Root elongation in various agronomic crops by the plant growth promoting rhizobacterium *Pseudomonas putide* GR 12-2. Isr. J. Plant Sci. 44: 37-42.

Hamdali H, Koriko M, Tchangbedji G, Ouhdouch Y, and Hafidi M. 2012. Isolation and characterization of rock phosphate solubilizing actinobacteria from a Togolese phosphate mine. African J. Biotech. 11(12): 312-320.

Khalid A, Arshsd M, and Zahir AZ. 2004. Screening plant growth-promoting rhizoacteria for improving growth and yield of wheat. J. Appl. Microbiol. 96: 473-480.

Kloepper JW, Lfshitz R, and Zablotowicz RM. 1989. Free living baterial inocula for enhancing crop productivity. Trends Biotech. 7: 39-44.

Kucey RMN. 1983. Phosphate solubilizing rhizobacteria and fungi in various cultivated and virgin Alberta soil. Can. J. Soil sci. 63: 671-678.

Lynch JP, and Brown KM. 1998. Regulation of root architecture by phosphorus availability. In: J.P. Lynch, J. Deikman (eds.), Phosphorus in Plant Biology: Regulatory Roles in Molecular, Cellular, Organismic and Ecosystem Processes. Amer. Soc. Plant Physiol., Rockville, MD, pp. 148-157.

Marschner H, Romheld V, Horst WJ, and Martin P. 1986. Root induced changes in the rhizosphere; importance for mineral nutrition of plants. Z. pflanzenernachr Bodenkd. 149: 441-456.

Mayak S, Tivosh T, and Glick BR. 1999. Effect of wild type and mutant plant growth-promoting rhizobacteria on the rooting of mung been cuttings. J. Plant Growth Regul. 18: 49-53.

Mehta S, and Nautiyal CS. 2001. An efficient method for qualitative Screening of phosphate solubilizing bacteria. Curr. Microbiol. 43: 57-58.

Morgan PW, Sarquis JL, He C, Jordan WR and Drew MC. 1993. Regulation of ethylene biosynthesis in maize root responses to stress. In: Kluwer Academic publishers, Dordecht, The Netherland. pp: 232-237.

Nautiyal CS, Bhadauria S, Kumar P, Lal H, Mondal R, and Verma D. 2000. Stress induced phosphate solubilization in bacteria isolated from alkaline soils. FEMS Microbiol. Lett. 182: 291-296.

Noel TC, Cheng C, Yost CK, Pharis RP, and Hynes MF. 1996. *Rhizobium leguminosarum* as a plant growth rhizobacterium: direct growth promotion of canola and lettuce. Can. J. Microbiol. 42: 279-283.

Okon Y and Vanderleyden J. 1997. Rootassociated *Azospirillum* sp. Can stimulates plants. Am. Soc. Microbiol. News. 63: 366-370.

Penrose DM, and Glick BR. 2003. Methods for isolating and characterizing ACC-deaminase containing plant growth-promoting rhizobacteria. Physiol. Plant 118: 10-15.

Penrose DM, Barbara M, and Glick BR. 2000. Determination of ACC to assess the effect of ACC-deaminase-containing bacteria on roots of canola seedlings. Can. J Microbial 47: 77-80.

Ponmurugan P, and Gopi C. 2006. In vitro production of growth regulators and phosphatase activity by phosphate solubilizing bacteria. Agri. J. Biotech. 4 (4): 348-350.

Puente ME, Bashan Y, Li CY, and Lebsky VK. 2004. Microbial populations and activities in the rhizoplane of rock- weathering desert plants. I. Root colonization and weathering of igneous rocks. Plant Boil. 6: 629-642.

Sarquis JI, Jorden WR, Morgan PW. 1991. Ethylene evaluation from maize (Zea mays L.) seedling roots and shoots in response to mechanical impedance. Plant Physiol. 96: 1171-1177.

Simons M, VanderBij AJ, Brand I, Wijffelman LA and Lugtenberg BJJ. 1996. Gnotobiotic system for studying rhizosphere colonization by plant growth-promoting *pseudomonas* bacteria. Mol. Plant Microbe Interact. 9: 600-607.

Steel RGD and Torrie JH. 1993. Principles and procedures for statistics. McGraw Hill Book Co., NY, USA.

Steenhoudt O and Vanderleyden J. 2000. Azospirillum, a free-living nitrogen-fixing bacterium closely associated with grasses: genetic, biochemical and ecological aspects. FEMS Microbiol. Rev. 24: 487-506.

Vazques P, Holguin G, Puente ME, Lopez-cortes A, and Bashan Y. 2000. Phosphate solubilizing microorganisms associated with the rhizosphere of mangroves in a semiarid oastal lagoon. Biol. Fert. Soil. 30: 460-468.

Zapata F and Zaharah AR. 2002. Phosphorus availability from phosphate rock and sewage sludge as influenced by addition of water soluble phosphate fertilizer. Nutr. Cycle Agroecosyst. 63: 43-48.

HERITABILITY AND RELATIONSHIP BETWEEN DROUGHT TOLERANCE TRAITS AND YIELD IN GROUNDNUTS (*ARACHIS HYPOGAEA* L.) UNDER DIFFERENT WATERING REGIMES

Mashamba Philipo[1,*], Clara Mollay[1] and Susan Nchimbi-Msolla[2]

[1]*The Nelson Mandela African Institution of Science and Technology (NM-AIST), Arusha, Tanzania*

[2]*Sokoine University of Agriculture (SUA), Morogoro, Tanzania*

ABSTRACT

Improvement of groundnuts for drought tolerance could increase production in drought prone areas. This study aimed at determining the heritability estimates and relationship of HI and SCMR with yield and yield components using 30 groundnut genotypes planted under different watering regimes, so as to speed up the selection and breeding of groundnut genotypes tolerant to drought. A completely randomized design arranged in a split plot experiment with four replications was conducted in screen house in Morogoro, Tanzania. Data collection was done on plant height, number of pod/plant and pod yield/plant. Drought tolerant traits measured included, harvest index (HI), SCMR at 40, 60 and 80 DAS. Broad-sense heritability was calculated for HI, pod yield, number of pod/plant and SCMR. Results showed that number of pods/plant, SCMR at 60 DAS and HI were significantly related to pod yield in all watering condition. Heritability of the traits ranged from 0.22 to 0.59 with HI having highest value and number of pods lowest in WW while in WS condition heritability was generally lower from 0.04 to 0.45. Due to SCMR at 60 DAS, number of pods/plant and HI having moderate heritability and significant correlation with pod yield under water stress condition, these could be useful criteria in drought tolerance selection.

Keywords: Groundnut, SCMR, Water stress, Heritability, Pod yield.

INTRODUCTION

Groundnut (*Arachis hypogaea* L.) is an important leguminous crop in the world grown for oilseed, food and animal feed (Jasem et al., 2013). The genus and species names *Arachis hypogaea* comes from Greek words (Holbrook and Stalker, 2003). Groundnut is tetraploid with 2n=4x=40 chromosomes (Mamadou, 2011). It has 44 to 50 % edible oil, 25 % easily digestible protein and 20 % carbohydrate (FAO, 2008). In Tanzania, total groundnuts production is 465,290 tons produced within the area of 482,311 ha, this gives a yield of 965 kg ha^{-1} (MAFC, 2010).

Groundnut is mostly cultivated as a rain fed crop by the resource poor farmers, hence affected by intermittent drought stress of different duration and intensities. Annually world-wide drought causes a loss of 6.7 million metric tons on groundnut production (Devi et al., 2010). Regardless of its importance in Tanzania, groundnut productivity is still low. Most of the groundnut growing areas in Tanzania are affected by intermittent drought of different duration and intensity.

**Corresponding author:* e-mail: lugendomas@gmail.com

Long and severe drought experienced in Dodoma, Tanzania in the year 2007/08, caused 100 % loss in groundnut production in Chamwino, Bahi and Kongwa (Monyo, 2009). In Southern Tanzania, drought accelerated rosette outbreak and thus reduced productivity and production. Breeding and selection of groundnut tolerant to drought have been conducted for a long time based on pod yield, but it has been unsuccessful in getting drought tolerant genotypes (Wunna et al., 2009), the genotypes which could maximize groundnut yield in drought prone areas where groundnut is grown under rain fed condition.

Since breeding approach utilizing selection for pod yield been unsuccessful for a long time (Jongrungklang et al., 2008), recently, other surrogate traits such as Soil Plant Analysis Development (SPAD) Chlorophyll Meter Reading (SCMR) and Harvest Index (HI) have been used during drought tolerance selection in breeding programme. Low genotype by environment interaction on Specific Leaf Area (SLA) and SCMR was obtained by Songsri et al. (2008). Wunna et al. (2009) reported the use of HI and DTI for yield and yield components in drought tolerance selection.

Drought stress causes wilting to groundnut genotypes, hence reduce the green leaf area and water uptake from soil profile to some extent. Early studies reported that drought stress increases SPAD chlorophyll meter reading (SCMR) in groundnut. Higher SCMR under water stress condition shows that, the groundnut genotypes could continue having high photosynthesis rate per unit leaf area regardless of water stress. In groundnut SCMR has strong relationship with pod yield and Transpiration Efficiency (TE) under water stress conditions (Bootang et al., 2010). In groundnut harvest index is the ratio of pod weight to total biomass and it varies depending on the severity and timing of water stress relative to pod set. Smartt (1994) reported that, differences in pod yield among groundnut genotypes during a terminal drought were due to differences in harvest index which was associated with variation in pods and the effective duration of pod filling developmental stage.

Heritability is the measure of relative importance of genetic and non-genetic factors in the expression of phenotypic differences among genotypes in a population. Heritability in the broad sense is the proportion of the phenotypic variance of family means that is due to all genetic effects. Heritability is used to estimate expected response to selection and to choose the best breeding approach to improve the target trait(s). Traits with high heritabilities can be selected on a single-plant basis (e.g. mass selection) faster, and in a low number of environments. In contrast, traits with low heritabilities require selection on a family basis and in a greater number of environments to determine breeding values of genotypes. A high heritability would likely result in high response to selection to advance the population in the desired direction of change (Acquaah, 2007).

This study aimed at determining the heritability estimates and relationship of HI and SCMR with pod yield and number of pods per plant from 30 groundnut genotypes under different watering regimes, so as to speed up the selection and breeding of groundnut genotypes tolerant to drought.

MATERIALS AND METHODS

The experiment was conducted at Sokoine University of Agriculture (SUA) farm which is located at S 06° 50' 24.7" and E 37° 38' 59.8", Morogoro, Tanzania. Morogoro has mean annual rainfall of 1031 mm and temperature range of 18.6 – 30.0 °C. A set of 30 groundnut genotypes were used in the study. The genotypes used were obtained from the International Crops Research Institute for the Semi-Arid Tropics (ICRISAT). The genotypes were planted in screen house to determine heritability and correlation among drought tolerant traits and yield under different watering regimes. The screen house condition during experimental period recorded temperature range of 25 - 31°C and the relative humidity was 85 – 95%

The genotypes were planted using a completely randomized design arranged in a split plot experiment in four replications with watering regimes being main plot treatments and genotypes sub plot treatments. Three seeds of each genotype were sown in a pot which had five drainage holes and filled with 12 kg of soil each. After sowing, 500 ml of water were applied to each pot on the sowing day, this was followed by applying 250 ml of water two times a day on alternate days until germination. Thinning was done 14 days after sowing (DAS), whereby one plant was left per every pot. The crops were irrigated to field capacity by 2 litres of water at 5 days intervals from sowing to about flowering time, by compensating evapotranspiration. From there onwards, irrigation for water stressed plants (WS) was 2 litres of water after every 10 days and for well watered plants (WW) 2 litres of water were given to each pot after every 5 days. The decision to irrigate was based on leaf wilting symptoms of WS plants, irrigation being supplied when the wilting score of a majority of WS plants reached a value of three and below score of two for WW plants (Hamidou et al., 2012). Scoring scale of wilting symptoms (recorded early afternoon) was used to monitor stress in the WS block, as illustrated in Table. 1. (Ratnakumar and Vadez, 2011).

Table. 1. Wilting symptoms scoring scale

Scale	Explanation
1	No wilting
2	Some leaves wilted ($\leq$ 50 % leaves)
3	Most leaves wilted ($\geq$ 50% leaves)
4	Permanent wilting on some leaves ($\leq$ 50 % leaves)
5	Permanent wilting on most leaves ($\geq$ 50% leaves)

After harvesting the groundnut pods from individual plant were air dried separately and weighed to get pod yield/plant. The roots were washed with water to remove the soil and then shoot and roots for each genotype were packed in paper bag and oven dried at 70 °C for 72 hours in order to obtain constant weight. After oven drying, shoot and root dry weight was recorded (Painawadee et al., 2009).
SCMR was recorded twice on each leaflet of the tetra foliate leaf along the mid-rib at 40, 60 and 80 days after sowing using SPAD chlorophyll meter (Figure 1.0). The third fully-expanded leaves from each plant were used for determination of SCMR, between 0830 and 1000 hours as during this time, there is high stomatal conductance to allow photosynthesis take place since evaporation demand is low particularly in stressed groundnut genotypes (Smartt, 1994).

Figure 1: Recording SCMR using SPAD Chlorophyll Meter at SUA

Harvest index was calculated based on the following relationship (Fageria et al., 2011):
HI = Pod yield / (Pod yield + Shoot and root dry weight)
The collected data were subjected to analysis of variance using GenStat 14th edition Statistical package for analysis. The estimate mean squares from ANOVA table for drought tolerance associated traits and pod yield were used to calculate genotypic variance and environmental variance. Broad-sense heritability estimates for the measured traits was calculated using the formula described by Ceccarelli et al. (2009) as follows:

$$h^2 = \sigma^2_G/(\sigma^2_G + \sigma^2_E)$$

h^2 = heritability in broad sense, σ^2_G = genotypic variance {(genotype mean square – residual mean square)/number of replication} and σ^2_E = residual mean square (environmental variance). The correlation coefficients (r) among all the measured traits were determined using GenStat 14th edition Statistical package.

RESULTS AND DISCUSSION

Heritability of yield and drought tolerant traits

Heritability or percentage of all differences between groundnut genotypes that is caused by gene effects and that can be transferred from generation to generation was determined in this study as summarised in Table 2. Results showed that broad-sense heritability estimates obtained from 30 groundnuts genotypes for pod yield and drought tolerant traits ranged from low (0.04) to medium (0.59) for the groundnut genotypes drought phenotyping pot experiment conducted in Morogoro at SUA. Under well watered condition the highest heritability value was rated as medium, and was observed from HI which had 0.59, followed by pod yield with 0.4, low heritability was observed from SCMR which had 0.25, plant height with 0.24 and number of pods per plant with 0.22.
Under water stress condition the highest broad-sense heritability estimates rated as medium was observed from HI which had 0.45 followed by SCMR with 0.35 and number of pods per plant with 0.32. Low heritability was obtained from pod yield which had 0.27 and plant height with 0.04. The variability in heritability estimates for the same trait under the two watering regimes was caused by the interaction between the genes that encode the trait and the environment in which the genes are being expressed (Acquaah, 2007). The heritability estimates in this experiment were not above 0.6 as the estimates were on plant basis and not on family basis, the heritability on family basis is known to be greater than that on plant basis (Ceccarelli et al., 2009)
In order for a trait to be more useful in selection of drought tolerance, the heritability estimates for the trait should be greater than the heritability for yield in the specific environment, so that selecting and advancing only the top few performers is likely to produce a greater genetic advance than selecting many moderate performers when heritability is low (Acquaah, 2007). In this study heritability estimates of HI, SCMR and number of pods per plant were higher than heritability estimates of pod yield per plant under water stress condition.

Therefore, HI, SCMR and number of pods per plant can be used as selection criteria in breeding programmes for drought tolerance.

Table 2: Heritability Estimates in Broad Sense for Harvest Index, SCMR, Plant Height, Number of Pods and Pod Yield per Plant

Characteristics	Heritability Well Watered	Heritability Water Stress
Harvest Index	0.59	0.45
Number of pods/plant	0.22	0.32
Pod yield/plant	0.40	0.27
SCMR	0.25	0.35
Plant height	0.24	0.04

Relationship between yield and yield components with drought tolerance traits (HI and SCMR)

Relationship among drought tolerant traits

The simple correlation coefficients between drought tolerant traits from the analysis are shown in Tables 3 and 4. No significance relationship ($P \leq 0.05$) between SCMR at 40, 60 and 80 DAS with HI was observed to the tested groundnut genotypes under the two watering regimes. The results from this study support the study conducted by Bootang et al. (2010) which found no association between SCMR and HI. Therefore from this study SCMR cannot be used as a surrogate trait for HI during selection of groundnut genotypes in breeding programmes for drought tolerance.

Table 3: Correlation among SCMR, HI, Number of Pods/Plant and Pod Yield in Well Watered Condition

	SCMR 40 DAS	SCMR 60 DAS	SCMR 80 DAS	Harvest Index	Number of pods/plant	Pod yield (g/plant)
SCMR 40 DAS	1.00					
SCMR 60 DAS	0.41***	1.00				
SCMR 80 DAS	0.41***	0.47***	1.00			
Harvest Index	0.14^{ns}	0.16^{ns}	-0.10^{ns}	1.00		
Number of pods per plant	-0.14^{ns}	-0.05^{ns}	-0.05^{ns}	0.24**	1.00	
Pod yield (g/plant)	-0.02^{ns}	0.20*	0.05^{ns}	0.63***	0.49***	1.00

ns = not significant, *, **, *** Significant at $P \leq 0.05$, $P \leq 0.01$ and $P \leq 0.001$, respectively, SCMR = SPAD Chlorophyll Meter Reading and DAS = Days after sowing

Table 4: Correlation among SCMR, HI, Number of Pods/Plant and Pod Yield in Water Stress Condition

	SCMR 40 DAS	SCMR 60 DAS	SCMR 80 DAS	Harvest Index	Number of pods/plant	Pod yield (g/plant)
SCMR 40 DAS	1.00					
SCMR 60 DAS	0.14^{ns}	1.00				
SCMR 80 DAS	0.20*	0.30***	1.00			
Harvest Index	-0.13^{ns}	0.08^{ns}	0.06^{ns}	1.00		
Number of pods per plant	-0.05^{ns}	0.24**	0.14^{ns}	0.39***	1.00	
Pod yield (g/plant)	0.06^{ns}	0.38***	0.19*	0.61***	0.64***	1.00

ns = not significant, *, **, *** Significant at $P \leq 0.05$, $P \leq 0.01$ and $P \leq 0.001$, respectively, SCMR = SPAD Chlorophyll Meter Reading and DAS = Days after sowing

Relationship between drought tolerant traits with yield and yield components

The relationship between yield and yield components with drought tolerance traits were from non-significant to highly significant ($P \leq 0.001$), positive and negative. Under well watered condition, moderate positive significant relationship ($r = 0.2$, $P \leq 0.01$) was observed between SCMR at 60 DAS and pod yield/plant, this means that knowing the value of SCMR at 60 DAS gives little information for predicting pod yield in groundnut under well watered condition. HI and number of pods/plant had strong positive and highly significant relationship ($r = 0.63$, $P \leq 0.001$ and $r = 0.5$, $P \leq 0.001$) with pod yield/plant, this means that knowing HI and number of pods/plant gives a greater accuracy prediction of pod yield/plant. This study supports the earlier study conducted by Jogloy et al. (2011) on strong and positive relationship between HI and number of pods/plant with pod yield.

Under water stress condition, SCMR at 60 DAS had moderate positive and highly significant relationship ($r = 0.38$, $P \leq 0.001$) with pod yield/plant while SCMR at 80 DAS had moderate positive and significant relationship ($r = 19$, $P \leq 0.05$) with pod yield/plant. This means that, understanding the values of SCMR at 60 and 80 DAS in groundnut genotypes gives some significant information about pod yield under water stress condition. SCMR at 60 DAS also had moderate positive and highly significant relationship ($r = 0.24$, $P \leq 0.01$) with number of pods/plant. Understanding the values of SCMR at 60 DAS in groundnuts genotypes gives some significant information about the number of pods/plant under water stress condition. HI and number of pods/plant had strong positive and highly significant relationship ($r = 0.61$, $P \leq 0.001$ and $r = 0.64$, $P \leq 0.001$) respectively with pod yield/plant. Understanding the values of HI and number of pods/plant gives greater accuracy information on pod yield/plant in groundnut.

This study supports the previous study which obtained significant relationship between SCMR and pod yield (Bootang et al., 2010). Positive and highly significant relationship between SCMR at 60 DAS, which is an indirect measure of chlorophyll density and pod yield under water stress condition, means groundnuts genotypes with high SCMR could maintain higher photosynthetic capacity and because of thicker leaves they have more leaf carbon exchange rate and chlorophyll content and therefore leads to high pod yield. The higher significant correlation between SCMR 60 DAS with number of pods/plant and pod yield/plant under water stress condition compared to well watered condition, means SCMR can be used in selection of drought tolerant genotypes in breeding programmes. Strong positive and highly significant relationship between HI and number of pods/plant with pod yield/plant implies that genotypes with high HI and number of pods/plant under water stress condition had high pod yield. Such results imply that, HI is the relative distribution of photosynthesis products between economical sinks and other existing sinks in the plant. Therefore this can be used in breeding program as a selection criterion in drought tolerance breeding programmes.

CONCLUSION

For effective selection of genotypes as drought tolerant genotypes, association of drought tolerance traits with pod yield and among themselves is important. In this study SCMR at 60 DAS, number of pods/plant and HI were found to be positively and significantly correlated to pod yield under all watering condition. A heritability estimate of a trait is also important due to the information on the variation of genotypes on certain trait on genetic bases. In this study high heritability estimates under water stress condition, was calculated from HI, SCMR and number of pods/plant. Therefore, these traits can be good criteria to be used in breeding and selection programmes in improving groundnut pod yield in drought prone areas.

ACKNOWLEDGEMENT

The authors express their sincere thanks to the International Crops Research Institute for the Semi-Arid Tropics (ICRISAT) and Tropical Legume Projects management team for granting the research of this study. Special thanks should be extended to the office of Faculty of Agriculture and the Department of Crop Science for providing screen house in which the experiment of this study was conducted.

REFERENCES

Acquaah G. 2007. Principles of Plant Genetics and Breeding. Blackwell Publishing Ltd., Oxford. 584pp.

Bootang S, Girdthai T, Jogloy S, Akkasaeng C, Vorasoot N, Patanothai A and Tantisuwichwong N. 2010. Response of Released Cultivars of Groundnut to Terminal Drought for Traits Related to Drought Tolerance. Asian J. Plant Sci. 9(7): 423 – 431.

Ceccarelli S, Guimarães EP and Weltzien E. 2009. Plant Breeding and Farmer Participation. Food and Agriculture Organization of the United Nations, Rome. 671pp.

Devi JM, Sinclair TR and Vadez V. 2010. Genotypic variation in Groundnut for transpiration response to vapor pressure. Crop Science. 50: 191 – 196.

Fageria NK, Moreira A and Coelho AM. 2011. Yield and Yield Components of Upland Rice as Influenced by Nitrogen sources. J. Plant Nutr. 34:361–370.

FAO, 2008. FAOSTAT Agricultural Statistical Data Base. [http://faostat.fao.org] site visited on 30 /3/ 2012.

Hamidou F, Ratnakumar P, Halilou O, Mponda O, Kapewa T, Monyo E, Faye I, Ntare B, Nigam SN, Upadhyaya HD and Vadez V. 2012. Selection of intermittent drought stress tolerant lines across years and locations in the reference collection of groundnut (*Arachis hypogaea* L.). Field Crops Research. 126: 189 – 199.

Holbrook CC and Stalker HT. 2003. Peanut breeding and genetic resources. Plant Breeding Reviews. 22: 297 – 356.

Jasem A, Mohsen M and Alireza S, 2013. Grain yield improvement of groundnut (*Arachis hypogaea* L.) under drought stress conditions. Intl J Agri Crop Sci. 6 (12); 819-824.

Jogloy C, Jaisil P, Akkasaeng C, Kesmala T, and Jogloy S. 2011. Heritability and Correlation for Maturity and Pod Yield in Peanut. Journal of Applied Sciences Research, 7(2): 134-140.

Jongrungklang N, Jogloy S, Vorasoot N, Kesmala T, Patanothai A and Toomsan B. 2008. Identification of Groundnut genotypes with higher water use efficiency under drought stress conditions from Groundnut germplasm of diverse origins. Asian J. Plant Sci. 7: 628 – 638.

MAFC. 2010. Agricultural Statistics, Agricultural Basic Data. [http://www.kilimo.go.tz] site visited on 2 /12/ 2016.

Mamadou AC. 2011.Utilization of Existing Germplasm Pool to Develop Groundnut Lines with Early Maturity and Improved Drought Tolerance. Thesis for Award of PhD Degree at College of Agriculture and Consumer Sciences, University of Ghana, 100pp.

Monyo ES (Ed.). 2009. Developing short-and medium-duration groundnut varieties with improved yield performance, acceptable market traits and tolerance to foliar diseases. Proceedings of McKnight Foundation Workshop, Lilongwe, Malawi. 10 September, 2008. 39.

Painawadee M, Jogloy S, Kesmala T, Akkasaeng C and Patanothai A. 2009. Heritability and Correlation of Drought Tolerance Traits and Agronomic Traits in Groundnut *(Arachis hypogaea* L*)*. Asian J. Plant Sci. 8(5): 325 – 334.

Ratnakumar P and Vadez V. 2011. Groundnut (*Arachis hypogaea*) Genotypes Tolerant to Intermittent Drought Maintain a High Harvest Index and have Small Leaf Canopy under Stress. Funct. Plant Biol. 38(12): 101 – 1023.

Smartt J, (Eds.). 1994. The Groundnut crop: A Scientific Basis for Improvement. Chapman and Hall, London. 756.

Songsri P, Jogloy S, Kesmala T, Vorasoot N, Akkasaeng C, Patanothai A and Holbrook CC. 2008. Heritability of drought tolerance traits and correlation of drought tolerance and agronomic traits in Groundnut. Crop Sciences. 48: 2245 – 2253.

Wunna H, Jogloy S, Toomsan B, Sanitchon J and Patanothai A. 2009. Inheritance of Traits Related to Biological Nitrogen Fixation and Genotypic Correlation of Traits Related to Nitrogen Fixation, Yield and Drought Tolerance in Groundnut (*Arachis hypogaea* L.) Under Early Drought Asian J. Plant Sci. 8: 265 – 275.

ESSENTIAL MINERAL AND TOXIC ELEMENTS IN OYSTER MUSHROOM (*Pleurotus florida*) CULTIVATED ON WATER HYACINTH AND RICE STRAW

S. Bandopadhyay *
Department of Botany, Sonamukhi College, Bankura, W.B., India

ABSTRACT

The present study investigated the concentration of three mineral elements (Fe, Zn, Cu) and three toxic elements (Pb, Cd, As) in the cultivation substrate (1:1 mixture of water hyacinth and rice straw) as well as in the cultivated oyster mushrooms *Pleurotus florida* grown on this substrate. The mean concentrations (mg kg^{-1} dry weight) obtained for Fe, Zn, Cu, Pb, Cd and As in the mushrooms were 172, 52.3, 10.5, 2.3, 0.95 and 0.7 whereas in the cultivation substrate were 201.5, 56.5, 11.9, 7.2, 2.7 and 2.4 respectively. The contents of Fe, Zn and Cu minerals are in consonance with the reports in literature. Pb and Cd contents are however slightly higher than some of the reports in literature but are within the of recommended safe dietary intake limits established by FAO/WHO expert committee. The coefficient of accumulation of these elements in the mushroom were also calculated by comparing with their concentration in the cultivation substrate and was found in the order of Zn > Cu > Fe > Cd > Pb > As indicating *P.florida* as good accumulator of the minerals. In terms of nutritional aspects, taking into account the recommended dietary allowances or provisional tolerable weekly intake recommended by FAO/WHO, optimum dietary intake of average 100 g day^{-1} fresh mushroom of *P.florida* grown on water hyacinth and rice straw is beneficial for health and poses no toxicological risk.

Keywords: Mineral, *Pleurotus florida*, rice straw, substrate, toxic elements, water hyacinth.

INTRODUCTION

Mushrooms are valuable health food, low in calories, in general high in vegetable proteins, minerals, vitamins and fiber. In most countries there is a significant consumer acceptance of cultivated mushrooms for their texture, flavor, nutritional properties as well as for their therapeutic uses. Oyster mushrooms or species of *Pleurotus* are popular mushrooms with high commercial value. In India this mushroom contribute to the maximum production over other due to its ease of cultivation on a wide variety of lignocellulosic materials beside straw of wheat or rice. As such, oyster mushrooms can play an important role in managing and recycling organic waste through biodegradation. Mushrooms are known to uptake and bio-accumulate essential mineral elements as well as some polluting toxic elements in high concentrations from growth medium (Kalac and Sovodon, 2001, Demirbas, 2001). Mineral and toxic metal (as lead, cadmium, mercury) concentration in higher mushrooms are often considerably higher than those in agricultural crop plants, vegetables and fruits (Zhu et al., 2010). The principal factors influencing the accumulation of these elements in macrofungi are environmental factors such as metal concentration in substrate, pH, organic matter, contamination by atmospheric deposition and fungal factors such as species and age of mushroom, development of mycelium and fruit bodies, morphological portion (Garcia et al., 1998). Growing up on a substrate with high concentration of various toxic elements, edible mushrooms can become toxic, accumulating a large amount of those elements.

Mineral content analysis allows the evaluation of nutritional quality of food. RDA (Recommended dietary allowances) are daily intake of essential nutrients judged to be adequate to meet the known nutrient needs of practically all healthy persons (NRC, 1989). Provisional Tolerable Weekly Intakes (PTWI) are set for substances such as heavy metals, that are contaminants in food and are known to accumulate in animals and humans (Codex Alimentarius, 2011). PTWIs have been set for cadmium, lead, mercury and tin (Luis et al., 2012). The determination of toxic element (As, Pb, Cd, Hg) concentration in the food allows for the evaluation of health risk and is thus part of every food safety program and also essential

**Corresponding author:* e-mail: sunita_babli@rediffmail.com in dietary intake studies. Food contains a wide range of elements such as potassium, sodium, iron, copper, zinc, magnesium which are essential elements since they play an important role in biological systems, whereas aluminum, lead, cadmium or arsenic are non-essential elements as they are toxic even though excessive levels of essential elements can be detrimental to organisms (Zhu et al., 2010, Akyuz and Kirbag, 2011). Therefore, it is worthwhile to evaluate the levels of essential mineral and toxic elements in mushrooms grown on any substrate (artificially / natural occurrence), to assess the contribution of the mushroom to daily intake of several toxic elements and to compare the results with the norms for these toxic elements in food stuff to report any possible contamination that would represent a health hazard.

Water hyacinth [*Eichhornia crassipes* (Mart.) Solms.] is one of the top ten problematic weeds of the world (Aboul-Enin et al., 2011) which adversely affect both human and animal due to its enormous rate of propagation. So, many workers (Nageswaran et al., 2003, Bandopadhyay, 2009) attempted to eradicate the vast quantity of this weed through its cost-effective and eco-friendly utilization as a substrate for cultivation of mushroom or as an ingredient in vermicompost. It is also used in waste water treatment for its unique property to absorb heavy metal on its roots (Dos Santos, 2001). In the present study water hyacinth along with rice straw (1:1) has been used as the substrate for cultivation of oyster mushroom *Pleurotus florida.*

In the light of above, the objective of this study is (i) to determine the levels of three essential mineral elements and three toxic elements in samples of cultivation substrate of water hyacinth (WH) mixed (1 : 1) with rice straw (RS) as well as in the oyster mushroom (*P.florida*) cultivated on this substrate and (ii) to estimate the contribution of this cost-effective oyster mushroom consumption to the RDA or PTWI set for mineral and toxic elements respectively by expert council or committee (NRC, 1989, EC No 1881/2006, Codex Alimentarius, 2011) of food safety program. Iron, copper and zinc (Fe, Cu, Zn) have been chosen as representative essential mineral and lead, cadmium, arsenic (Pb, Cd, As) as representative toxic elements whose levels in environment represent a reliable index of environment pollution.

MATERIALS AND METHODS

The samples used in this study were obtained from previous cultivation work (Bandopadhyay 2009). Pure culture of *Pleurotus florida nomen nudum* (Eger, strain P-1) was procured from NCMRT, Solan, H.P., India. Rice straw and sundried water hyacinth (without root) was mixed in 1:1 proportion to use as substrate for cultivation of *P. florida.* Preparation of substrate and cultivation of *Pleurotus* sp. were carried out following standard polythene bag method (Upadhyay, 1997). Spawns of *P. florida* were prepared on wheat grains in polypropylene packets following Suman and Sharma (2005) and spawned @5% wet substrate. Temperature and relative humidity regimes were maintained between 14 – 22°C and 75 - 80% respectively. Mushrooms were harvested upto 3rd flush to prepare mushroom samples for determination of essential mineral and toxic element contents.

Mushrooms were cut into slices, air dried at 60°C for 24 h. and homogenized to fine powder. Sample of substrates (WH+RS 1 : 1) was dried at 70°C for 72 h , then homogenized to fine powder and weighed for acid digestion.

Samples from each of dried mushroom powder (500 mg) and substrate powder (1 gm) were digested at 100°C for 30 min. in acid solution containing mixture of HNO_3 : H_2SO_4 : H_2O_2 (10 : 1 : 1, 12 ml for 1g sample). After cooling to room temperature, the volume of clear digested solution was made upto 50 ml using deionized water. A blank was also prepared using similar experimental procedure (Anonymous, 1990). The concentrations of Cu, Zn, Fe, Pb and Cd were determined using Atomic Absorption Spectrometer (AAS) [Perkin Elmer 5100 PC] in Bose Institute, Kolkata, while Arsenic (As) was determined by AVANTA GBC AAS with flamed hollow cathode lamp (HCL).

The samples after acid digestion were passed through the AAS system using different lamps, and calibrated with different minerals in different concentrations for different micronutrients (Anonymous, 1990). The concentrations of all the essential and toxic metals are the mean of three replicates and are expressed in mg related with kg of dry mushroom or substrate. The minimum detection limits of the device ranged according to analyzed elements and were found to be (mg kg^{-1} dry weight) 0.077 for Cu, 0.018 for Zn, 0.11 for Fe, 0.028 for Cd and 0.45 for Pb.

Table 1: Content (mg kg^{-1} dw*) of mineral and toxic elements in cultivated *P.florida* in the present study and the range of contents of these elements in selective references on other cultivated spp. of *Pleurotus*

Mineral/ Toxic element	Contents of elements in the substrate of water hyacinth and rice straw (1:1)in the present study (mg kg^{-1} dw of substrate)	Contents of elements in *P.florida* in the present study (mg kg^{-1} dw of mushroom)	Range of contents in cultivated spp. of *Pleurotus* in published literature (mg kg^{-1} dw of mushroom)	References
Iron (Fe)	201.5 ± 104.3	172 ± 61.5	67-1524	Chang et al., 1981; Bisaria et al., 1992; Caglarirmak, 2007; Akyuz and Kirbag, 2010; Zahid et al., 2010.
Zinc (Zn)	56.5 ± 6.2	52.3 ± 3.7	54-180	Wang, 2001; Rasad and Abdon, 2002; Gong et al., 2003; Guo et al., 2007; Caglarirmak, 2007; USDA, 2010; Akyuz and Kirbag, 2010.
Copper (Cu)	11.9 ± 2.6	10.5 ± 1.1	11-36	Rasad and Abdon, 2002; Dogan et al., 2006; Akyuz and Kirbag, 2010; Zahid et al., 2010; USDA, 2010.
Lead (Pb)	7.2 ± 1.5	2.3 ± 1.6	n.d. – 4.4	Dogan et al., 2006, Regula and Siwulski 2007, Garcia et al. 2009.
Cadmium (Cd)	2.7 ± 1.8	0.95 ± 1.3	0.3 – 2.9	Dogan et al. 2006; Regula and Siwulski, 2007; Garcia et al., 2009.
Arsenic (As)	2.4 ± 1.2	0.7 ± 0.3		No reference in literature for As content in cultivated mushroom

*dw =Dry weight

Table 2: Coefficient of accumulation (K_a) of the mineral and toxic elements in *P.florida* cultivated on water hyacinth and rice straw (1:1).

Mineral/ Toxic element	Average yield* of *P.florida* (g fresh weight per kg dw of substrate (WH+RS)	Mineral/ Toxic element content (g) in mushroom (dw) produced per kg dw of substrate (considering 82% moisture content of mushroom)	Coefficient of accumulation (K_a)
Iron (Fe)	1522	47.1	0.23
Zinc (Zn)	1522	14.3	0.25
Copper (Cu)	1522	2.9	0.24
Lead (Pb)	1522	0.63	0.09
Cadmium (Cd)	1522	0.26	0.1
Arsenic (As)	1522	0.2	0.08

* Referred to Bandopadhyay (2013).

Table 3: Recommended daily dietary intake values (mg day^{-1} person^{-1})*

Mineral/ Toxic element	Recommended values of daily dietary intake (mg day^{-1} person^{-1} derived from RDA[a]/PTWI[b]/)	Mineral/ Toxic element content(mg) per kg fresh weight of experimental mushroom of *P. florida*	Mineral/ Toxic element content (mg) per 100g fresh weight of experimental mushroom of *P. florida*
Iron (Fe)	18[a]	31	3.1
Zinc (Zn)	15[a]	9.4	0.9
Copper (Cu)	2[a]	1.9	0.2
Lead (Pb)	0.2[b]	0.4	0.04
Cadmium (Cd)	0.06[b]	0.17	0.02
Arsenic (As)	0.12[b]	0.13	0.01

* Referred to [a]NRC, 1989, [b]Codex Alimentarius, 2011.
[a]RDA: Recommended dietary allowances.
[b]PTWI: Provisional Tolerable Weekly Intake.

RESULTS AND DISCUSSION

The concentrations (mg kg^{-1}) of representative essential and toxic elements in substrate and in mushrooms are detailed in Table 1, which reveal gradual decrease of the element contents in the descending order of Fe > Zn > Cu > Pb > Cd > As. Among all the analyzed elements in the mushrooms, iron was the most important trace element with highest concentration

(172mg kg^{-1} dry weight of mushroom) followed by zinc (52.3 mg kg^{-1}) and copper (10.5 mg kg^{-1}). The table also summarizes the references of a selection of studies on the range of content of these elements in other cultivated oyster mushrooms published in literature.

The contents of all the essential minerals considered in this study fall within the range in literature and consistent with previous studies on cultivated mushrooms of *Pleurotus* spp. As the micronutrients (both mineral and toxic elements) flows from substrate to mushroom and accumulate, the co-efficient of accumulation of essential minerals and toxic elements are also calculated using relation:

$$K_a = \frac{C_m}{C_s}$$

Where C_m is the concentration of elements in mushrooms and C_s is their concentration in initial substrate of cultivation. Table 2 represents the coefficient of accumulation in mushroom indicating highest accumulation for Zn followed by Cu and Fe indicating *P. florida* as good accumulator of these elements.

According to Demirbas (2000) the level of mineral and heavy metal accumulation depends on the ability of the species to extract the elements from the substrate and on their selective uptake and deposition of such elements in the tissues. In the present study higher concentrations of toxic elements have been detected in initial substrates of water hyacinth and common paddy straw though they are present in much lower concentration in the mushrooms. Therefore, they are not getting accumulated in toxic level in the mushrooms which supports cultivation of mushroom on water hyacinth supplemented rice straw.

In order to contribute to the safe consumption of the experimental mushroom (grown on WH+RS) to the dietary intake of essential minerals and to evaluate the risk of dietary exposure to the toxic elements, the concentration of the mineral and toxic elements obtained in this study were compared with RDA (for Fe, Cu, Zn, As) or PTWI (for Pb, Cd) for an average adult of 60 kg body weight (b.w) as established by the Joint National Research Council (NRC, 1989), FAO / WHO (2010) , Codex Alimentarius (2011) Expert Committee on Food Additives and European Commission (toxicological fact sheet series, 2009 EC No 1881/2006) (Table 3). The recommended dietary values are on the fresh weight of foodstuff as such the obtained values for the mineral and toxic element contents on dry weight basis were recalculated on fresh weight basis considering 82% moisture content of experimental mushroom of *P. florida* (Bandopadhyay, 2013). Mean concentrations of Fe, Zn, Cu, Pb and Cd obtained in the experimental mushrooms were approximately equivalent to 31 mg kg^{-1}, 9.4 mg kg^{-1}, 1.9 mg kg^{-1}, 0.4 mg kg^{-1} and 0.17 mg kg^{-1} fresh weight of mushroom. Arsenic was obtained in trace amount of 0.7 mg kg^{-1} dry weight or 0.13 mg kg^{-1} fresh weight of mushroom. Recommended allowances for iron, zinc and copper for average adult person are 15, 12-15 and 1.5-3 mg day^{-1} respectively (NRC, 1989) while their provisional maximum tolerable daily intake (PMTDI) are 45, 40 and 10 mg day^{-1} respectively (Codex Alimentarius, 2011). The PTWI recommended by the Joint FAO/WHO Expert Committee on Food Additives for cadmium was previously 7 µg Cd kg^{-1} body weight (JECFA, 2003). In 2010 the JECFA (Joint FAO / WHO Expert Committee on Food Additives) reviewed the evaluation on cadmium and established a new provisional tolerable monthly intake (PTMI) of 25 µg kg^{-1} body weight which corresponds to a weekly intake of 5.8 µg kg^{-1} body weight or 0.3 mg for an adult. The PTWI for Pb, set by JECFA (25 µg Pb kg^{-1} body weight) was withdrawn in 2010. Regulation (EC) of European commission has set maximum levels for certain contaminants in vegetables establishing a maximum level of 0.2 mg kg^{-1} and 0.3 mg kg^{-1} for Cd and Pb respectively in oyster mushrooms like *Pleurotus ostreatus* (Toxicological factsheet series, 2009). For arsenic, inorganic As is more toxic than organic As. The safe range in toxicological guidance value is 2-7 µg kg^{-1} body weight per day (Codex Alimentarius, 2011) which is equivalent to 0.12 mg day^{-1} for a 60kg adult (Toxicological factsheet series, 2009). Considering the above recommended values, an adult person (of average 60kg b.w.) can therefore safely consume average 100g fresh mushroom of *P.florida* per day. The estimated intake of referred elements from dietary intake of this fresh mushroom will be 3.1mg day^{-1} Fe, 0.9mg day^{-1} Zn and 0.02mg day^{-1} Cu while 0.04 mg day^{-1} Pb , 0.02 mg day^{-1} Cd and 0.01 mg day^{-1} As, which are well below the permissible levels set by standard expert committee or commission for human consumption.

In the present study, Fe and Zn were found as the dominant essential mineral elements in comparison to other minerals. Iron is vital in a variety of metabolic processes. High intake of iron is desirable to compensate excretion and unutilization (Lynch and Baynes, 1996) and to reduce absorption, utilization and retention of most toxic metals particularly Pb and Cd with its antagonistic nature. Sufficient Zn in food is also essential to neutralize the toxic effects of Cd.

Comparison of the levels of essential minerals and toxic elements in the studied mushroom of *P. florida* indicated that permissible levels must be followed for safe food. Regular monitoring of toxic elements from mushrooms cultivated on substrates collected from roadside fields or even from rural areas is required to do vegetables safety. Monitoring accompanied by regular reevaluation of acceptable levels must continue but with the realization that some toxic heavy metals will probably always be found in very low quantities and they are considered to be unavoidable contaminants. It can therefore be concluded that cultivation of *P.florida* on mixture of water hyacinth and rice straw is a cost-effective, eco-friendly project and controlled consumption of this mushrooms daily does not pose any toxicological risk.

ACKNOWLEDGEMENT

Author sincerely thanks Department of Science and Technology, New Delhi, Govt. of India for providing financial assistance under “Women Scientists Scheme for Societal Programmes”.

REFERENCES

Aboul-Enein AM, Al-Abd AM, Shalaby EA , Abul-Ela F, Nasr Allah AA , Ahmoud AM and El-Shemy HA, 2011. *Eichhornia crassipes* (Mart.) Solms: from water parasite to potential medicinal remedy. Plant Signalling & Behavior. 6: 834-836.

Akyuz M and Kirbag S, 2010. Element Contents of *Pleurotus eryngii* (DC. ex Fr.) Quel. var. *eryngii* Grown on Some Various Agro-Wastes. Ekoloji. 19, 74: 10-14.

Anonymous, 1990. Official Methods of Analysis of Association of the Official Analytical Chemists. In: Helrich K (eds.), Fifteenth edition. Virginia 22201 USA, 1213, ISBN: 0-995584-42-0, ISSN: 0066-96IX.

Bandopadhyay (Mukhopadhyay) Sunita and Chatterjee NC, 2009. Water hyacinth, a lost-cost supplement for oyster mushroom (*Pleurotus florida*) cultivation. Mushroom Res. 18, 1: 5-9.

Bandopadhyay S, 2013. Effect of supplementing rice straw with water hyacinth on yield and nutritional qualities of oyster mushroom (*Pleurotus* spp.). Micologica Applicada Internatl. 25, 2: 15-21.

Bisaria R, Madan M and Bisaria, VS 1987. Biological efficiency and nutritive value of *P. sajor-caju* cultivated on different agro-wastes. Biol. Wastes. 19: 239-255.

Çaglarirmak N, 2007. The Nutrients of Exotic Mushrooms (*Lentinula edodes* and *Pleurotus* species) and an Estimated Approach to the Volatile Compounds. Food Chem. 105: 1188-1194.

Chang ST, Lau DW and Cho KY, 1981. The cultivation and nutritional value of *Pleurotus sajor-caju*. Eur. J. Appl. Microb. Biotech. 12: 58-62.

Codex Alimentarius (2011) Joint FAO/WHO Food Standards Programme Codex committee on contaminants in foods. Fifth Session. CF/5 INF/1. The Hague, The Netherlands, 21 - 25 March 2011.

Demirbas A, 2001. Concentrations of 21 metals in 18 species of mushrooms growing in the East Black Sea region. Food Chem. 75: 453–457.

Demirbas A, 2000. Accumulation of heavy metals in some edible mushrooms from Turkey. Food Chem. 68:415–419.

Dogan HH, Sanda MA, Uyanoz R, Ozturk C and Çetin U, 2006. Contents of metals in some wild mushrooms: Its impact in human health. Biol. Trace Element Res. 110:79-94.

Dos Santos MC and Lenzi E, 2000. The use of aquatic macrophytes (*Eichhornia crassipes*) as a biological filter in the treatment of lead contaminated effluents. Env. Tech. 21: 615-622.

FAO/WHO, 2010. Joint FAO/WHO expert committee on food additives. Seventy-third Meeting, Geneva, June 8-17, 2010. Summery and conclusions. Issued 24th june 2010. Available at: http://www.fao.org/ag/agn/jecfa/jecfa73sc.pdf.

Garcia MA, Alonso J, Fernandez MI and Melgar MJ, 1998. Lead content in edible wild mushrooms in Northwest Spain as indicator of environmental contamination. Arch. Env. Contam. Toxicol. 34: 330–335.

García MÁ, Alonso J and Melgar MJ, 2009 Lead in Edible Mushrooms: Levels and Bioaccumulation Factors. J. Hazardous Materials. 167: 777-783.

García-Gabarra A, 2006. Ingesta de Nutrientes: Conceptos y Recomendaciones Internacionales (2ª Parte). Nutr. Hosp. 21,4: 437-47.

JECFA (Joint FAO/WHO Expert Committee on Food Additives), 2003. Summary and conclusions of the sixtyfirst meeting of the Joint FAO/WHO Expert Committee on Food Additives (JECFA), pp. 18-22. Available on http://www.who.int/pcs/jecfa/Summary61.pdf

Kalac P and Svoboda L, 2001. A review of trace element concentrations in edible mushrooms. Food Chem. 69: 273–281.

Luis G, Hernández C, Rubio C, González-Weller D, Gutiérrez Á, Revert C and Hardisson A, 2012. Trace elements and toxic metals in intensively produced tomatoes *(Lycopersicon esculentum)*. Nutr. Hosp. 27, 5:1605-1609.

Lynch SR and Baynes RD, 1996. Deliberations and evaluations of the approaches, endpoints and paradigms for iron dietary recommendations. J. Nutri. 126:2404-2409.

Nageswaran M, Gopalakrishnan M, Ganeshan M, Vedhamurthy A and Selvaganapadhy E, 2003. Evaluation of Water hyacinth and paddy straw waste for culture of oyster mushroom. J. Aquat. Plant Manage. 41: 122-123.

NRC, 1989. Recommended Dietary Allowances (RDA), 10th Edition. Subcommittee on the Tenth Edition of the recommended Dietary Allowances, Food and Nutrition Board , Commission on Life Sciences, National Research Council, National Academy Press, Washington, D.C.20001. ISBN: 0-309-53606-5, 302 pages, 6 x 9, (1989).

Obodai M and Apertorgbor M, 2008. Proximate Composition and Nutrient Content of some Wild and Cultivated Mushrooms of Ghana. J. Ghana Sc. Assn. 10: 139-144.

Rashad MM and Abdou HM, 2002. Production and Evaluation of *Pleurotus ostreatus* Mushroom Cultivated on some Food Processing Wastes. Adv. Food Sc. 24:79-84.

Reguła J and Siwulski M, 2007. Dried Shiitake (*Lentinulla edodes*) and Oyster (*Pleurotus ostreatus*) Mushrooms as a Good Source of Nutrients. Acta Sci. Pol. Technol. Aliment. 6:135-142.

Suman, B.C. and V.P. Sharma, 2005. Steps in mushroom growing. pp. 70-74. In: Suman BC, Sharma VP (eds), Mushroom cultivation and uses, Agrobios, India.

Toxicology Factsheet Series, 2009. Mercury, lead, cadmium, tin and arsenic in food. Commission Regulation (EC) No 1881/2006 of 19 December 2006 setting maximum levels for certain contaminants in foodstuffs. Food Safety. 1:1-13.

Upadhyay RC, 1997. Cultivation of oyster mushroom – complete technology package. pp. 53-64. Proceedings of National Training Course on Mushroom Production Technology for entrepreneurs, 21st April – 30th April, NCMRT, Solan.

USDA, 2010. USDA National Nutrient Database for Standard Reference. Release 23. Nutrient Data Laboratory Homepage (database), United States Department of Agriculture,Washington D.C.

Zhu F, Qu Li, Fan W, Qiao M, Hao H and Wang X, 2010. Assessment of heavy metals in some wild edible mushrooms collected from Yunnan Province, China. Environ. Monit. Assess. DOI 10.1007/s10661-010-1728-5.

INFLUENCE OF SOIL APPLIED MORINGA LEAF EXTRACT ON VEGETATIVE GROWTH OF *CYPERUS ROTUNDUS*

Anser Ali[1], Nasir Abbas[1], M.M. Maqbool[1], Tanveerul Haq[1], Muhammad Ahmad*[2], and Rashid Mahmood[3]

[1] *Faculty of Agriculture, Ghazi University, Dera Ghazi Khan, Pakistan*

[2] *Agricultural Training Institute, Karor, Layyah, Pakistan*

[3] *Institute of Agricultural Sciences, Punjab University, Lahore, Pakistan*

ABSTRACT

Allelopathy refers to the beneficial or harmful effects of one plant on another plant through the release of secondary metabolites from plant parts in both natural and agricultural systems. Moringa leaf extract (MLE) is generally considered to have enhancing or suppressing effect on the plant growth in a dose dependent manner. Therefore, the current study was planned to evaluate the effects of MLE on the growth of purple nutsedge (*Cyperus rotundus*). MLE was added into pots three days after transplanting with 25, 50, 75 and 100% concentrations (v/v) whereas distilled water was used as control. Application of MLE at 100% concentration notably improved the root and shoots lengths, shoot fresh and dry weights. MLE application had non-significant effect on root fresh and dry weights, number of nodes and node weights of purple nutsedge. It was concluded that MLE improved the growth of purple nutsedge even higher concentrations rather to suppress it.

Keywords: *Moringa oleifera*, leaf extract, purple nutsedge, root shoot weight

INTRODUCTION

Purple nutsedge (*Cyperus rotundus* L.) is amongst the most serious and troublesome weeds occurring in many parts of the world (Travlos et al., 2009; Shabana et al., 2010) including Pakistan (Riaz et al., 2009). It reproduces mostly through tubers (Sharma and Gupta, 2007; Lati et al., 2011). The spread of the plant in the short term is accomplished through rhizomes, which may extend upward, downward or horizontally (Stoller and Sweets, 1987). Purple nutsedge is difficult to control with herbicides (Gilreath and Santos, 2004).

Allelopathy plays an important role in regulating plant diversity (Chou and Lee, 1991). In the beginning most of the allelopathic research was conducted to investigate the effect of weeds on crops and one crop on another. In crop science, mechanism of allelopathy is exploited successfully by exogenous application of plant extracts for weeds suppression and crop growth enhancement (Farooq et al., 2011). Cheema et al. (1997) found that aqueous extracts of sorghum and sunflower have the potential to suppress the weed infestation in wheat crop. Moradshahi et al. (2003) found that aqueous extracts of *Eucalyptus camaldulensis* (Schlecht.) has the potential to suppress growth of *Echinochloa crus-galli* (L.) Beauv., *Avena fatua* L. and *Rumex acetosella.*

Moringa (*Moringa oleifera*) is an important plant of Moringaceae family having tremendous allelopathic potential either to suppress or stimulate the neighboring crop. *M. oleifera* is widely cultivated plant of tropical as well as subtropical areas and is native to sub-Himalayan tracts of India, Pakistan, Bangladesh and Afghanistan (Nikkon et al., 2003). Having medicinal value, it has been used by all civilizations as a source of medicines since ancient times. It is used as human food, medicine and in oil production (Anwer et al., 2007). The plant growth is controlled by certain growth substances such as auxins, gibberellins, and cytokinins, which are organic in nature and produced in minute concentrations (Taiz and Zeiger, 2010). MLE is a rich source of growth regulators such as zeatin (a cytokinin), ascorbate, phenolics, and many essential plant minerals such as Ca, K, Mg, Mn, P, B, Zn, and Fe (Fuglie, 1999; Anwar et al., 2007; Basra et al., 2011; Hussain et al., 2013) and therefore can be used either as foliar spray or seed priming agent for growth promotion (Hussain et al., 2013; Nouman et al., 2012). Moringa leaf extract application also

**Corresponding author:* e-mail: mahmada2003@yahoo.co.uk

enhanced the seed germination of sorghum crop (Phiri, 2010).

It is documented that diluted MLE (20 times with water) has the potential to perk up the growth and grain yield of crops and vice versa (Basra et al., 2009b). Keeping in view the above studies, it was imperative to conduct a study that that how the soil applied MLE application at different concentrations affects the growth of *C. rotandus.*

MATERIALS AND METHODS

Well-nourished nursery of purple nutsedge (*C. rotundus*) was raised by using bulbs as a seed. Ten days after sowing it was transplanted into the plastic pots filled with soil. MLE was added into pots three days after transplanting with 25, 50, 75 and 100% concentrations (v/v), while distilled water (0% MLE) was used in control one. To prepare MLE, fresh leaves of *M. oleifera* were collected and ground in a blender for juice extraction. The extracted juice was filtered with sieve and then mixed with distilled water to prepare 25, 50, 75 and 100% concentrations (v/v) of MLE. The pots were arranged in completely randomized design having three replications. Data regarding root and shoot length, root and shoot fresh and dry weight, and number of nodes (seeds) and weight of nodes (seeds) per pot were collected 30 days after transplanting (DAT) by using standard procedures. The collected data were analyzed statistically at 5% level of significance and Bartlett test was used to compare its means at 5% level of probability (Steel et al., 1997).

RESULTS

Application of 100% MLE significantly improved the shoot length of purple nutsedge compared with all other treatments including control, however the minimum shoot length was observed in control (Table 1). Maximum root length of purple nutsedge was recorded at 50% MLE application but it was at par with all other treatments except control which recorded the minimum root length. It was also at par with all other MLE applications except of 50% MLE (Table 1). Moreover, application of MLE had non-significant effect on root fresh and dry weight of purple nutsedge (Table 1). MLE application at 75 and 100% concentrations enhanced the shoot fresh and dry weight against the minimum shoot fresh and dry weight as observed in control (Table 1). Nonetheless, MLE application had non-significant effect on the number of nodes and weight of nodes per pot of purple nutsedge (Table 1).

Table 1.Effect of various doses of moringa leaf extract on vegetative growth of *Cyperus rotundus*

MLE levels No. of pods/pot	SL	RL	RF wt.	SF wt.	RD wt.	SD wt.	Wt. of nodes per pot	No. of nodes/pot
Control	12.00c	15.00b	0.71^{NS}	0.40d	0.37^{NS}	0.25b	2.55^{NS}	7.25 ^{NS}
25%	16.75b	21.50ab	0.67	0.71cd	0.37	0.29b	3.37	7.75
50%	19.00b	26.75a	0.75	0.91bc	0.53	0.30b	3.78	8.50
75%	19.50b	21.50ab	0.80	3.81	8.75	1.28ab	0.51	0.41ab
100%	23.75a	19.75ab	0.71	1.44a	0.47	0.53a	3.18	9.75

SL, RL, RF Wt., SF Wt., RD Wt. and SD Wt. represent shoot length, root length, root fresh weight, shoot fresh weight, root dry weight and shoot dry weight, respectively.

DISCUSSION

MLE application has proven its worth as an excellent source of plant growth-promoting substances. MLE is either used as foliar spray or seed priming agent for growth promotion (Mehboob et al., 2011; Nouman et al., 2012) and behaves differently with change in its dilution (Basra et al., 2011). The higher concentrations of growth promoters including MLE are generally considered to suppress the plant growth. The present work indicated that MLE at higher concentrations had the promoting effect on *C. rotundus* rather to suppress its growth. In the current study, it was observed that MLE promoted the shoot growth more in comparison to root. MLE applied at higher doses significantly increased the shoot length and its weight. It had no prominent effect on root length and its weight. The similar trend was observed for number of nodes and their weights. In a recent study Nouman et al. (2012) reported that seed priming with MLE (1:30) substantially increased the number of leaves, roots and tillers, shoot vigor, and root length and weight in three rangeland grasses including barnyard grass, blue panic grass, and buffalo grass. Seed priming with MLE diluted 20 times with water improved root length in sunflower (Basra et al., 2009b).

The growth promotion of plants owing to MLE application is due to containing the growth regulators and essential minerals (Fuglie, 1999; Anwar et al., 2007; Basra et al., 2011; Hussain et al., 2013). In conclusion, the exogenously applied moringa leaf extract may have promoting effect depending upon the nature of plant, even if its higher doses are applied.

REFERENCES

Anwer F, Latif S, Ashraf M, and Gilani AH, 2007. *Moringa oleifera*: A food plant with multiple medicinal uses. Phytother. Res. 21: 7–25.

Basra SMA, 2011. Moringa leaf extract: a natural crop growth enhancer. In: Proc. Intl. Seminar on ''Crop Management: Issues and Options'', Univ. Agri., Faisalabad, Pakistan, June 1–2: 21.

Basra SMA, Zahoor R, Rehman H, Afzal I and Farooq M, 2009b. Response of root applied brassica and moringa (*Moringa oleifera*) leaf water extracts on seedling growth in sunflower. In: Proc. Intl. Conf. on Sustainable Food Grain Production: Challenges and Opportunities. Univ. Agri. Faisalabad, Pakistan: 24.

Basra SMA, Iftikhar MN and Afzal I, 2011. Potential of Moringa *(Moringa oleifera*) leaf extract as priming agent for hybrid maize seeds. Int. J. Agr. Biol. 13: 1006–1010.

Farooq M, Jabran K, Cheema ZA, Wahid A and Siddique KHM, 2011. The role of allelopathy in agricultural pest management. Pest Manag. Sci., 67: 493–506.

Fuglie LJ, 1999. The miracle tree: *Moringa oleifera*: natural nutrition for the tropics. Church World Service, Dakar: 68.

Gilreath JP and Santos BM, 2004. Herbicide dose and incorporation depth in combination with 1,3-dichloropropene plus chloropicrin for *Cyperus rotundus* control in tomato and pepper. Crop Prot. 23: 205-210.

Hussain M, Farooq M, Basra SMA and Lee D-J, 2013. Application of moringaallelopathy in crop sciences. In: Allelopathy; current trends and future applications. Cheema et al. Eds., 1st ed., Springer Heidelberg, New York and Dordrecht, London.

Lati RN, Filin S and Eizenberg H, 2011. Temperature- and radiation-based models for predicting spatial growth of purple nutsedge (*Cyperus rotundus*). Weed Sci. 59: 476-482.

Mehboob W, Rehman H, Basra SMA and Afzal I, 2011. Role of seed priming in improving performance of spring maize, In: Proc. Intl. Seminar on Crop Management: Issues and Options June 01-02, 2011, Univ. Agri., Faisalabad, Pakistan: 55.

Makkar HPS and Becker K, 1996. Nutritional value and anti-nutritional components of whole and ethanol extracted *Moringa oleifera* leaves. Anim. Feed Sci. Tech. 63: 211–228.

Nikkon F, Saud A, Rahman MH and Haque ME, 2003. In vitro antimicrobial activity of the compound isolated from chloroform extract of *Moringa oleifera* Lam., Pak. J. Biol. Sci. 6: 1888-1890.

Phiri C, 2010. Influence of *Moringa oleifera* leaf extracts on germination and early seedling development of major cereals. Agri. Biol. J. N. Am. 5: 774–777.

Riaz T, Khan SN, and Javaid A, 2009. Weed flora of Gladiolus fields in district Kasur, Pakistan. J. Anim. Plant Sci.19: 144-148.

Shabana, YM, Charudattan R, Abou-Tabl, AH, Morales-Payan JP, Rosskopf EN and Klassen W, 2010. Production and application of the bioherbicide agent Dactylariahigginsii on organic solid substrates. Biol. Cont. 54: 159-165.

Sharma R and Gupta R. 2007. *Cyperus rotundus* extract inhibits acetyl-cholinesterase activity from animal and plants as well as inhibits germination and seedling growth in wheat and tomato. Life Sci. 80: 24-25.

Steel, R.G.D., J.H. Torrie and D.A. Deekey.1997. Principles and Procedures of Statistics.A Biometrical Approach.3rd Ed. McGraw Hill Book. Int. Co. New York. 14: 400-428.

Stoller EW and Sweet RD, 1987. Biology and life cycle of purple and yellow nutsedge

(*Cyperus rotundus* and *C. esculentus*). Weed Tech. 1: 66–73.

Taiz L, Zeiger E, 2010. Plant physiology, 5th edn.,Sinauer Associates, Sunderland.

Travlos IS, Economou G, Kotoulas VE, Kanatas PJ, Kontogeorgos AN, and Karamanos AI, 2009. Potential effects of diurnally alternating temperatures and solarization on purple nutsedge (*Cyperus rotundus*) tuber sprouting. J. Arid Environ. 73: 22-25.

Yasmeen A, 2011. Exploring the potential of moringa (*Moringa oleifera*) leaf extract as natural plant growth enhancer, Ph.D. thesis, Dept. Agron., Univ. Agri., Faisalabad, Pakistan.

THERMAL HARDENING IMPROVES GERMINATION AND EARLY SEEDLING GROWTH OF CHICKPEA

Atique ur Rehman[1*], Muhammad Farooq[2], Hakoomat Ali[1], Naeem Sarwar[1], Rafi Qamar[3]

[1]*Department of Agronomy, Bahauddin Zakariya University, Multan.*

[2]*Department of Agronomy, University of Agriculture, Faisalabad.*

[3]*Department of Agronomy, University of Sargodha, Sargodha.*

ABSTRACT

Poor quality seeds of chickpea have germination problem and issue is serious in chickpea growing areas where quality seed is not available. This study was an attempt to solve the problem. Experiment was done in Seed Technology Laboratory of Department of Agronomy, University of Agriculture, Faisalabad. Two chickpea cultivars viz. Kabli type cultivar Noor-91 and desi type cultivar Punjab-91 were treated with different temperatures of 40, 50 and 60 °C at 24, 48 and 72 hrs. Different thermal hardening treatments significantly improved different germination related traits including germination percentage, germination index, germination energy and reduced time taken for 50% germination. Similarly heating of chickpea seeds resulted in more number of secondary roots, and improved fresh and dry weights. Although germination was improved with thermal treatments at 40 and 50 °C for different time durations, however, an increased temperature (60 °C) was better only at 24 hrs and keeping seeds at this temperature for more time resulted in reduced germination and other traits. It is thus concluded that thermal hardening at 40 and 50 °C for up to 72 hrs can be used to improve germination of chickpea.

Keywords: Chickpea, Germination Index, Germination Energy, Dry weight, Temperature.

INTRODUCTION

Chickpea (*Cicer arietinum* L.) is a chief protein rich source of cool-season usually grown under rainfed environments of the world. It is an important human consumption which is roughly alienated into desi with dark brown coat (Microsperma) and kabli with light yellow covering (Macrosperma) (Muehlbauer and Singh, 1987). Both are used as different food stuff due to difference in seed size, color, water uptake and cooking properties, crude fiber and caloric value (Khan et al., 1995). As grown mostly under rainfed environments, a good stand is possible in good dew precipitation.

Seed hardening is a technique which has been used successfully for vigor enhancement in several seeds like rice, wheat (Basra et al., 2003, 2004; Lee and Kim, 2000). In this technique, seeds may be exposed to different conditions like alternate wetting and drying, chilling treatment or high temperature for different durations (Farooq et al., 2005). The practice of heating, chilling or hydration may be repeated several times according to the seed structure or composition. (Lee and Kim, 2000). It is possible to carry out thermal treatments using dry heat by keeping seeds in an oven at varying controlled temperatures with a series of temperatures imitating the conditions that may exist in the soil with different exposure times (Perez and Pita, 1997).

Several researchers worked on hardening and other techniques for increase in vigor of seeds and found hardening better than other. For example, in a study on rice seeds, Farooq et al. (2004) exposed seeds to dry heat (at 40 °C for 72 h, at 60 °C for 24 h) and chilling (at -19 °C for 72 h). The seeds treated with dry heat at 40 °C for 72 h found with more vigor as compared to chilling treatments. Earlier, Lee and Kim (2000) compared osmoconditioning and hardening for germination of standard and physically aged seeds with analysis of α-amylase activity and total sugars. Although total sugar content and α-amylase activity increased with both techniques but hardening was found to be more effective than osmoconditioning.

Many plant species in different climatic regions of the world are exposed to the passage of fire and it is found that rather a

Corresponding author: e-mail: atiqjugg@gmail.com

destructive influence, it triggers and facilitates in germination and seedling establishment (Whelan, 1995). Seeds of such species not only tolerate high temperatures more than 100 °C, but also frequently respond with high germination rate following the heat shock. It is argued that these seeds may have a dormancy imposed by an impervious seed coat (Cavanagh, 1987) or alterations to cell structure of seed (Tran and Cavanagh, 1980). The exposure to high temperature and heat shock to such seeds is considered to scarify the seed coat and allow water and gas exchange resulting in smashed seed dormancy (Thanos and Georghiou, 1988). Duration of exposure for seed hardening is also counted important for better results. Improving germination by thermal treatment is proved to be important tool and may use for seeds which have poor quality. In our present study, attempt was made to achieve a better germination of chickpea with different temperatures.

MATERIAL AND METHODS

Seeds of the chickpea (*Cicer arietinum* L.) Kabli cultivar Noor-91 and desi cultivar Punjab-91 were used as the experimental material. Seeds of both cultivars were obtained from the Pulses Research Institute, Ayub Agriculture Research Institute (Faisalabad, Pakistan).

Thermal hardening treatments include dry heating 25 seeds of each cultivar incubated at 40, 50 and 60 °C for 24, 48 and 72 hrs each in an oven. Seeds were incubated in glass jars tightly covered with lids.

After treating seeds with different temperatures for different durations, these were sown in Petri plates between layer of filter paper which were moisturized with water and kept at 27 °C in an incubator. To record germination characteristics, seeds were observed daily according to the methods of the Association of Official Seed Analysis (1990).

The time taken to complete 50% germination (T_{50}) was computed according to the formula of Coolbear et al. (1984) and modified by Farooq et al. (2004):

$$T_{50} = t_i + \left[\frac{N/2 - n_i}{n_j - n_i}\right]$$

Where N is the final number of germinated seeds and nj and ni are the cumulative number of seeds germinated by adjacent counts at times tj and ti, respectively, when ni < N/2 < nj.

Mean germination time (MGT) was calculated according to the equation of Ellis and Roberts (1981):

$$MGT = \frac{\sum Dn}{\sum n}$$

Where n is the number of seeds that had germinated on day D and D is the number of days counted from the beginning of germination.

The germination index (GI) was determined as described by the Association of Official Seed Analysis (1983) using the following formula:

Germination Index

$$(GI) = \frac{\text{No. of germinated seeds}}{\text{Days of first count}} + \text{-------} + \frac{\text{No. of germinated seeds}}{\text{Days of final count}}$$

The energy of germination (GE) was recorded on the 4th day after planting of seeds. This is the percentage of germinating seeds 4 d after planting relative to the total number of seeds tested (Ruan et al., 2002).

Secondary roots were counted daily to determine the rate of secondary root emergence. While on 10th day after sowing, the seedlings were tested for vigor after careful removal. Radicle and plumule lengths were recorded of 5 randomly selected seedlings per replicate and averaged. Seedling fresh weight was noted by weighing seedlings with digital balance and dry weight was taken after drying the seedlings at 70 °C for 7 days.

Graphical presentation of the data was made in Microsoft Excel. Standard error bars were computed to compare the real difference amongst the treatments.

RESULTS

Heating chickpea seeds improved germination and stand establishment of seedlings. In both chick pea cultivars, treated seeds were germinated earlier compared with untreated (control) seeds. However, both cultivars responded different for different temperatures with different durations. In Kabli, seeds subjected to temperature of 50 °C for 48 or 72 h germinated earlier than seeds subjected to other treatments while desi type chickpea responded best at lower temperature i.e. 40 °C for 24 and 48 h. The highest T_{50} was noted in seeds treated with high temperature of 60 °C for all the durations (Fig 1). A similar trend was observed in case of mean germination time in both chickpea cultivars. Kabli chickpea responded best at 50 °C for 48 and 72 h while desi chickpea seeds at 40 °C for 24 and 48 h. Seeds treated with high temperature of 60 °C could not differ significantly from untreated seeds or control (Fig 1). In case of final germination percentage, seed treatments did not significantly affect; however, in temperature of 60 °C for all duration, significantly less germination percentage was observed as compared with control. The energy of germination was improved significantly by temperature treatments up to a certain limit. Both cultivars responded differently for energy of germination with different temperatures. Kabli chickpea seeds were observed with significantly more germination energy at temperature of 40 °C for 72 h and 50 °C for 24 h. In all other treatments of 40 °C or 50 °C, kabli seeds remained at par with control while temperature more than 50 °C significantly reduced the energy of germination. In desi chickpea seeds, germination energy was significantly high with treatments of 40 °C for 24 and 48 h. All treatments of 40 and 50 °C were at par with that of control while temperature of 60 °C significantly reduced the energy of germination (Fig 1). Similar trend was in case of germination index in both chickpea cultivars. In kabli chickpea, maximum germination index was recorded in seeds subjected to temperature of 40 °C for 72 h and 50 °C for 24 h, then followed by the remaining treatments, except control and 60 °C for all durations where germination index was significantly lower than control. Germination Index of desi chickpea seeds was high in temperature of 40 °C for 24 and 48 h. All other temperature treatments remained at par with control except 60 °C which could not improve the germination index (Fig 1).

Radicle length was drastically improved by pre sowing heating treatments (Fig 2); however, minimum radicle length was noted in untreated seeds. Although different thermal treatments improved radicle length of chickpea, however, seeds of kabli heated at 40 °C were noted with longer radicle than all other. Whereas radicle length was decreased with increasing temperature and duration and at 60 °C radicle length of kabli was improved when seeds were heated for 24 hrs while in case of desi cultivar, radicle length was less than that of control. The plumule length of seeds undergoing the heating treatment protocols was higher at 40 and 50 °C while it was reduced when seeds were treated at 60 °C especially for a long time (Fig 2). However, among different treatments, heating seeds for 24 hrs at 40 and at 50 °C were found better than others.

Number of secondry roots were significantly improved with thermal hardening of chickpea seeds (Fig 3). Different hadening treatments were found effective in improvement of secondary roots, however, heating seeds at 40 and 50 °C for any time duration was most effective for improvement than control. Although heating at 60 °C for 24 hrs was also improved number of secondary roots but a longer duration viz. 48 and 50 hrs could found effective rather these were found harmful (Fig 3).

Fresh and dry weights of chickpea seedlings were improved with thermal hardening (Fig 4). Heating seeds at 40 and 50 °C for 24 hrs to 72 hrs were found effective in improving both fresh and dry weights of both chickpea cultivars and all these treatments were found at par. However, heating at 60 °C could not increase fresh or dry weight of any cultivar (Fig 4).

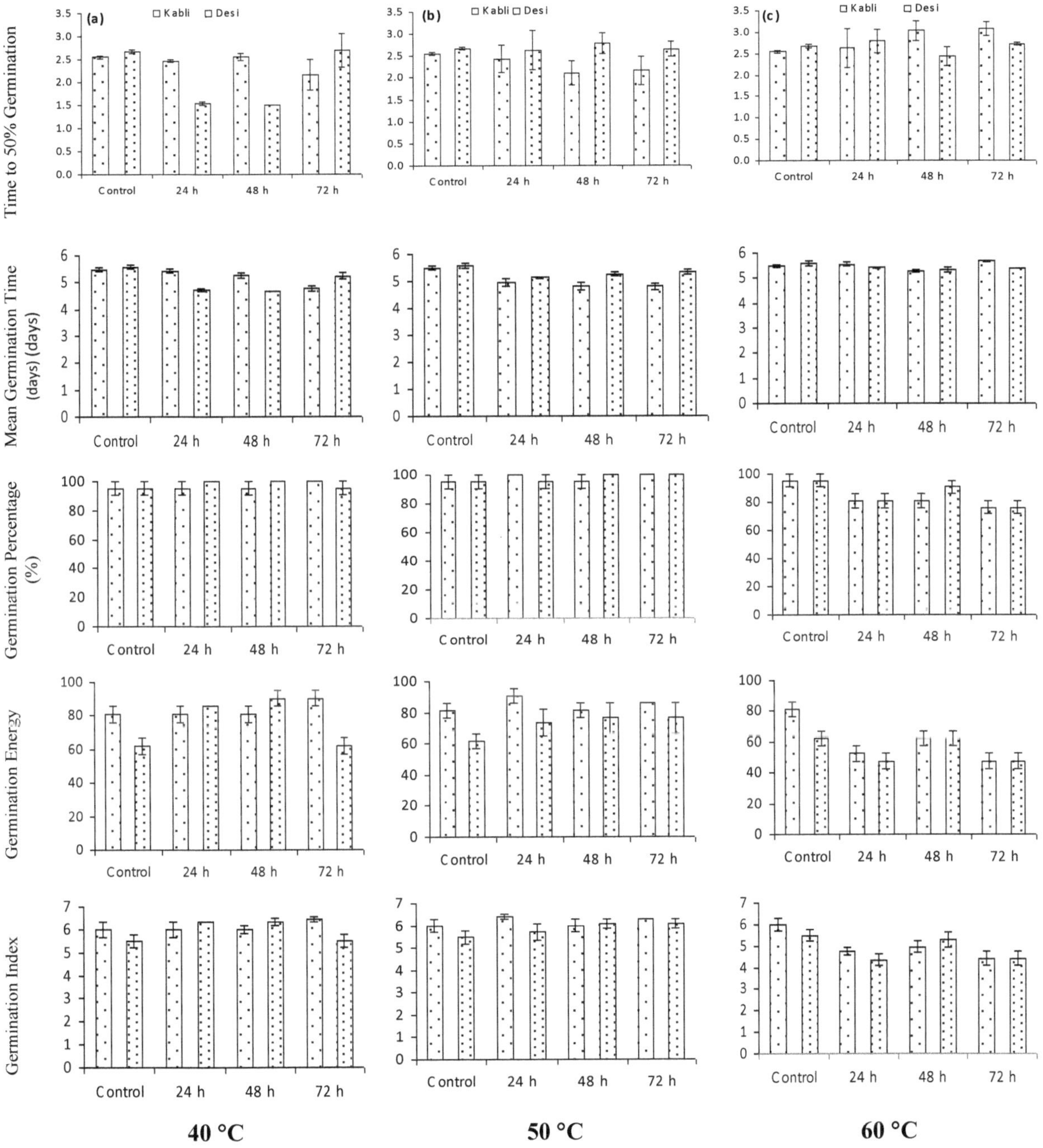

Fig. 1: Influence of dry heat treatments at (a) 40 (b) 50 and (c) 60 °C on the germination traits of chickpea cultivars

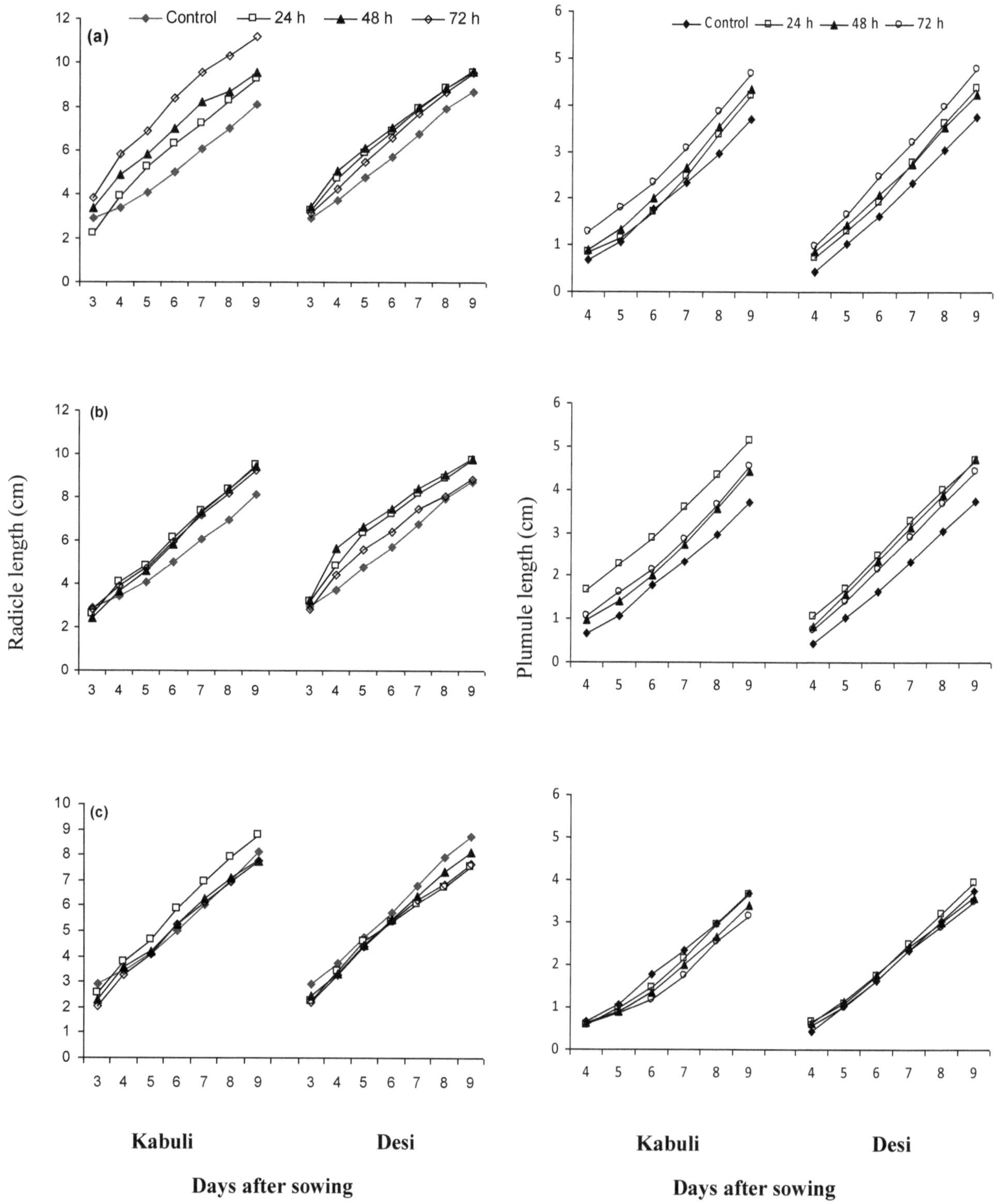

Fig 2. Influence of dry heat treatments at (a) 40 (b) 50 and (c) 60 °C on the radicle and plumule length of chickpea

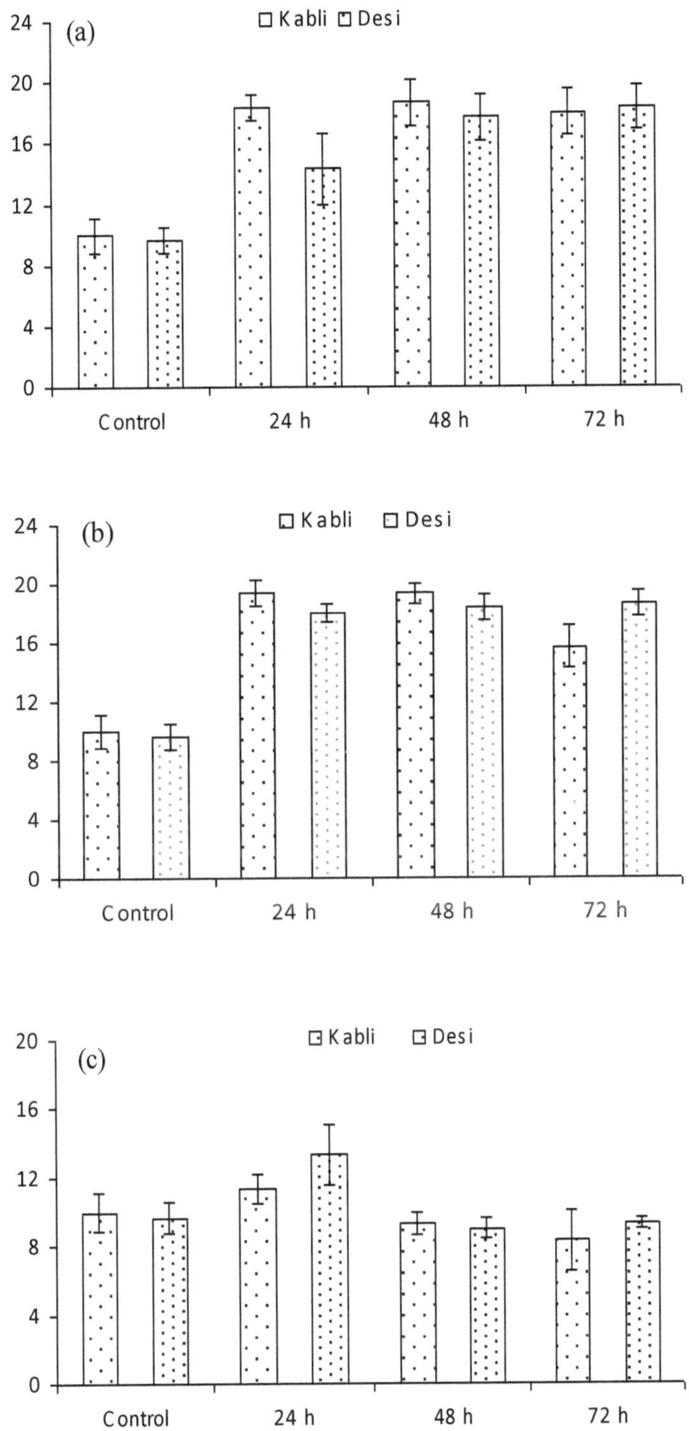

Fig 3. Influence of dry heat treatments at (a) 40 (b) 50 and (c) 60 °C on number of secondary roots of chickpea

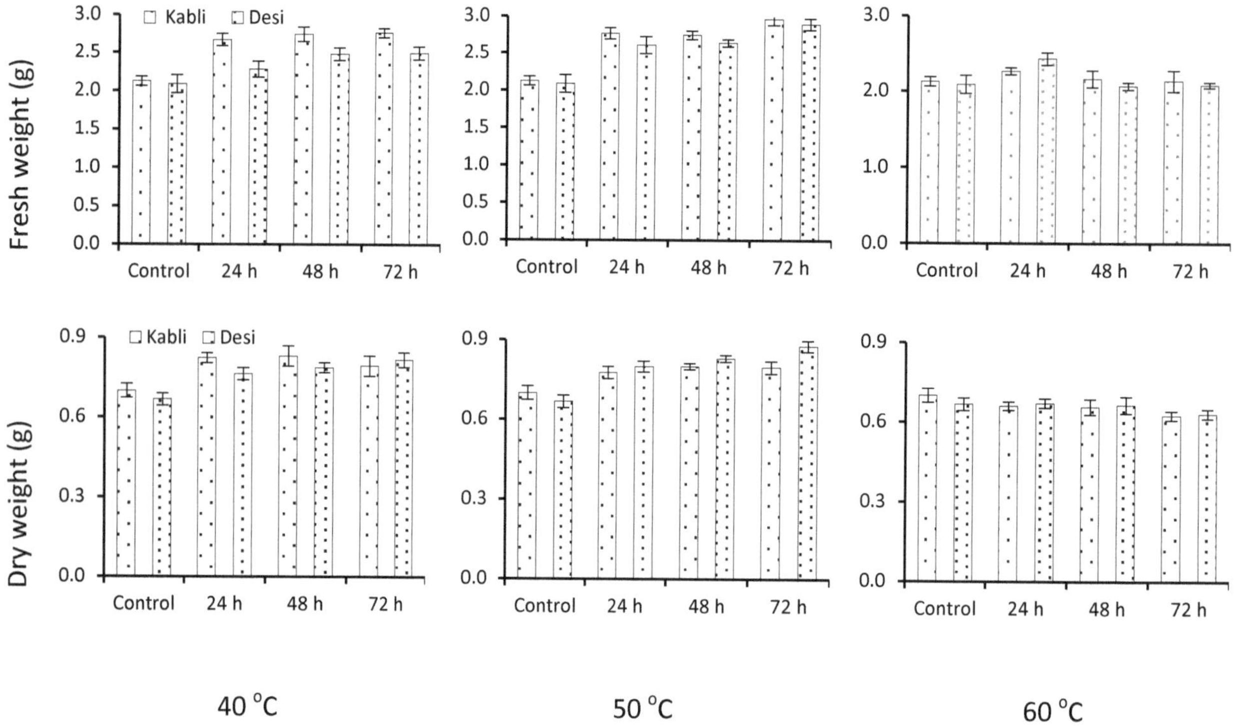

Fig 4. Influence of dry heat treatments at (a) 40 (b) 50 and (c) 60 °C on the fresh and dry weight of chickpea

DISCUSSION

Thermal hardening found to have a significant effect on the germination and seedling vigor of both chickpea cultivars (Fig 1, 2). However, the responses of both cultivars of chickpea to different dry heat treatments were not similar (Fig 1, 2). In both cultivars, earlier and synchronized germination was observed in treated seeds as compared to control seeds, as evidenced by a lower time to start germination, T_{50}, MGT and a higher GI, GE, FGP, lengths of radical and plumule, number of secondary roots as well as seedling fresh and dry weight (Fig 1, 2). Germination percentage and radicle length remained unaffected, but higher germination energy was recorded in treated seeds. It is well postulated that several seed treatments have to improve different germination traits of different crop species. Ruan et al. (2002) have also reported that seed treatments were not able to invigorate rice seeds, but resulted in a higher energy of germination and germination index compared with untreated seeds. Improvement in germination was observed due to some beneficial aspects primarily are due to pre-enlargement of the embryo (Austin et al., 1969), improvement of the germination rate (Gray and Steckle, 1977), and effects that are attributed to alternating wetting and drying processes (Basra et al., 2003). Greater enhancement was noted in seeds subjected to dry heat treatment as compared to those seeds treated with chilling (Farooq et al., 2005). Farooq et al. (2004) also reported vigor enhancement in indica rice seeds subjected to dry heat treatment. However, these improvements was reported mainly in rice seeds while our study was on chickpea seeds. In our study is evident from the results that heat treatments significantly improved not only germination traits but also further improved number of secondary roots, radicle and plumule length and fresh and dry weights of both chickpea cultivars (Fig 1,2,3). Treating seeds with a higher temperature (60 °C) for long duration (72 hrs) was observed harmful indicating that chickpea seeds cannot tolerate a high temperature and increasing temperature from 60 °C will definitely damage the seeds. However, treating at high temperature for short duration improved different traits than control. Earlier Mott et al. (1982) reported that treating seeds with very high temperature 160 °C for seconds or a few minutes improved germination.

Improved fresh and dry weights are a result of earlier and healthy germination, which contributed to further growth of chickpea cultivars. There was a little difference in chickpea cultivars in response to heat treatments, which is due to their genetic makeup. Genetic difference is found a reason for differential response to different treatments in some other species like rice (Lee et al., 2002).

In conclusion, heat treatment (upto 50 °C for not more than 72 hrs) are helpful in improving the germination of chickpea seeds.

REFERENCES

Association of Official Seed Analysis, 1983. Seed Vigor Testing Handbook. Contribution No. 32 to the Handbook on Seed Testing. Association of Official Seed Analysis, Springfield, IL.

Association of Official Seed Analysis, 1990. Rules for testing seeds. J. Seed Technol., 12: 1–112.

Austin RB, Longden PC and Hutchinson J, 1969. Some effects of "hardening" carrot seed. Ann Bot 33, 883–895.

Basra SMA, Farooq M and Khaliq A, 2003. Comparative study of pre-sowing seed enhancement treatments in fine rice (*Oryza sativa* L.). Pak. J. Life Soc. Sci. 1: 5-9.

Basra SMA, Farooq M and Tabassum R, 2004. Physiological and biochemical aspects of seed vigor enhancement treatments in fine rice (*Oryza sativa* L.). Seed Sci. Technol., 33: 21.

Cavanagh T, 1987. Germination of hard seeded species (Order Fabales). *In*: Langkamp, E.P. (Ed.). Germination of Australian Plant Seed, Inkata Press, Melbourne, pp. 58–70.

Coolbear P, Francis A and Grierson D, 1984. The effect of low temperature pre-sowing treatment under the germination performance and membrane integrity of artificially aged tomato seeds. J. Exp. Bot., 35: 1609-1617.

Ellis RA and Roberts EH, 1981. The quantification of ageing and survival in orthodox seeds. Seed Sci Technol 9: 373–409.

Farooq M, Basra SMA, Hafeez K and Warriach EA, 2004. Influence of high and low temperature treatments on the seed germination and seedling vigor of coarse and fine rice. Int. Rice Res. Notes, 29: 69–71.

Farooq M, Basra SMA, Ahmad N and Hafeez K, 2005. Thermal Hardening: A New Seed Vigor Enhancement Tool in Rice. J. Integ. Plant Biol., 47: 187−193.

Gray D and Steckel JRA, 1977. Effects of pre-sowing treatments on the germination and establishment of parsnips. *J Hort Sci* 52, 525–534.

Khan MA, Akhtar N, Ullah I and Jaffery S, 1995. Nutritional evaluation of desi and kabli chickpeas and their products, commonly consumed in Pakistan. Int. J. Food Sci. Nutr., 46: 215–223.

Lee SS and Kim JH, 2000. Total sugars, α-amylase activity, and germination after priming of normal and aged rice seeds. Kor. J. Crop Sci., 45: 108–111.

Lee SY, Lee JH and Kwon TO, 2002. Varietal differences in seed germination and seedling vigor of Korean rice varieties following dry heat treatments. *Seed Sci Technol* 30, 311–321.

Mott, JJ, Cook SJ and Williams RJ, 1982. Influence of short duration high temperature seed treatment on the germination of some tropical and temperate legumes. Tropical Grasslands, 16: 50-55.

Muehlbauer FJ and KB Singh, 1987. Genetics of chickpea. *In*: Saxena, M.C. and K. B. Singh (Eds.). The Chickpea. Wallingford, OX: CAB International. pp. 99–125.

Pe´rez F and Pita JM, 1997. Ecofisiologı´a de la germinacio´ n de las jaras (Cistus ssp.). Lecciones hipertextuales de bota´ nica (revista@quercus.es).

Ruan S, Xue Q and Tylkowska K, 2002. The influence of priming on germination of rice (*Oryza sativa* L.) seeds and seedling emergence and performance in flooded soils. *Seed Sci Technol* 30, 61–67.

Thanos CA and Georghiou K, 1988. Ecophysiology of fire stimulated seed germination in *Cistus incanus* ssp. *creticus* (L.) and *C. salvifolius* (L.). Plant Cell Env., 11: 841–849.

Tran VN and Cavanagh AK, 1980. Taxonomic implication of fracture load and deformation histograms and the effects of treatment on the impermeable seed coats of Acacia species. Aust. J. Bot., 28: 39–51.

Whelan RJ, 1995. The Ecology of Fire. Cambridge University Press, Cambridge.

A REVIEW PAPER ON POTATO LEAF ROLL VIRUS (PLRV) OF POTATO IN PAKISTAN

Aqleem Abbas[1]*, Muhammad Arif[1], Murtaza Ali[2]
[1]*Department of Plant Pathology, The University of Agriculture Peshawar Pakistan*
[2]*Department of Plant Breeding and Genetics, The University of Agriculture Peshawar Pakistan*

ABSTRACT

Potato leaf roll virus (PLRV) is one of the most important viruses infecting members of Solanaceae family. Among the members of Solanaceae family, potato crop is the most significant host of PLRV. The PLRV belongs to genus *Polerovirus* in the family *Luteoviridae*. It causes variety of symptoms depending on the viral strain, time of infection, host tolerance and environmental factors. The symptoms of PLRV are stunting and dwarfing of infected potato crops, reddening or yellowing and rolling of their leaves. It also reduces the size as well as number of potato tubers with annual global yield losses up to 20 million tonnes. In Pakistan PLRV has caused severe yield losses. It contains positive sense single stranded RNA ((+)ssRNA) of 5.9 kb genome. The virus can be transmitted by aphid vectors in circulative, non-propagative manner and experimentally by grafting. Among the aphid vectors, *Myzus persicae* is known to be its most efficient vector. There are various management strategies but most economic and environmentally satisfactory way of managing is breeding of resistant potato cultivars with effective vector control. Other approaches are thermotherapy, tissue culture, pathogen derived resistance and seed potato certification programs. Since PLRV is responsible for significant yield loss in potato crops of Pakistan, so understanding its biology and developing an efficient management strategy is very important.

Keywords: PLRV, occurrence, distribution, symptoms, transmission, control

INTRODUCTION

Potato (*Solanum tuberosum* L.) is the most significant food and vegetable crop of the world and rank fourth in production after cereals such as maize, wheat and rice. Potato is having high nutritive value which contains water (78 %), starch (18 %), protein (2%), vitamins (1%) and several trace elements (Hasse, 2008). The origin of potato is mountainous regions of South America and then it spreads to other countries of the world (Beukema and Eanderzaag, 1990).

Potato is infected by various viruses along with other pathogens (Khan *et al.,* 2008). More than forty viruses are known to infect potato crop (Valkonen, 2007). In Pakistan potato is grown in spring and autumn season in plain areas and as summer crop in hilly areas (Abbas *et al.,* 2012). Plant viruses such as *Potato virus* Y, *Potato leaf roll virus* (PLRV), *Potato virus* A (PVA), *Potato virus* S (PVS), *Potato virus* M (PVM), and *Potato virus* X (PVX) had been reported in these potato growing areas. Among potato viruses PLRV has been considered as major threat to potato production (Peter *et al.,* 2000; Zhang et al., 2010). PLRV is the type species of genus *Polerovirus* in the family *Luteoviridae* (Robert and Lemaire, 1999). The virus particles are isometric, 24nm in diameter (Harrison, 1984). PLRV is single stranded positive sense RNA of about 5.9 kb genome (Miller *et al.,* 1997).

Various aphid species are responsible for natural transmission of PLRV. Among these aphid species green peach aphid (*Myzus persicae*) is the most efficient vector which transmits PLRV through persistent, circulative and non-propagative manner. The virus is a phloem limited virus and is also experimentally transmissible by grafting (Harrison, 1984). The prevalence of *M. persicae* in Pakistan was first reported in 1978 (Mirza, 1978). PLRV has been an emerging problem and widely prevalent in all potato growing areas of Pakistan (Gul *et al.,* 2013). PLRV causes rolling and yellowing of potato leaves that later become dry, stiff, leathery, crisp and papery to touch. PLRV also causes net necrosis in potato tubers and inferior crop quality (Harrison, 1984; Radcliffe and Ragsdale, 2002). In Pakistan the yield reduction as high as 90% have been reported for PLRV (Bhutta and Bhatti, 2002). PLRV can be managed by procuring virus free certified seeds, eradicating volunteers and weed hosts and early roughing of infected plants. The management of aphids is also an important strategy to minimize the spread of PLRV

**Corresponding author:* e-mail: aqlpath@gmail.com

(Alani *et al.*, 2002). Efforts have been made to defend potato crop against viruses by using thermotherapy, tissue culture and micro-propagation. Implementation of strict quarantine measures for seed certification schemes and field spray of insecticides to control vectors were used successfully to manage crop against PLRV (Oosterveld, 1987). This review will focus on occurrence and distribution, management strategies as well as biological, molecular, serological and physical properties of PLRV.

OCCURRENCE AND DISTRIBUTION

PLRV damages potato tubers, fresh markets and processing potatoes of the world. PLRV cause stunting of potato crop and reduction in tuber number as well as tuber size consequently farmers get low yield. PLRV is also responsible for internal net necrosis that makes the tubers unfit for processing and consumption (Radcliffe and Ragsdale, 2002). PLRV was primarily reported from Germany and Denmark in 1905 and now its distribution is worldwide. In Pakistan, PLRV is the most disparaging virus of potato crop (Ahmad and Ahmad, 1995). It was first reported from Punjab followed by Khyber Pakhtunkhwa, Sindh and Baluchistan (Khan and Abbas, 2008). PLRV has severely infected potato growing districts of Khyber Pakhtunkhawa province with the percent incidence as 9.44%, 9.45%, 14.33%, 14.43% and 17.68% in Swat, Dir, Abbotabad and Mansehra respectively. Halim, (1999) reported 60% reduction in yield of potatoes grown from PLRV infected tubers (Bhutta and Bhatti, 2002). Lack of resistance in potato cultivars against PLRV indicate that inoculums level of the PLRV virus is building up and may cause serious effects on yield of potato crop (Khan and Abbas, 2008). PLRV has severely devastated potato crops in plains of Punjab and Sindh with incidence 20-60% (Batool *et al.*, 2011).

Abbas *et al.,* (2012) recorded 52.3 % average incidence of *Potato virus* A (PVA), *Potato virus* X (PVX), *Potato virus* S (PVS), *Potato virus* M (PVM) and *Potato leaf roll virus* (PLRV) from Rawalpindi, Faisalabad, Islamabad and Sahiwal

BIOLOGICAL PROPERTIES

Characteristics symptoms caused by PLRV include chlorosis, reddening, rolling and leathering of leaves, stunting of potato crop and net necrosis of potato tubers (Alani *et al.,* 2002). Symptoms caused by PLRV are categorized into two major types i.e. primary symptoms and secondary symptoms. Primary symptoms are also considered as primary infection which is caused by current season aphid transmitted viruses. The primary symptoms include pallor or reddening of leaf tips of upper or young leaves, after ward these young leaves become roll and erect. Secondary symptoms appear on the potato crops which are grown from the infected tubers. These symptoms consist of stunting of shoots and rolling of oldest or lower leaves upward. Secondary symptoms turned out to be more severe as compared to primary symptoms with leathery texture and overall roll of leaves and with stunted growth and net necrosis. Plants exhibiting primary infection the virus can be transmitted through a variable portion of tubers but in plants with secondary infection all tubers will remain viruliferous (Harrison, 1984). The plants which are infected early in the growing season will be dwarfed and if virus infects the plants late in the growing season even foliar symptoms may not be exhibited. With the age the potato plants are shown resistance to foliar infection consequently no foliar symptoms but there would be virus particles within the host cell (Erik *et al.,* 1993). Infection caused by PLRV can be observed in a circular pattern in the potato fields, generally surrounding the source of virus inoculums, an infected seed piece. Aphids are responsible for direct damage and even kill potato plants technically called aphid's hole in the field (Robert *et al.,* 2000). PLRV translocated through the phloem of the plant into tubers. Afterward the virus reduces size of potato tubers and causes net necrosis. Net necrosis or tuber necrosis is actually manifestations of darkening or browning of the vascular bundles which are extended throughout potato tubers. Potato tubers with symptoms like net necrosis are undesirable for processing into chips and fries (Scagliusi, 2000).

The members of Solanaceae family are the major hosts of PLRV Non-*solanaceous* hosts belong to nine plant families have also been considered as hosts of PLRV. These nine plant families are as;

Chenopodiaceae, Brassicaceae, Malvaceae, Asteraceae, Cucurbitaceae, Lamiaceae, and

Portulacracea (Tamada *et al.,* 1984). *Datura spp* and *Physalis floridana* are considered most excellent diagnostic and propagative host respectively (Harrison, 1984).

PLRV acquired and transmitted through aphid vectors and mechanisms associated with PLRV are circulative, non-propagative and persistent manner (Thomas, 1987). PLRV is restricted in phloem cells therefore it required more time to acquire by aphids (Ragsdale *et al.,* 2001).

PHYSICAL AND MOLECULAR PROPERTIES

PLRV particles are isometric and 25 nanometer in diameter (Harrison, 1984). The genome of PLRV is a single-stranded, positive-sense RNA(5.8kb) with a molecular weight of 2 × 10^6 daltons (, with a small protein covalently linked to the 5' end of no 3' poly (A) tail (Casper, 1988). The positive sense RNA of PLRV has neither a 5'cap nor 3' poly (A) tail but carries a special protein called VPg and an OH group at the 3' end. The protein VPg stands for viral genome linked protein at the 5' end of positive sense RNA of PLRV (Taliansky *et al.,* 2003). Nucleic acid of PLRV is 30% of particle weight and its molecular weight is 2.0 x 10^6 daltons (Harrison, 1984). Genome sequencing has revealed the presence of six open reading frames (ORFs) alienated by a small non coding RNA (Priifer *et al.* 1992). These ORFs are responsible for encoding of various important viral proteins (Prfifer *et al.* 1992). Small ORFs is followed by large ORFs 2a and 2b at the 5'end which may code for 72 and 68 kDa proteins and also contains motifs which are characteristics of enzymes such as helicases and RNA polymerases (Prfifer *et al.* 1992). The genes which are present in highly conserved region in 3' half are translated from a 2.4kb sub genomic RNA (Tacke *et al.* 1990). The subgenomic RNA is not encapsidated into proteins (Smith and Harris 1990). One of the genes in the sub-genomic RNA encodes the coat protein of approximately 23 kDa (Tacke *et al.* 1991).

There is also a single ORF for 17 kDa protein which has a regulatory role during replication of PLRV (Tacke *et al.* 1991). The coat protein of luteoviruses is responsible for transmission specificity (Vanden Heuvel *et al.* 1990) and serological properties (Massalski and Harrison 1987). It is assumed that the read through protein coupled with the coat protein at its C terminal plays a significant role in the transmission of virus (Mayo *et al.* 1993). PLRV consist of single stranded non-polyadnylated RNA molecule with plus orientation that encodes six open reading frames (ORFs). Among these six ORFs the three (ORFs) are located near the 3' ends and encode sub genomic RNA molecules, the 23 kDa coat protein (CP) (Vanderwilk *et al.,* 1989). Sokolova *et al.* (1997) concluded that the given three open reading frames encode a 17 kDa fluolimetric movement protein. Chayca *et al.,* (1996) reported that a 56 kDa protein plays major role in the virus or virus vector interaction. Awan *et al.* (2010) demonstrated molecular detection of PLRV through reverse transcription polymerase chain reactions (RT-PCR) in potato tubers. It was concluded that expected nucleotide sequence of amplified PLRV-CP gene shows homology of 94 to 97 % when compared to the sequences already reported in Gen-Bank database. Jeevalatha *et al.,* (2013) reported that genome of PLRV harbors nine open reading frames (ORFs). These nine ORFs are numbered from 0 to 8 coding for proteins (P0-P7) and Rap1 respectively. Three 5' proximal ORFs encode the proteins (P0, P1 and P2). Five other ORFs are expressed from two subgenomic RNAs (sgRNAs) namely sgRNA1 and sgRNA2. The proteins P3, P5 and P4 are encoded by sgRNA1 where as P6 and P7 which are 3' –proximal proteins are encoded by the sgRNA2. The proteins P3 and P5 are the structural proteins. The remaining ORFs, the ORF1 contains a small ORF whereas ORF 8 encodes Rap 1. The functions of various proteins harbored by ORFs of PLRV are shown in Table 1

Table 1: Functions of various proteins of PLRV

Proteins	Functions
P0	a. Symptom development, suppressor of RNA silencing b. F-Box motifs to overcome PTGS
P1	a. VPg
P2	a. Conserved motifs typical of RNA dependent RNA polymerases (RdRp)
P3, P4 and P5	a. Capsid protein (CP) b. Movement protein (MP) c. Read through domain (RTD)
P5	a. Interaction between virus and vector b. Vector specificity
P6	a. Functions are not known
P7	a. Nucleic acid (NA) binding properties
Rap1	a. Virus replication

SEROLOGICAL PROPERTIES

PLRV virus is serologically associated with *Tobacco necrotic dwarf virus* (TNDV), *Pepper vein yellows virus* (PeVYV), *Barley yellow dwarf virus(*BYDV) and *Beet mild yellowing virus* (BMYV) or *Beet western yellows virus* (BWYV) (Harrison, 1984). Relationships to several members of this group have been detected by gel-diffusion serological tests, immune electron microscopy and by density gradient zone-depletion serological tests The closest relationships were found to be with tobacco necrotic dwarf (SDI = 2), beet western yellows (SDI = 2-5) and bean leaf roll viruses, however, antisera to PLRV react more powerfully with particles of the beet viruses than do antisera to the beet viruses with particles of PLRV (Harrison, 1984).

CROP MANAGEMENT STRATEGIES AGAINST POTATO LEAF ROLL VIRUS

Thermotherapy

Thermotherapy is one of the components of physical method that could be used for the management of PLRV in infected potato tubers. Kassanis, (1950) demonstrated the efficacy of thermotherapy in freeing tubers of PLRV. PLRV was the only virus listed as being inactivated form potato tubers by thermotherapy (Kaiser, 1980). Kaiser, (1980) demonstrated the effectiveness of hot air treatments for eliminating three viruses i.e. alfalfa mosaic virus (AMV), PLRV and tomato black ring viruses (TBRV) from potato tubers. These viruses were eliminated from cultivars of potato by hot air treatment at 37 ^{0}C at various intervals of time. This treatment also eradicated viruses in tubers dually infected with PLRV and TBRV. Viruses could not be detected from these tubers following repeated indexing by mechanical means and by serological methods. Kaiser (1980) report was the first report about thermotherapy on the African continents to free potato tubers of virus infection. Since its introduction in 1949 thermotherapy has been widely used to inactivate PLRV from the potato tubers. However thermotherapy did not inactivate rod-shaped viruses. The reported control of PLRV in tubers of potato by hot water treatment at 50-52 ^{0}C for 17-20 minutes would noticeably reduce the time and space required to liberate tubers of virus infection. Abbas *et al.,* (2016) conducted an experiment to evaluate the effectiveness of hot water treatment to inactivate PLRV from potato tubers. Potato tubers were treated at average 37 °C for various intervals of time. Treatment of potato tubers with hot water at 2 ½ hours were found to be effective in fully or partially elimination of PLRV from potato tubers.

RESISTANT POTATO CULTIVARS AND TRANSGENIC POTATOES

Cultivation of resistant potato cultivars is sustainable strategy for control of PLRV (Beekman, 1987). Unluckily partial resistance against PLRV is present in most of the potato cultivars. Breeders are attempting to produce resistant varieties but hampered by its polygenic inheritance. The remarkable resistance against PLRV so far identified in wild species such as *Solanum brevidens, Solanum demissum, Solanum acaule* or in complex hybrids (Baker *et al.,* 1992). It can be transferred into potato cultivars using the fusion of protoplast. These attempts to incorporate the resistant genes from wild species into potato crop resulted in the release of resistant cultivars but are not immune to PLRV. Since the success of using genes of wild species to increase resistance to PLRV is limited because the resistance is a polygenetic trait and is controlled by many genes. The trait is also associated with shape and size of potato tubers and yield. Recently transgenic plants have been produced which confers resistance to PLRV. The concept that the expression of viral gene sequences in transgenic plants (Pathogen derived resistance) can protect potato crop against virus infection. The gene sequences include sequences which encode viral coat protein (Coat protein mediated resistance) and replicase related protein (Replicase protein mediated resistance). The other pathogen derived resistance is; movement protein mediated resistance, polymerase mediated resistance and ribozyme mediated resistance (Ruth and Barker, 2001). The other phenomenon to manage PLRV is genetically engineered cross protection which is expression of mild strains in potato crops against severe strains of viruses. The results obtained so far with transgenic potatoes having viral sequences are promising but not fully satisfactory (Syller, 1996).

VECTOR CONTROL THROUGH INSECTICIDES

Vector control consists of several practices (Milosevic, 1996) which often include insecticide treatments as fundamental part since aphids transmit viruses in field during a growing season. Robert *et al.* (2000) applied some epidemiological methods to manage aphid born virus diseases i.e. PLRV has been managed in seed potato crop in Northern Europe. The research has shown that mineral oils interfere with virus retention in the vector (Aphids) mouth parts. Similarly *Cucumber mosaic virus* (CMV) transmission was reduced by foliar applications of mineral oils on pepper plants. All these studies demonstrate that mineral oils are effective tools to control non persistent viruses (Loebstein and Raccah, 1980). Ali *et al.* (2011) evaluated some chemicals against aphids, jassids and white flies in potatoes. Six insecticides were applied against these pests of potato. All these insecticides showed 86 % mortality. Sharp 25WP and 10EC caused the highest 96.5% mortality in aphids. Milosevic *et al.*(2012) study was to analyze a potential use of insecticides in preventing the transmission and spread of PLRV during seed potato production, by controlling aphid vectors. Their studies revealed that Confidor showed the highest mortality in aphids as compared to other insecticides. Olubyo *et al.* (2010) recommended that insecticides in combination with mineral oils could play most important role in reduction of aphids whereas synthetic insecticides were regarded more effective in managing PLRV as compared to mineral oils. Furthermore, the effect of insecticides used to control PLRV transmission has also been reported in the literature (Van Tor and Teulon, 2006). Insecticides cause specific effects in preventing the transmission of PLRV from infected to healthy plants (Van Toor and Teulon, 2006).

Mowry, (2005) reported differences in the efficacy of insecticides in controlling PLRV, emphasizing that the most effective are Imidacloprid and Thiamethoxam. The trial involving experimental transmission of PLRV by the *Myzus persicae* aphid showed high efficacy of insecticides such as Pymetrozin, Imidacloprid and Thiametoxam.

Kotzampigiikis *et al.* (2008) reported higher efficacy of insecticides in preventing the transmission of PLRV by aphids. The researcher further suggested that the most effective insecticides are Imidacloprid and Thiamethoxam in controlling aphids species. Further he concluded that systemic insecticides are efficient foliar insecticide to reduce within field spread of PLRV, especially if colonizing aphids are virus-free on arrival. Ragsdale *et al.* (2001) recommended insecticides Carbamate and Aldicarb to control wingless aphids which are responsible for spread of PLRV within field but doesn't prevent transmission by winged aphids. He also studied that Chloronicotinyl class of insecticides i.e. Imidacloprid and Pymetrozine and revealed that insecticides are appearing to be effective in controlling within-field spread of PLRV. Boiteau and Singh, (1999) studied Confidor and found effective in reducing PLRV spread in the potato crop. *M. persicae* is the most important vector of PLRV so its control is usually focused (Ragsdale *et al.,* 1994). Ragsdale *et al.* (2001) suggested timely application of insecticides for managing PLRV in potato tubers is usually affected by aphid biology and their arrival time relative to age of potato crop. Mowry, (1994) assigned ten aphids per hundred leaves as action thresholds for seed potato crop but such action thresholds are region specific. Weekly monitoring of aphids is essential which provide rapid results to apply control measures. However intensive practice of using insecticide is expected to contribute resistance in aphids. Cranshaw and Baxendale, (2005) reported that repellants such as azadirachtin(neem) slowed down the spread of PLRV but these may be more effective when applied as oil formulations.

REFERENCES

Abbas A, Arif M and Ali A, 2016. Use of Hot Water-Thermotherapy to Free Potato Tubers of Potato Leaf Roll Virus. Int. J. Life. Sci. Scientific. Res. 2: 155-162.

Abbas FM, Hammed S, Rauf A, Nosheen Q, Ghani A, Qadir A and Zakia S, 2012. Incidence of six viruses in potato growing areas of Pakistan. Pak. J. Phytopathol. 24: 44-47.

Ahmad N, Khan MA, Ali S, Khan NA, Binyamin R, Sandu AF and Rahman AU, 2011. Epidemiological studies and Management of Potato Germplasm against PVX and PVY. Pak. J. Phytopathol. 23: 159-165.

Ahmed M and Ahmed W, 1995. Detection of major potato viruses from different potato growing localities of Punjab. Natl. Seminar. R and D. potato. Prod. Pakistan PSPDP/PARC, Islamabad. 175-179.

Alani RA, Alessawi UN and Almashaikhy SA, 2002. .Isolation of proteins from *Datura stramonium* has ability to inhibition the multiplication of potato virus Y (PVY^n). Jerash J. Res. Studies. 7: 9-21.

Ali H, Shah RA, Zeb RQ, Badshah H and Rehaman M, 2011. Evaluation of some chemicals against the aphids jassids and white flies in potato Sci. Intl. (Lahore). 23: 67-69

Awan AR, Haq I, Babar ME and Nasir IA, 2010. Molecular detection of potato leaf roll polerovirus through reverse transcription polymerase chain reaction in dormant potato tubers. Pak. J. Bot. 42: 3299-3306.

Barker H and Woodford JAT, 1992. Spread of potato leaf roll virus is decreased from plants of potato clones in which virus accumulation is restricted. Ann. Appl. Biol. 121:345-354.

Batool A, Khan MA, Farooq J, Mughal SM and Iftikhar Y, 2011. Elisa based screening of potato germplasm against potato leaf roll virus. J. Agric. Res. 49:57-63.

Beekman AGB, 1987. Breeding for resistance. In J.A. de Bokx and J.P.H. van derv ants (eds). Viruses of potatoes and seed potato production Wageningen Netherlands: Pudoc. 162-170.

Beukama HP and Eanderzaag DE, 1990. Introduction to Potato Production, Center for Agricultural Publishing and Documentation (PUDOC), Wageningen. 8-13.

Bhutta AR and Bhatti MFJ, 2002. Seed potato certification in Pakistan. Federal Seed Certification and Registration Department Ministry of Food Agriculture and Livestock, Islamabad, Pakistan. 60-66.

Boiteau G and Singh RP, 1999. Field assessment of imidacloprid to reduce the spread of PVY and PLRV in potato. Amer. J. Potato Res. 76: 31-36.

Casper R, 1988. The plant viruses, polyhedral virions Luteoviruses with monopartite RNA genomes New York: Plenum. 235-258.

Chayca UB, Gunasinge, Kumar SP, Dinesh, Mille RWA and Gray SM, 1996. Aphid transmission and systemic plant infection determinants of barley yellow dwarf Luteovirus-PLRV are contained in the coat protein read through and 19Kda, respectively. Virology. 219: 57-65.

Cranshaw WS and Baxendale B, 2005. Insect control through oils. J. Gen. Virol. 33: 470-481.

Erik J, Marianne JH and Ben JCC, 1993. Agronomic performance and field resistance of genetically modified, virus-resistant potato plants. 4: 407-416.

Gul Z, Khan AA, Khan AUR and Khan Z, 2013. Incidence of potato viruses in different districts of Khyber pakhtunkhawa, Pak. J. Plant Pathol. 2: 32-36.

Haase NU, 2008. The nutritional value of potatoes in Canada. J. Potatoes Res. 50: 415-417.

Halim MA, 1999. Serological detection of PLR V and PVY from certified potato seeds and their impact on yield. Thesis of MS in Plant Path, BSMRAU. 80.

Harrison BD, 1984. Descriptions of Plant Viruses CMI/AAB. Potato Leafroll virus (revised).www.dpvweb.net/dpv/showdpv.p hpdpvno 36: 198- 291.

Jeevalatha A, Kaundal P, Shandil RK, Sharma NN, Chakrabati SK and Singh BP, 2013.

Complete Genome Sequence of *Potato leafroll virus* Isolates Infecting Potato in the Different Geographical Areas of India Shows Low Level Genetic Diversity. Indian J. Virol. 24:199-204.

Kaiser WJ, 1980. Use of thermotherapy to free potato tubers of alfalfa mosaic, potato leaf roll and tomato black ring viruses. J. Phytopathol. 70:1119-1122.

Kassanis B, 1950. Heat inactivation of leaf roll virus in potato tubers. Ann. Appl. Biol. 37(3):339-41.

Khan MA and Abbas W, 2008. Multiple regression models based upon epidemiological factors to predict *M.persicae* population and PLRV disease incidence. Int. Nat. Conf. Plant Scientists, ICPS, 155-168.

Kotzampigikis A, Hristova D and Tosheva-Terzieva E, 2008. Distribution of potato leafroll virus (PLRV) and potato virus Y – (PVYN) in a field experiment. Bulg. J. Agric. Sci. 14: 56-67.

Lobenstein G and Raccah B, 1980. Control of non-persistently transmitted aphid borne viruses. Phytoparasitica. 8: 221-235.

Massalski PR and Harrison BD, 1987. Properties of monoclonal antibodies to potato leafroll luteovirus and their use to distinguish virus isolates differing in aphid transmissibility. J Gen. Virol. 68: 18-21.

Mayo MA, Jolly CA, Duncan GH, Lamb JW and Hay RT, 1993. Experimental approaches to determining roles for the proteins in the particles of PLRV. Scottish Crop Research Institute. 97-100.

Miller WA, Brown CM and Wang S, 1997. New punctuation for the genetic code: Luteovirus gene expression. J. Gen. Virol. 8: 3-13.

Milosevic D, 1996. Efficacy of oil and insecticides in potato plant protection against infection by potato virus y and potato leaf roll virus (PVY and PLRV), Zastita Bilja. 47: 333-342.

Milosevic D, Stamenkovic S and Peric P, 2012. Potential use of insecticides and mineral oils for the contol of transmission of major Aphid-Transmitted potato viruses. Pestic. Phytomed. (Belgrade). 27: 97-106.

Mirza MS, 1978. The role of aphids in spreading potato virus diseases in the plains of Pakistan. Potato Research in Pakistan. Shah M.A (ed.), Pak. Agric. Res. Council. 29-32.

Mowry TM, 2005. Insecticidal reduction of Potato leafroll virus transmission by *Myzus persicae*. Ann. appl. Biol. 146: 81-88.

Mowry TM, 1994. Potato leaf roll virus management in the pacific north west (USA). pp. 111-123. In Zehnder GW, Powelson ML, Jansson RK and Raman KV (eds). Advances in Potato Pest Biology and Management. APS Press, Saint Paul.

Mughal SM and Khalid S, 1985. Virus diseases in relation to potato production in Pakistan. In National Seminar on Potato in Pakistan, Islamabad (Pakistan), 2-4 Apr. 1985. PARC.

Olubyo F, Kibaru A, Nderitu J, Njeru R and Kasina M, 2010. Management of Aphids and their vectored diseases on seed potatoes in Kenya using synthetic insecticides, mineral oils and plant extracts. J. Inov. Dev. Strtegy. 4: 1-5.

Oosterveld P, 1987. Inspection and grading. In J.A. de Bokx and J.P.H. van der Want (eds) Viruses of potatoes and seed-potato production, Wageningen, the Netherlands: Pudoc. 204-214.

Peter ET, Cliff EL, James CZ, Gary LR and Wojciech KK, 2000. Extreme resistance to potato leaf roll virus in potato cv. Russet Burbank mediated by the viral replicase gene. Vir. Res. J. 71: 49-62.

Prfifer D, Tacke E, Schmitz J, Kull D, Kaufmann A and Rohde W, 1992. Ribosomal frameshifting in plants: a novel signal directs the frameshift in the synthesis of the putative viral replicase of potato leafroll luteovirus. Embo. J. 11: 11-17.

Radliffe, EB and Ragsdale DW, 2002. Aphid-transmitted potato viruses: The Importance of understanding vector biology. American Journal of Potato Research. 79: 352-286.

Ragsdale DW, Radcliffe EB and Difonzo CD, 2001. Epidemiology and field control of PVY and PLRV. Pp. 237-270. In Loebenstein G, Berger PH, Brunt AA and Lawson RH (eds.), Virus and Virus-like diseases of potatoes and production of seed-potatoes. Kluwer Academic Publishers, Dordrecht.

Ragsdale DW, Radcliffe EB, Difonzo CD and Connelly MS, 1994. Action thresholds for and aphid vector of potato leaf roll virus, - in advances in potato pest biology and management. American phytopathological Society, Minnesota, USA. 99-110

Robert Y, Wood ford JAT and Ducray-Bourdin DG, 2000. Some epidemiological approaches of aphid borne virus diseases in seed potato crops in northern Europe. Virus Res. 71:33-41.

Robert Y and Lemaire O, 1999. Epidemiology and control strategies. pp. 211-279. In The Luteoviridae Smith, H. G. and H. Barker (eds.), CAB Inter., Wallingford U.K.

Ruth MSB and Barker H, 2001. Breeding virus resistant potatoes (*Solanum tuberosum*): a review of traditional and molecular approaches. Heredity 86:17-35.

Scagliusi SM and Lockhart BEL, 2000.Transmission, characterization and serology of a luteovirus associated with yellow leaf syndrome of sugarcane. J. Phytopathol. 90: 120-124.

Sokolova M, Prufer D, Tcacke E and Rhode W, 1997. The potato leaf roll virus 17k movement protein is phosphorylated by a membrane associated protein kinase from potato with biochemical features of protein kinase C. FEBS let. 400: 201-205.

Syller, J, 1996. Potato leaf roll virus; its transmission and control. Integrated pest management reviews. 1: 217-227.

Tacke E, Prfifer D, Schmitz J and Rohde W, 1991. The potato leafroll luteovirus 17k protein is a single-stranded nucleic acid binding protein. J. Gen. Virol. 72: 20-35.

Tacke E, Priifer D, Salamini E and Rohde W, 1990. Characterization of a potato leafroll luteovirus subgenomic RNA: differential expression by internal translation initiation and UAG suppression. J. Gen. Virol. 71: 2265-72.

Taliansky M, Mayo MA and Barker H, 2003. Potato leaf roll virus: a classic pathogen shows some new tricks. Mol. Plant. Pathol. 1:81-90.

Tamara T, Harrison BD and Roberts IM, 1984. Portulacraceae as PLRV hosts Variation among British isolates of potato leaf roll virus. Ann. App. Biol. 104: 107-116

Thomas JE, 1987. Potato leaf rolls Luteovirus, Plant viruses online. Descriptions and lists from the vide database http://micronet.im.ac.cn/vide/descr644.htm

Van Tor RF and Teulon DAJ, 2006. Insecticide practice for aphid control in potatoes. New Zealand Plant Protection. 59: 235-241.

Vanden Heuvel, DeBlank CM, Goldbach RW and Peters D, 1990. A characterization of epitopes on potato leafroll virus coat protein. Arch. Virol. 115: 185-97.

Vanderwilk, Huisman E, Cornelissen MJ, Huttinga BJCH and Goldbach R, 1989. Nucleotide sequence and organization of potato leafroll virus genomic RNA. FEBS Lett. 6: 51-245.

Valkonen JPT, 2007. Viruses: economic losses and biotechnological potential. In Vreugdenhil, D. Potato Biology and Biotechnology Advances: a Perspectives. Elsevier, Amsterdam, 619-641

Zhang W, Bai YQ, Gao YL, Sheng Y, Fan GQ, Gen HW and Meng XX, 2010. A survey on occurrence frequencies of potato viruses in major potato-producing provinces in China. Heilongjiang Agric. Sci. 4: 71-73.

PREVALENCE OF EARLY BLIGHT OF TOMATO AND DIFFERENCES AMONG ISOLATES OF *ALTERNARIA SOLANI* IN PESHAWAR DIVISION

Afaq Ahmad, Ihsan Ul Khaliq* and Maid Zaman
Department of Plant Pathology,The University of Agriculture,Peshawar-Pakistan.

ABSTRACT

A survey was conducted in Peshawar division to determine the incidence of early blight of tomato caused by *Alternari asolani*. High disease incidence (100%) was recorded in Pathwarbala, Sufaid sang, kanderysadin and Shahibala. Regiaftezai and Malakandhir had the lowest (50%) incidence of the disease. *Alternaria solani* was successfully isolated on PDA medium from the infected tomato leaves of three locations. Comparison of the isolates showed Regilalma isolate to be the most vigorous. The conidia of this isolate were found to be wider than those of the other isolates. All the three isolates grew better at pH 6.0 and 5.5 as compared to pH4.5. Regilalma and Nasir bagh isolates registered more abundant growth than the Malakandhir isolate. These result shows that isolates of *Alternaria solani* may differ in their growth and virulence over short distances.

Keywords: *Alternaria solani,* disease incidence, PDA medium, pH, isolates.

INTRODUCTION

Tomato (*Lycopersicum esculentum*) which belongs to the family *Solanaceae* is believed to be first domesticated in Mexico (anonymous 1983). It is the second most important vegetable crop after potato. Present world production is about 126 million tons (FAO statistics database, 2009). Tomato fruit is rich in vitamins A and C and contains an antioxidant, lycopene (Jones 1999). Statistical data for the year 2008-2009 shows that total area under tomato cultivation in Pakistan is 53400 hectares, with a total production of 561900 tonnes and 10.5 tonnes/hectare, whereas tomato crop in KPK occupied 16500 hectares with a total production of 161800 tonnes and 9.8 tonnes/hectares (MINFAL,2008-2009).

Tomato plants are subject to a large number of pests and diseases from the time of emergence to harvest. Among these, early blight disease induced by *Alternariasolani* is one of the most important limiting factors in tomato production (Rich, 1983). It is widespread in the tropics, subtropics and temperate zones and can attack the plants at any stage of development causing a significant risk to crop productivity in the field and quality in the market (Anonymous 1983). Early blight, caused by *A.solani* is a common and serious fungal pathogen of tomato as it results in yield loss and reduced fruit quality (Jones et al. 1991).

**Corresponding author:* e-mail: hsnkhaliq@gmail.com

Early blight inflicts a wide range of symptoms at all stage of plant growth. It can cause damping-off, collar rot, stem cankers, leaf blight and fruit rot. The classic symptoms occur on the leaves where circular lesions up to 1/2" in diameter are produced. Within these lesions, dark concentric circles can be seen. The leaf blight phase usually begins on the lower, older leaves and progresses upwards. Infected leaves eventually wither, die, and fall from the plant (Sherf and Macnab, 1986).

All *Alternaria* species are highly resistant to adverse weather. They develop in a wide range of temperatures and use the locally available source of moisture. They sporulate best on necrotic and dead tissues and produce a relatively small number of spores, mainly at the end of the season, and they are decisively affected by the age condition and susceptibility of host plants (Rotem, 1998). The best way to manage the disease is to use preventive measures by combining many control measures together. Once early blight is established in the crop, it is very difficult to be controlled (Smith and kotcon, 2002). Management of early blight disease in tomato currently relies on fungicides such as Ridomil Gold, Sulphur and copper compounds (Markham 1990). The widespread use of these fungicides have significant drawbacks including increased cost, handling hazards, pesticide residues in agricultural commodities, and threat to human health and environment (Paranagamaetal., 2003).

Alternaria genus is cosmopolitan in occurrence. The members of this species like

*Alternaria alternata, A. solani, A. porri, A. dauci, A. helianthi, A. carthami*and *A. macrospore* causing different diseases in their respective host (Rotem, 1998). Among them,*A.solani causing* early blight of potato (*Solanum tuberosum*) and tomato (*L.esculentum*) is the most destructive (Rotem, 1998) disease of field crops. It causes diseases on foliage (blight), basal stems if seedlings (collar rot and damping off), stems of adult plants (stem lesions), fruits (fruit rot) of tomato and may also infect egg plant and pepper. This disease can be very destructive if left uncontrolled, often resulting in complete defoliation of plants (Dater and Mayee, 1985).

Therefore, there is dire need for advance planning and research to increase food production and improve quality in order to meet the needs of ever increasing population.

Considering the importance of tomato, the costs of diseases in terms of yield reduction, expenditure on their control, the present study is designed to investigate the prevalence of early blight in tomato in Peshawar division, variation among different isolates and their responses to different pH levels.

MATERIALS AND METHODS

Discased survey and collection of specimen:

The survey was conducted at different locations of Peshawar division. Early blight infected leaves were collected in polythene bags and brought to the laboratory of department of plant pathology, The University of Agriculture, Peshawar-Pakistan, where they were stored at 4^0c until isolation of *Alternaria solani*was made.

Disease Incident (%):

Diseases incidence in tomato fields was recorded by randomly throwing a 1m diameter ring. The total and infected plants inside the ring were than counted and % disease incidence was calculated by the formula:

Disease Incident (%):

Diseases incidence in tomato fields was recorded by randomly throwing a 1m diameter ring. The total and infected plants inside the ring were than counted and % disease incidence was calculated by the formula:

$$\frac{\text{No. of infected plants}}{\text{Total no. of plants}} \times 100$$

Isolation of *Alternariasolani*:

Alternaria solani was isolated from diseased leaves on PDA medium. PDA medium was prepared using standard procedure (for one liter medium, 250g potato, 20g dextrose and 20g agar). The medium was sterilized at 121^0c for 20 minutes. After sterilization streptomycin was added for the inhibition of bacteria growth while pouring media into Petri dishes.

The infected leaves were cut into small pieces, surface sterilized with mercuric chloride (0.1%) for 15-30 seconds and then rinsed with sterilized distilled water thrice. The small pieces were placed on PDA medium (4-5 pieces/petri dish). The plates were then incubated at $25C^0$ for growth of *Alternaria solani*. Pure cultures of *A. solani* were prepared on fresh PDA medium by sub culturing.

Comparison of*Alternariasolani* isolates:

Alternaria solani were inoculated at the centre if fresh PDA medium in petridishes under aseptic conditions; for each isolates four petridishes were used in completely randomized (CR) design. The plates were incubated at 25^0c for the growth of the isolates. Colony diameters of the isolates were measured periodically along two perpendicular lines. Slides were prepared for determining the size of conidia through micrometry. Average conidium size was calculated from 20 individual measurements.

Response of *Alternariasolani* isolates to changes in pH of PDA medium:

Alternaria solani isolates were culture on PDA medium adjusted to pH levels of 4.5,5.5 and 6.0 pH was lowered or raised by adding hydrochloric acid (HCL) or Sodium Hydroxide (NaOH) to PDA medium, respectively, prior to sterilization. The factorial experiment (isolates and pH levels) was replicated four times. After incubation at 25^0Cdata on colony diameters were recorded periodically as described above. At the end of the experiment, the petridishes were flooded with sterile distilled water (10 ml each) conidia were detached from the cultures by rubbing gently with a glass rod. The spore concentration was determined with the help of Haemocytometer. The data were analyzed statistically by ANOVA and LSD test at 5% probability level.

RESULTS

Incidence (%) of early blight diseases of tomato

Survey conducted on diseases incidence at different locations of Peshawar division (Table 1). The highest diseases incidence (100%) was recorded in areas of Pathwarbala, Sufaid sang, kandarysadin and Shahibala. Areas of Malkandhir and RegiAftezai had the lowest diseases incidences (50%).

Table-1: Disease incident of early blight disease in Peshawar Division

S. No.	Location	Incidence (%)
1	PathwarBala	100
2	RegiAftezai	50
3	Malkandhir	50
4	Sufaid Sang	100
5	Kandarysadin	100
6	Regilalma	80
7	Nasir bagh	70
8	Amangharh	60
9	Jamma khan kalay	60
10	Shahibala	100
11	Palosi	60
12	Safdarabad	70
13	Azakhailbala	75
14	AzakhailPayan	80
15	Peer Bhai	60

Comparison among *Alternaria solani* isolates in terms of colony diameter and conidia size

Data presented in table-2 shows comparison of three isolates in terms of colony diameter and conidia size. After 10 days of incubation, the highest colony diameter (55.53) was recorded for isolate Regilalma. Isolate Malakandhere and Nasir Bagh were statistically at par with each other. However, the three isolates did not differ significantly in conidium length and width (Table-2).

Table-2: Comparison among *Alternariasolani*isolates in terms of colony diameter and conidia size

Location	Colony diameter (mm) after 3 days	Colony diameter (mm) after 10 days	Spore width (µm)	Spore length (µm)
Malakandhir	17.00 B	43.66 B	0.453	1.17
RegiLalma	20.33 A	55.33 A	0.463	1.14
Nasir Bagh	18.44 AB	45.11 B	0.46	1.13

Response of *Alternaria solani* isolates to changes in pH:

After 3 days of incubation, colony diameters recorded for the three isolates (Table-3) revealed that isolates of Regilalma and Nasir Bagh had grown significantly more than Malakandhir isolate. On the other hand, maximum growth (19.83mm) was recorded at pH 6.0, which reduced significantly as the pH level was decreased to 5.5 and 4.5. However, isolates did not interact significantly with the pH levels.

After 10 days of incubation, Regilalma isolate maintained its superiority in growth recorded (Table-3), followed by Nasir Bagh isolate. Malkandhir isolate registered the lowest growth. Growth on pH 6.0 and 5.5 was observed to be similar but higher than at pH 4.5. The isolates interacted significantly with pH levels. While Regi Lalma isolate had more abundant growth at pH 6.0 And 5.5 than 4.5, Nasir Bagh isolate grew as much as at ph 4.5 as at either 6.0 or 5.5. On the other hand,Malakandhir isolate had similar growth at all pH levels.

After 10 days of incubation, colony diameters of RegiLalma and Nasir Bagh Isolates of *Alternaria solani* were statistically at par (Table.3) and significantly higher than that of Malakandhir isolate, similarly, growth at pH 6.0 and 5.5 was similar but higher than that of pH 4.5. The interaction between isolates and pH levels was also significant. Isolates of Regi Lalma and Malakandhir had statistically similar colony diameters at pH 6.0 and 5.5 but Nasir Bagh isolate had significantly more growth at pH 6.0 than at pH 5.5.

Table-3: Colony diameters (mm) after three, seven and ten days of incubation

Location	Colony diameter	pH=4.5	pH=5.5	pH=6.0	Mean
Malakandhir	Colony diameter after 3 days of incubation	15.25	17	19	17.08 B
RegiLalma		17.37	19.62	20.5	19.17 A
Nasir Bagh		17.62	18.25	20	18.63 A
Mean		16.75	18.29 B	19.83 A	
Malakandhir	Colony diameter after 7 days of incubation	27.62 C	45.00 C	45.00 C	38.83 C
RegiLalma		44.00 C	56.62 A	56.00 A	52.21 A
Nasir Bagh		46.12 BC	43.87 C	50.00 B	46.83 B
Mean		39.25 B	48.13 A	50.50 A	
Malakandhir	Colony diameter after 10 days of incubation	33.75 D	66.12 BC	65.25 ABC	55.04 B
RegiLalma		56.00 C	70.12 AB	67.62 AB	64.58 A
Nasir Bagh		61.37 BC	56.62 C	75.12 A	64.37 A
Mean		50.38 B	64.29 A	69.33 A	

Spores concentration of *Alternaria solani* isolates:

Regarding spore concentration, neither the isolates of *A.solani*nor the pH levels differed significantly from one another. The interaction was also non-significant.

Spore concentration of *A. solani* isolates

Table-4 Spore Concentration of *A. solani*isolates

Location	pH=4.5	pH=5.5	pH=6.0	Mean
Malakandhir	12.99	10.66	10.32	11.32
RegiLalma	10.33	10.66	8.32	9.77
Nasir Bagh	10.33	9.82	5.99	8.71
Means	11.21	10.38	8.21	

DISCUSSION

This study found variation in disease incidence of early blight of tomato among different locations of Peshawar division. The variation in disease incidence may indicate the existence of isolates differing in virulence. *In vitro* differences in growth among selected isolates in this study confirmed this assumption. Isolates growing more vigorously on PDA medium caused higher disease incidence in the field. Another reason for variation in disease incidence may be the prevalence of different environmental conditions, such as temperature and humidity, at different locations which influence the final disease outcome. The use of different cultural practices by farmers may also result in different levels of the disease such practices may include the use of resistant varieties, proper time of sowing, sanitations and the use of appropriate fungicides in the proper dose, frequency and time. The choice and availability of different cultural practices may in turn be an indication of the socioeconomic and educational levels of the farmer.

The isolates of *Alternariasolani* grew best at slightly acidic medium. Growth decrease as pH of the medium was reduced. Again, isolates showed differences in their response to pH levels. This finding is similar with the results of Chaerani*et al.* (2007)

The results of the study show the need for further research to elaborate mechanisms of variation among isolates of *Alternaria solani* so that effective strategies could be devised for the management of this economically important disease of tomato in Peshawar and other areas of the province.

CONCLUSION

The prevalence of early blight disease of tomato varied from location to location in Peshawar division indicating pathogenic variation among isolates of *A. solani.*Isolates of *A. solani* showed variation in *In vitro* growth reflecting their filed variation. It is concluded that isolates of *A. solani* grow best at slightly acidic PDA medium and they respond differently to changes in pH levels.

REFERENCES

Anonymous, 1983. Pest control in tropical tomatoes. Center for Overseas Pest Research. Overseas Development Administration. London.

Chaerani R, R Groenwold, P Stam and RE Voorrips, 2007. Assessment of early blight (*Alternariasolani)* resistance in tomato using a droplet inoculation method. J. Gen PlantPathol. 72 (2): 96-103

Dater VV and Mayee CD, 1985. Chemical management of early blight of tomato. J. MahaGri. Uni. 10 (3): 278-280.

FAO STAT, 2009. Food and Agriculture Organization Corporate Statistical Database.

Jones, J Benton, 1999. Tomato plant culture in the field, greenhouse and home garden. CRC press LLC.

Jones JB, Jones JPS Stall and Zitter TA, 1991. Compendium of Tomato.

Markham LJ, 1999. Biological activity of tea tree oil in tea tree the genus *Melaleuca* (medicinal and aromatic plant: industrial profiles). Harwood Academic Publisher 2006.

MINFAL, 2009. Agriculture Statistic of Pakistan, Ministry Of Food, Agriculture and Livestock (Economic Wing) Islamabad. Pp.71-72.

Paranagama PA, Abeysekera KHT, Abeywickrama K and Nugaliyadde L, 2003. Fungicidal and anti-aflatoxigenic effects of the essential oil of *Cymbopogon citrus*. (Lemongrass) against *Aspergillusflavus*link. Isolated from stored rices. Letters in AppMicrobiol 37:86-90.

Rotem J, 1998. The genus *Alternaria* biology, epidemiology, and pathogenicity. American Phytopath. Soc. Press, St. Paul, Minnesota.

Sherf AF and AA Macnab, 1986. Vegetable diseases and their control. 2nd Ed. John Wiley & Sons, New York, NY. Pp.728

Singh BM, 1965. Inducing sporulation in different strains of *Alternariasolani*, Ph.D thesis department of botany and plant pathology, Punjab AgriculturalUniversity, Hissar, Punjab, India.

Smith LJ and KotconJ, 2002. Intercropping with tomato resistant variety ‘Juliet’ reduces early blight on susceptible variety ‘brandywine’. Phytopathology. 92(supplement): S77. Tomato diseases APS,St. Paul.

EVALUATION OF EXTRACTION METHODS FOR HOUSEHOLD HERBAL PRODUCTS

Sumera Javad[1], Sana Khalid[1], Shagufta Naz[2], Kanwal Naz[1]
[1]Department of Botany, [2]Department of Biotechnology, LCWU, Lahore

ABSTRACT

Microwave assisted extraction is a modern and novel approach for rapid and efficient extraction of the plant material which is of prime importance in herbal medicinal industry. For household herbal products (different tea samples, qehwa (Peshawari Qehwa), coffee (Nescafe Classic) and joshanda (Qarshi), their qualitative and quantitative analysis of phytochemicals were compared. Microwave assisted extraction gave the maximum amount of extracts than that of stove heating with significant difference of amount of phenolics and flavonoids. Maximum gallic acid (standard phenolics) was observed from microwave assisted extract of Tapal Tez dam. Stove extracts produced maximum phenolics from Lipton tea. Maximum catechin (Standard of flavonoids) was observed from microwave assisted extract of Johar Joshanda while stove extracts produced maximum catechin from Coffee sample.

Keywords: Microwave assisted extraction, stove heating, herbal products, phenolics

INTRODUCTION

Coffee and Tea are considered as the trendiest beverages for centuries, basically due to their pleasing taste and stimulant effects (Alan and Iris, 2004). Throughout the world, a cup of tea or coffee in the morning is considered as the sign of wakefulness; these are also in use to reduce the effects of wholeday work and resultant mental and physical exertion. These herbal products have a number of biologically active secondary metabolites such as phenolics, alkaloids, tannins and flavonoids. These all above mentioned plant secondary metabolites are beneficial to human health as antioxidants. There are scientific proofs in favor of tea and coffee which show that their regular use protect us from development of many chronic diseases and their acute effects (Al- Rasbi and Khan, 2013). Joshanda, herbal product which is in use as tea, is also considered as a treatment of some diseases. It was tested for the presence of acetyl salicylic acid and antioxidant character and positive results were obtained. It is obvious that the presence of these biochemicals in any extract make it much beneficial for human health (Soomro et al. 2011).

Scientific methods turned out to be more superior and preferred techniques in early 19th century, after this revolution people started to call the practice as quackery. But with increasing knowledge of side effects of conventional medicines as well as the increased resistance in pathogens, once again the trend of people began to transfer from conventional medicines to herbal medicines (Winslow and Kroll, 1998). Recent advances in functional foods also support the use of such products like tea, coffee and joshanda.

Extraction is a very basic step in formulation of herbal products and it involves the partition of medicinally activecomponentsof plant from the inactive, inert or undesired portion by using solvents of different polarity. The extract attained in this way may be prepared for use as a medicinal agent directly in the form of fluid extracts and tinctures and the most common example is tea, coffee and traditional Qehwa etc. These are the primary plant extracts which are in daily use, in further steps their extraction method may be checked for its efficiency to give more and more plant metabolites with its maximum capacity per unit of plant material (Handa et al. 2008).

After World War II, microwaves have been in use and now commercially used as domestic ovens. While use of microwave energy in science laboratories started in the late 1970s and for first time these were used in acid digestion. Microwaves are part of electromagnetic spectrum and have a frequency range of 0.3 to 300 GHz (Camel, 2000; Javad et al., 2014). Microwave assisted extraction (MAE) is going to be an efficient replacement of the conventional extraction methods like cold maceration, simple heating and boiling, Soxhlet, digestion, percolation and sonication. The majorbenefit of MAE is hidden in its

*Corresponding author's email: zif_4@yahoo.com

heating source. The high temperature attained by microwave heating in turn reduces the extraction time and ultimately the volume of solvent essential (Camel, 2000).

In present study, extraction of tea, coffee and Joshanda was done by using simple boiling method on hot flame and by using microwave assisted extraction. Then a comparison for the amount of extract, phenolics and flavonoids was made for all of above mentioned products.

MATERIALS AND METHODS

Ten different samples were taken including tea samples (Green tea, Vital tea, Supreme tea, Danedar tea, Tez Dam tea, Tetley and Lipton tea), Qehwa (Peshawari Qehwa), coffee (Nescafe Classic) and Joshanda (Qarshi Joshanda). For each extraction, 10g of the samples were weighed accurately and placed in the beakers. Measured quantity of distilled H_2O (50ml) was added in beaker as an extracting solvent along with the plant material and then heated for extraction. Two different methods i.e., microwave heating and stove heating were then opted to extract the bioactive compounds from plant material. For microwave assisted extraction, each sample was heated in the microwave at full power level, described 1000W (Model HGN-45100EB) for 2 minutes for each sample each time. For obtaining the extract by stove heating extraction, each sample was heated at normal heat of stove for 2 minutes for each sample each time. For both extraction methods, each plant sample was strained after heating by using common household strainer and poured into pre-weighed glass vials and was subjected to shade drying. After drying, extracts were weighed and tabulated. Phytochemical analysis (Qualitative and Quantitative) of the extracts were done for phenolics and Flavonoids.

Phenolics: For1 ml sample (1mg/ml), there was added 1 ml of Folin and Ciocalteu's phenol reagent. After 3 minutes saturated Na_2CO_3 (1ml) was added. After that, by adding distilled water, the total volume of reaction mixture was made upto 10 ml. The reaction mixture was then kept in dark for about 90 minutes. Absorbance was checked at 725nm. Gallic acid was used as standard phenolics to construct the standard curve (100-200μg/mL) (Arabshahi and Urooj, 2007).

Flavonoids: 250μL extract (1mg/ml) was added with 1.25ml distilled water and 75μL of 5% $NaNO_2$. After 5 minutes, 150μL of 10% $AlCl_3$ were added. After 6 minutes of this addition, 500μL of 1M NaOH and 275μL of distilled water were added. Then the solutions were mixed well. Absorbance was measured at 510nm. Catechin was used to construct the standard curve (20-120μg/ml) (Wierong et al., 2004). Statistical analysis was done by comparing the means by applying the ANOVA and each value is representative of five replicates (Mean ± SD). Meansnot sharing a common superscript significantly differ at 5% level of significance by applying Duncan's new multiple range test. Statistical software (COSTAT) was used for this purpose.

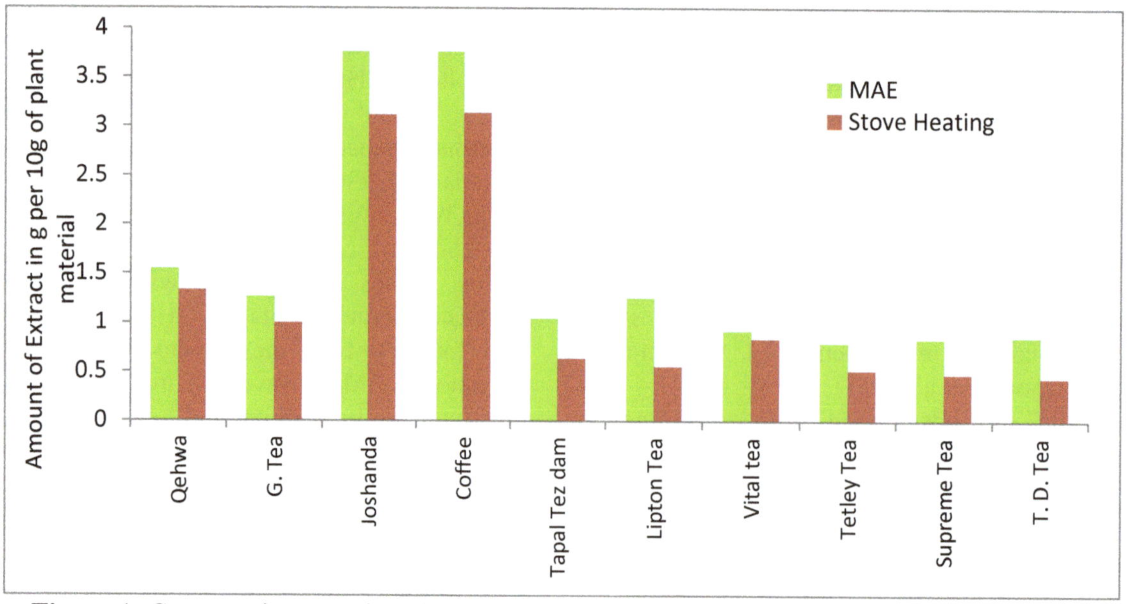

Figure 1: Comparative quantity of extract from ten different herbal products by two different methods of Extraction

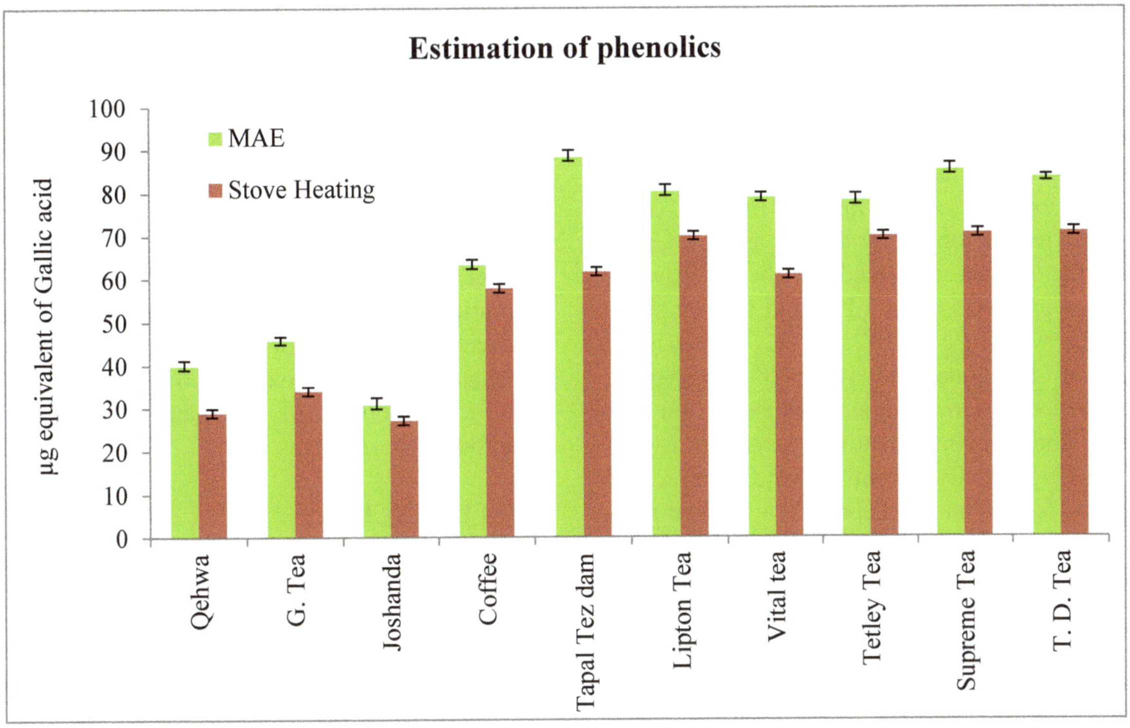

Figure 2: Comparative quantity of phenolics in extract from ten different herbal products by two different methods of extraction

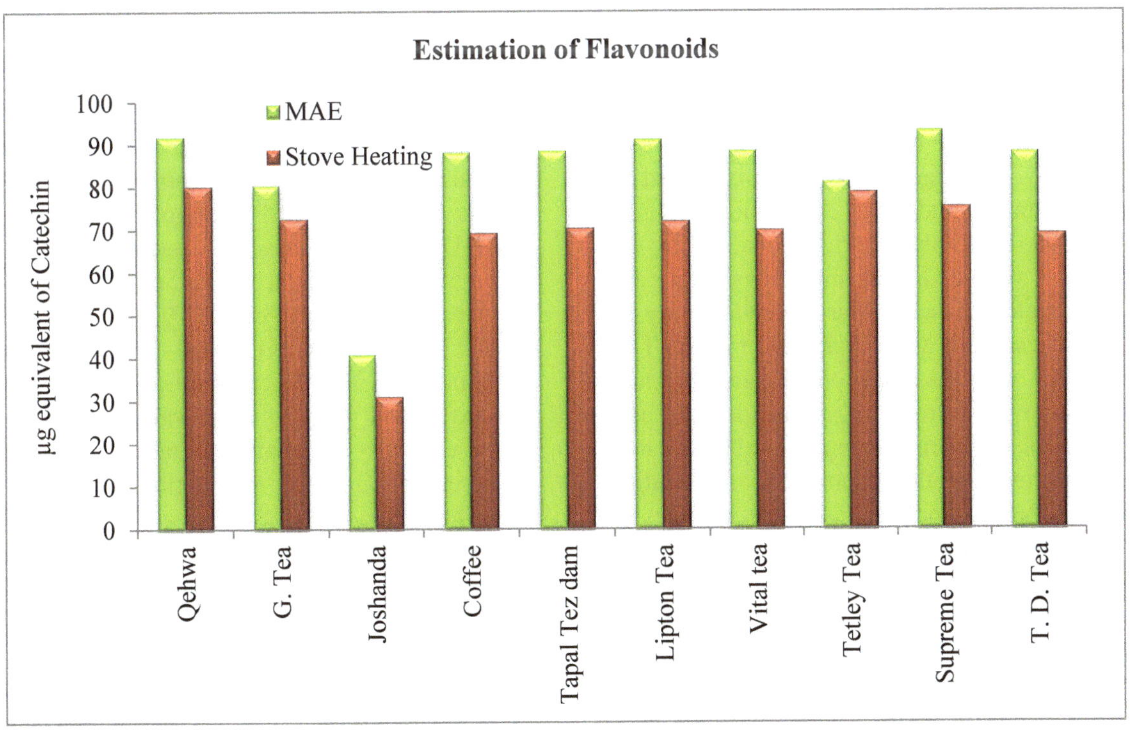

Figure 3: Comparative quantity of flavonoid extracts from ten different herbal products by two different methods of extraction

RESULTS AND DISCUSSION

In using the samples of Green Tea, Joshanda, Coffee, Tapal Tez Dam Tea, Tetley Tea, Supreme Tea and Tapal Danedar Tea, microwave assisted extraction method gave the higher extracts as compared to the stove heating extraction method. In case of Green Tea, the extract produced by the microwave assisted extraction was 1.27g and that by stove heating was 1.0g/5g of sample. Similarly in case of Joshanda, the value of the extract obtained by microwave assisted extraction was 3.76g and by stove heating extraction it was 3.12g (Fig. 1). On the other hand, samples of Qehwa, Vital Tea and Lipton Tea showed non-significant differences between the extracts obtained by microwave assisted extraction and stove heating extraction method. Similarly, significantly higher quantities of phenolics and flavonoids were found in microwave assisted extracts as compared to that in stove heated extract.

Microwave assisted extraction proved to be a rapid, efficient and modern method of extraction. When it is compared with conventional extraction methods, it showed targeted extraction, lesser by products, high extraction selectivity in a very short time and less labor. Microwaves are in fact electromagnetic radiations (Camel, 2000). The electric field of these waves causes heating with help of two mechanisms which are named as dipolar rotation and ionic conduction (Sinquin et al., 1993). Dipolar rotation is caused by the alignment of the dipolar molecules along the electric field. These dipolar molecules may be from both solvent as well as plant matrix. This alignment and resultant oscillations produce collisions with surrounding molecules and as a result a lot of heating energy is released. It is also reported that this phenomenon of heating occurs 4.91times per second if microwaves of 2.45 GHz are used (Onuska and Terry, 1995) and the resulting heating is very much fast. So it is clear that if a solvent with larger dielectric constant like water is used, there will be more resultant heating. Indeed, water is the common solvent used for household herbal products, showing more efficiency in microwaves as compared to conventional stove heating (Jassie et al., 1997). In such polar solvents, microwaves heat the whole sample at a time as compared to conductive heating method of stove heating. Another advantage of microwave heating is the breakage of weak hydrogen bonds in solvent as well as plant matrix, adding to its extraction efficiency (Camel and Bermond, 1999). Thirdly, Ionic currents are produced by electric field of microwaves in the solution, and the medium usually resists these currents. Due to these resistant effects once again heat is produced by a Joule effect. In fact, water content is very vital in MAE due to its polarity, heating effect and extraction efficiency which favor making teas, coffees and joshanda in microwaves. In addition, control of the water content of the matrix allows more reproducible results.

REFERENCES

Alan M and Iris Microwaves, 2004 .The Empire of tea.The overlook press. 32.

Al-Rasbi and Khan SA, 2013. Isolation and quantitative estimation of caffeine content in different brands of coffee and tea leaves. Sch. Aca. J. Biosci. 1(3): 67-68.

Arabshahi S and Urooj A, 2007. Antioxidant properties of various solvent extracts of mulberry (*Morus indica* L.) leaves. Food Chem. 102: 1233-1240.

Camel V, 2000. Microwave assisted solvent extraction of environmental samples. J Trends Anal. Chem. 19: 229-248.

Camel V and Bermond A, 1999. Microwave solvent extraction of environmental samples. J. Anal. Exp. Chem. 92: 117-135.

Handa SS, Khanuja SP, Longo G and Rakesh DD, 2008. Extraction technologies for medicinal and aromatic plants. J. Intl. Centre Sci. High Tech. 1-10.

Jassie L, Revesz R, Kierstead T, Hasty E and Matz S, 1997. Microwave assisted solvent extraction in microwave enhanced chemistry: fundamentals, sample preparation and applications. J. Amer. Chem. Soc. 569-609.

Javad S, Shagufta N, Saiqa I, Amna T and Farah A, 2014. Optimization of the Microwave assisted extraction and its comparison with different conventional extraction methods for isolation of stevioside from *Stevia rebaudiana*. Asian J. Chem. 26(23): 8043-8048.

Onuska FI and Terry KA, 1993. Extraction of pesticides from sediments using a microwave technique. J. Chrom. 36:191-194.

Sinquin A, Gorner T and Dellacheie E, 1993. Lutilisation des micro-ondesenchinie. Analytique Analusis. 21: 1-10.

Soomro TA, Mohammad M and Aqeel Z, 2011. Acetyl salicylic acid (aspirin) and antioxidative agents in Joshanda- a herbal (medicine) tea. J. Electrochem. Soc. 33(26): 1-8.

Weirong C, Xiaohong GU and Tang J, 2010. Extraction, purification, and characterization of the flavonoids from *Opuntia milpa alta* skin. Czech J. of Food Sci. 28(2): 108-116.

Winslow LC and Kroll DJ, 1998. Herbs as medicine. Arch. Internal Med. 158:2192-2199.

CONCENTRATION OF MACRO NUTRIENTS IN DRY FOLIAGE OF RUSSIAN OLIVE (*ELAEAGNUS ANGUSTIFOLIA*)

Aliya Noreen[1], Ghulam Raza[1], Haibat Ali[1], Khadim Hussain[2], Saif uddin[1], Qandeel Zehra[2], Babar Hussain[2*], Syed Waqar Hussain[2], Yawar Abbas[2], Rashid Alam[3], Nadia Ali[1]

[1]*Department of Environmental Sciences, Karakoram International University Gilgit-Pakistan.*
[2]*Gilgit-Baltistan Environmental Protection Agency, Gilgit-Pakistan (GB-EPA).*
[4]*Department of Food Science and Nutrition, Anhui Agriculture University China.*

ABSTRACT

To be acquainted with the essential nutrients for plants for the sake of their growth is an important aspect of successful production of crops. These macro nutrients are Nitrogen (N), Phosphorus (P), Potassium (K), Calcium (C), and Magnesium (Mg). *Elaeagnus angustifolia* is a small tree and it is also called a large shrub. It has silvery leaves and thorny body. This species is tolerant to saline as well as alkaline soils and is hosted from Russia. This essential plant nutrient research will definitely help to solve such nutrient deficiency problems. This aim of present study was to explore the importance of essential nutrients for livestock and to assist farmers in the selection of fodder and forage. The results showed that concentrations of N, P, K, Ca and Mg in the dry foliage of Russian olive were 3.75, 0.558, 4.37, 2.459 and 0.769 % respectively. It may be concluded from the results that the range of N, P, K, Ca and Mg in the leaves of *E. angustifolia* was very close to sufficient range of plant N, P, K, Ca and Mg content.

Keywords: Essential nutrients, *Elaeagnus angustifolia*, Fodder and forage.

INTRODUCTION

Currently the animal protein deficiency is an alarming issue in most of the developing countries. This deficiency is typically due to dearth of forage. In such developing countries mostly the available grazing is not sufficient for the livestock due to their harsh environmental conditions; in such condition trees and shrubs play a vital role by providing fodder to the livestock. These trees and shrubs are mostly known as a provider of proteins to animals (Paterson et al., 1998; Azim et al., 2011). Russian olive is an alien tree that is progressively common in riparian environments and is highly resistant to extreme environmental condition like floods and droughts. It has also the ability to resist different soil textures and moisture levels. This is actually the reason behind its successful occurrence in all types of environments (Katz and Shafroth, 2003). *E. angustifolia* adds substantial nitrogen to ecosystems as it has a good actinorrhizal nature, deciduous routine and high leaf nitrogen content (Castro et al., 1993; Katz and Shafroth, 2003). Leaves of *E. angustifolia* grown in the laboratory enclosed 3.3% N, which was related with a high photosynthetic rate (Cote et al., 1988; Katz and Shafroth, 2003). Due to this possession of high nitrogen content *E. angustifolia* is used in silvicultural activities and it is also used as a biofertilizer (Dawson and Sambeek 1993, Domenach et al., 1994; Katz and Shafroth, 2003). According to the author, 60% of cultivated soils suffer from different types of minerals and nutrient deficiencies and many types of toxicities; which can cause growth-limiting problems. It is also estimated that around 50% of the world population undergoes micronutrient deficiencies (Cakmak, 2002).

According to USDA (2015), the physical and environmental condition like high winds, extreme temperatures (- 50 °F to 115 °F), saline or alkaline soil settings, flooding, drought and race from other trees and shrubs leave little or no impact on the growth of the species *E. angustifolia*. This species was introduced in United States as an ornamental tree. Soon after its invasion the species become dominant due to its properties like aggressive reproduction, adaptability, and rapid growth rate. The components of growth of *E. angustifolia* include stem and foliage density, canopy cover etc. *Elaeagnus angustifolia* can grow in groups as well as it can also grow individually. This species is part of mature riparian vegetation (Kelly et al., 1999;

**Corresponding author:* e-mail: babar.ses@gmail.com

Muldavin et al., 2000; Zouhar and Kris, 2005). The height of the Canopy of Russian olive is normally from 16 to 33 feet which is 5-10 m but the density of the canopy is quite high. Live foliage of the tree is often present in the upper portion of the tree; in the lower 2 meters of the Russian olive tree dense branches are present which are often dead (Zouhar and Kris, 2005). Leaves of the species are generally 2 to 3 inches long and are arranged in an alternative way. Small and plane margins are present on the leaves. Color of upper leave surface of the species is green-gray and the lower superficial as well as leaf stalks are silvery gray in color. Foliage of the species has generally silvery look with tiny scales that cover the leaves (Creech et al., 2008). As reported by Coconino, Kaibab and Prescott national forest Russian olive is consider as noxious weed. This species is included in B class weed. Smaller groups of the species could be easily controlled but larger groups are quite problematic to control effectively. They are commonly found at the riparian zones and residential areas due to wide plantation of these species (Shafroth et al., 1995; Knopf and Olson, 1984) According to the author's personal observation, domesticated animals used to feed on young russian-olive trees but adult russian-olives prevent most livestock with the help of their piercing thorns and protective compounds in the leaves. Mostly the barks of the tree are eaten by rabbits. Poultry for example chickens, ducks, turkeys, and pigeons) may also eat the leaves from freshly implanted trees (Borell, 1962; Zouhar and Kris, 2005).

The purpose of the experiment was to find out the macro nutrients in the leaves of *E. angustifolia* and to recommend the specie as a source of foliage and fodder in Gilgit Baltistan and to compare the nutrients with the palatable plants macro nutrient standards.

MATERIALS AND METHODS

The macro-nutrients tested in the leaves of the species *E. angustifolia* were N, P, K, Ca and Mg.

Leaf Nutrient Analysis

The leaves of plants are the major and crucial components for analysis of various nutrients because leaves are indicators of essential nutrients accumulation and show changes in nutrients concentration. The amount of macronutrients in the dry part of leaves and its comparison with standard values helps to understand nutrients status. The data obtained from nutrients analysis of the species was used to find out the difference between the results and optimal nutrient concentration (Motsara and Roy, 2008).

Table. 1: General sufficiency or optimal range of nutrients in plants

Nutrients (Macronutrients)	Sufficiency or optimal range (%)
N	2.0–5.0
P	0.2–0.5
K	1.0–5.0
Ca	0.1–1.0
Mg	0.1–0.4
S	0.1–1.3

Source: FAO (Motsara and Roy, 2008)

Methods

The methods involve in this research is totally based on laboratory work. Before the laboratory processes sampling of the species was done.

Sampling

From the surrounding of KIU campus the trees were selected through simple random sampling technique. Sampling was done during the 1st week of March where the already dried leaves were collected from the trees. After collection the leaves were placed in a grinder and crushed into fine powder.

Sampling size

4 trees were used as replicates in the sampling technique because in laboratory work minimum 3 to 4 replicates were used for accurate results (ITRC, 2012).

Laboratory processing

Laboratory work was done at Pakistan Agriculture Research Council Jaglote (Gilgit). The work properly started with the formation of solutions of the powder of the species and leaving it in the oven overnight. The process properly started next day to find out the readings of N, P, K, Ca, and Mg with the help of following procedures.

Kjeldahl Nitrogen and Phosphorous

Reagents and Apparatus

1. Concentrated Sulfuric Acid (H_2SO_4) reagent grade.
2. Digestion tablets (Kjeltab).
3. Glass boiling beads.
4. Aluminum digestion block with temperature controller.
5. Kimax glass tubes with 50-ml graduation mark.
6. TKN Standards.
7. Carries Solution

Procedure

The weight of 0.150 g ground plant material was put into 50 ml digestion tube. After that one digestion tablet (Kjeltab 1.72 g) and two glass boiling beads were added in each tube. Then in the underneath hood 3.5 ml concentrated sulfuric acid (H_2SO_4) was pour through 5 ml repipet device. For each run there was at least one blank and one check was included. Preheat aluminum digester until it reach at the temperature of 180 °C. The tubes were then put into the block digester which was preheated and continue to heat till 390 +/- 5 °C then keep it for digestion up to two hours. After that the tubes were taken out from heating then cooled it for around 20 minutes. Add 10-15 ml distilled water in the digestion tubes despite of they were warm. The samples were dilute with distilled water to the mark of 50 ml then mix completely after capping. After that the solution was filter by # 2 filter paper. Finally the analysis of ammonium for TKN on the flow injection analyzer based on the Lachat method No. 13-107-06-2-D was done.

Calculations

For plant TKN %:
N % = reading (mg/l) x 50 / 0.150 / 10000

Total K, Na, Ca and Mg

Equipment
1. Adjustable pipette
2. 20-ml glass beaker
3. Atomic Absorption Spectrophotometer

Reagents
1. 1.2 N HCl solution Add 103.7 ml of hydrochloric acid (sp. Gr. 1.19, 37.5%) into 1 liter bottle. Dilute with distilled water and mix well.
2. 0.105 % Lanthanum diluent Place 1.2314 g lanthanum oxide (La_2O_3), low calcium grade, in a one liter volumetric flask. Add 4 ml of 6 N HCl to dissolve the La_2O_3 and then dilute to one liter with demineralized water.

Procedure
The filtrate was transfer from the dry ashing procedure into 20 ml beaker. Then 1.0 ml of filtrate was diluted by 9.0 ml of the 1.2 N HCl solution. The dilution was carried out 10 times. The dilute filtrate was dilute by 0.5 ml with 9.5 ml Lanthanum diluent. The reading of samples was taken at atomic absorption spectrophotometer using appropriate standards and instrument was set for Ca and Mg. For determination of K flame emission spectrometers may be used by using diluted filtrate and it is also used for determination of P.

Calibration and Standards

1. 1000 ppm K, Na, Ca and Mg stock solution
2. Working standards

Pipette the following volumes of 1000 ppm stock solution into 1000 ml volumetric flasks and dilute to volume with 1.2 N HCl solutions for Ca and Mg, and with 0.12 N HCl solutions for K and Na:

1000 ppm stock (ml)				Final working Standards			
K	Na	Ca	Mg	K	Na	Ca	Mg
0	0	0	0	0	0	0	0
20	15	50	5	20	15	50	5
40	30	100	10	40	30	100	10

Store in plastic bottles and keep in the refrigerator until ready to use.

Calculations
Ca and Mg % in plant = ppm in reading x 10 x 50 / 0.500 / 10000
K and Na % in plant = ppm in reading x 10 x 50 / 0.500 / 10000

RESULTS

Concentration of Nitrogen:

The leaves of *E. angustifolia* show the concentration of nitrogen as 3.75 % (Fig 1) in dry foliage collected from dry temperate mountainous range of Gilgit. The range of nitrogen in the leaves of *E. angustifolia* was very close to sufficient range of plant nitrogen.

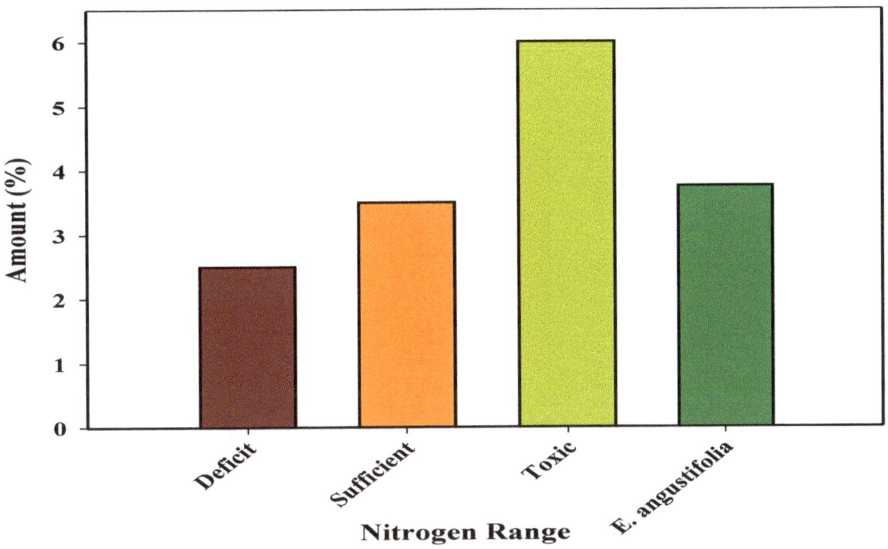

Figure 1: Percent value of Nitrogen in *Elaeagnus angustifolia* and its comparison with deficit, sufficient and toxic levels of Nitrogen in plants required for livestock.

Concentration of phosphorus

The leaves of *E. angustifolia* show the concentration of phosphorus as 0.558 % (Fig 4.2) in dry foliage collected from dry temperate mountainous range of Gilgit. The range of phosphorus in the leaves of *E. angustifolia* was also very close to sufficient range of plant phosphorus

.

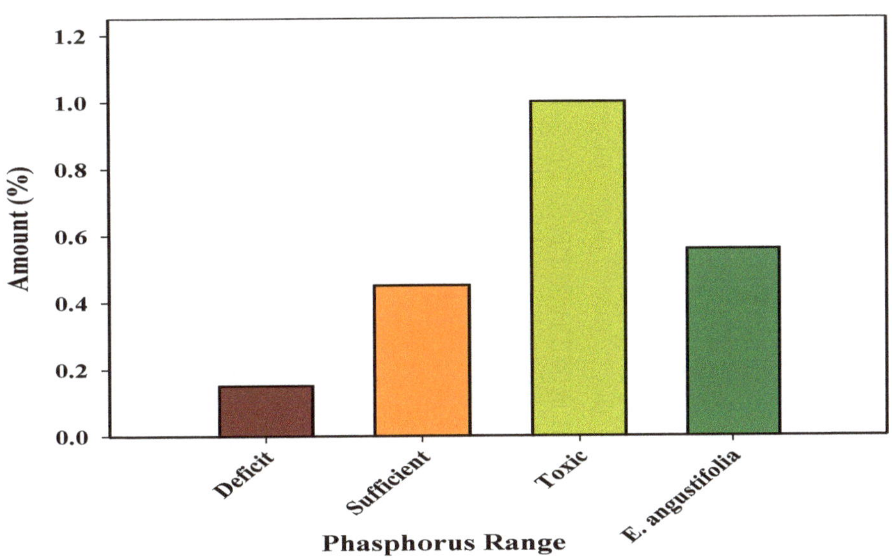

Figure 2: Percent value of Phosphorus in *Elaeagnus angustifolia* and its comparison with deficit, sufficient and toxic levels of Phosphorus in plants required for livestock.

Concentration of Potassium:

The concentration of potassium in the leaves of *E. angustifolia* was 4.37 % (Fig 4.3) in dry foliage collected from dry temperate mountainous range of Gilgit which is quite more than the sufficient range of plant potassium.

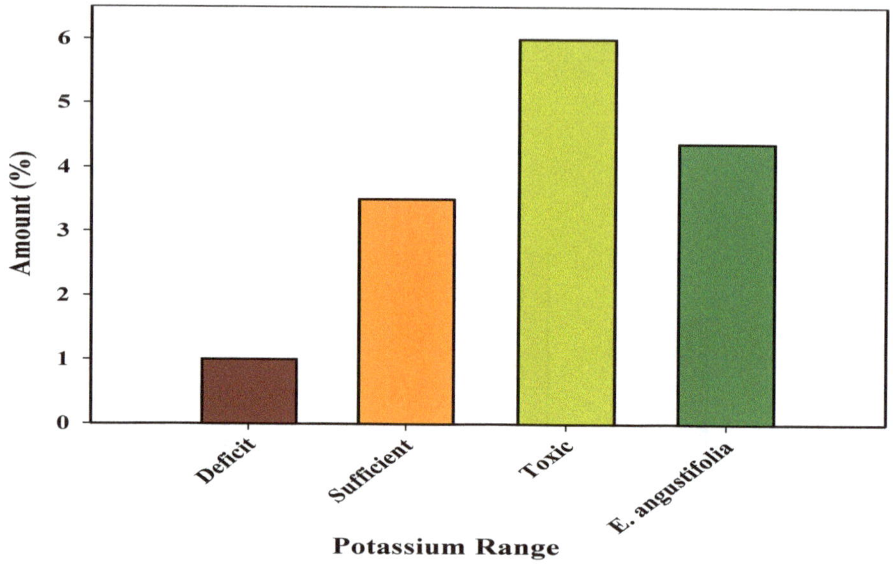

Figure 3: Percent value of Potassium in *Elaeagnus angustifolia* and its comparison with deficit, sufficient and toxic levels of Potassium in plants required for livestock.

Concentration of calcium:

The results given by the analysis of calcium content in the leaves of *E. angustifolia* was 2.459 % (Fig 4.4) in dry foliage collected from dry temperate mountainous range of Gilgit. This range of calcium in the leaves of *E. angustifolia* was very close to sufficient range of plant calcium content.

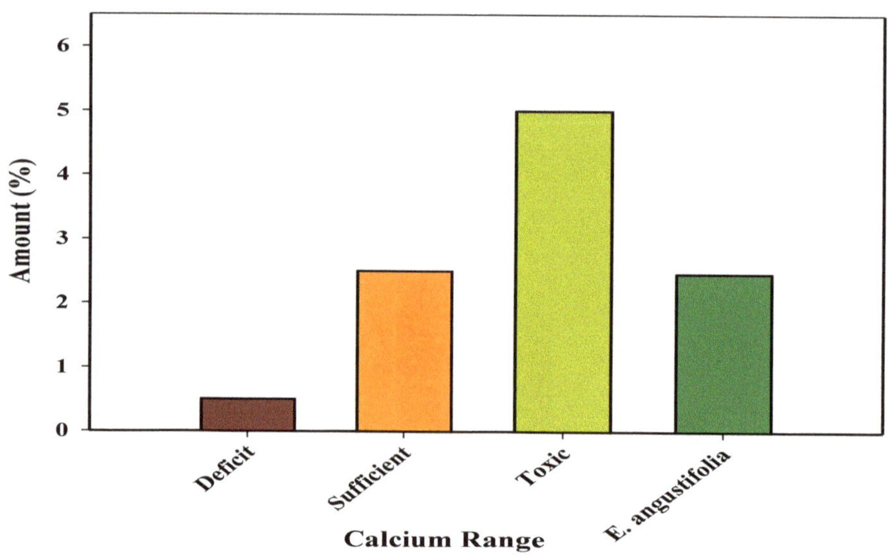

Figure 4: Percent value of Calcium in *Elaeagnus angustifolia* and its comparison with deficit, sufficient and toxic levels of Calcium in plants required for livestock.

Concentration of magnesium:

The concentration of magnesium in the leaves of *E. angustifolia*is 0.769 % in dry foliage of collected from dry temperate mountainous range of Gilgit (Fig 4.5). The range of magnesium in the leaves of *E. angustifolia* was also very close to sufficient range of plant magnesium content.

.

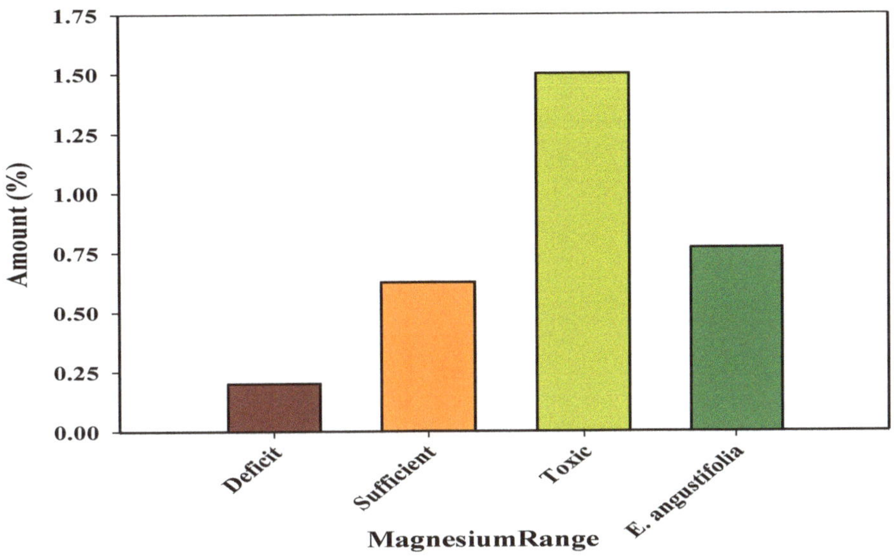

Figure 5: Percent value of Magnesium in *Elaeagnus angustifolia* and its comparison with deficit, sufficient and toxic levels of Magnesium in plants required for livestock.

A pie chart of the analyzed values

A pie chart showing the analyzed values of macronutrient in the dry foliage of Russian olive.

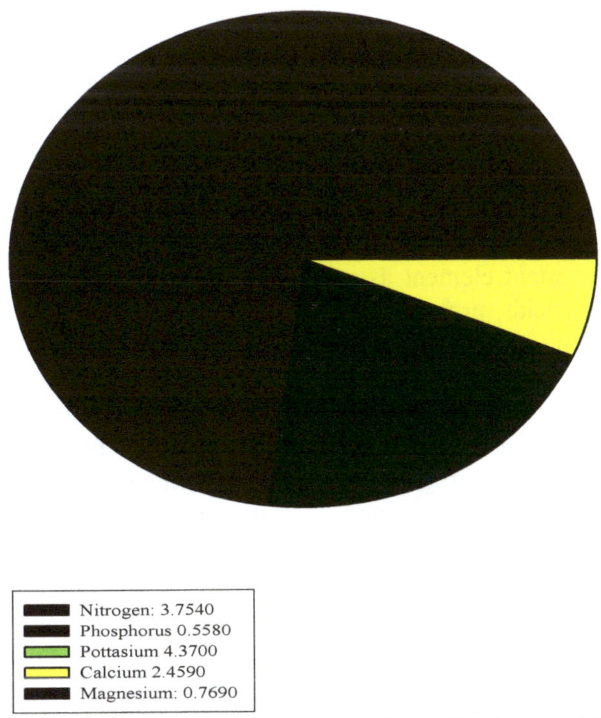

Figure 6: Concentration of Nitrogen, Phosphorus, Potassium, Calcium and Magnesium in dry foliage of *Elaeagnus angustifolia.*

Table 2: Comparison between the macronutrients present in the dry foliage of Russian olive along with their deficient, sufficient, toxic and analyzed ranges.

	Nitrogen	Phosphorus	Potassium	Calcium	Magnesium
Deficient	2.5000	0.1500	1.0000	0.5000	0.2000
Sufficient	3.5000	0.4500	3.5000	2.5000	0.6250
Toxic	6.0000	1.0000	6.0000	5.0000	1.5000
E. Angustifolia	3.7540	0.5580	4.3700	2.4590	0.7690

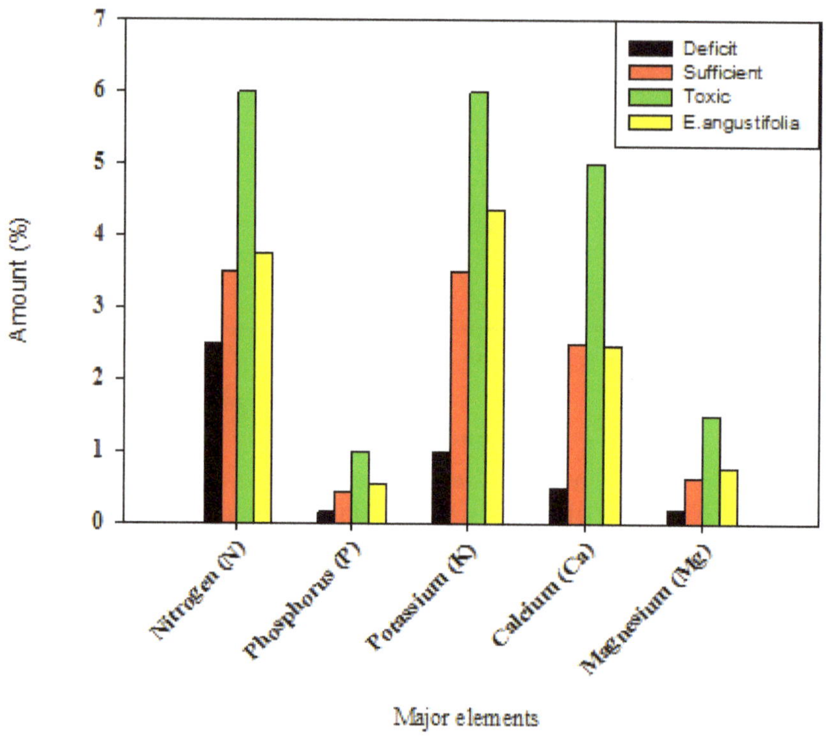

Figure 7: Comparison of Percent value of Nitrogen, Phosphorus, Potassium, Calcium and Magnesium in the dry foliage of *Elaeagnus angustifolia* with their deficit, sufficient and toxic levels in general plants required for livestock.

DISCUSSION

Nitrogen content is an important element as it is building block for amino acids, proteins and protoplasm in plants. Nitrogen behaves as a catalyst for other nutrients in plant. It is very essential for flowering differentiation. It is also important for the growth of shoots in trees (Abid et al., 2005). Deficiency of nitrogen can cause growth abnormalities in plants and yellowing of foliage (Saliva and Uchida, 2000). The current level of nitrogen in Russian olive is very close to general plant nitrogen level. This shows that the plant have sufficient level of nitrogen in the mountain region of Gilgit. The leaves of plant indicate an average N level of 3.23% in *E. angustifolia* analyzed by Llinares, el al. (1992) in San Pablo. In the Leaves of the species there is 3.1–3.3% N in Spain (Llinares et al., 1992; Katzl et al., 2003), 2.9% N on the Rio Grande in New Mexico (Johnson 1995), and 1.6% N in southern Idaho (Katz et al., 2003). Some of these values are quite similar to my analyzed value of nitrogen concentration in Russian olive.

Similarly in case of Phosphorus; it assists in respiration processes as well as in photosynthesis (Ayub et al., 2002). Besides this if there is any deficiency of phosphorus in the plants, abnormalities related to growth, death of tissues decreased fruits quantity and leaves quality may occur (Assuero et al., 2004). Phosphorus concentration in the analyzed results is much close to the general plant phosphorus standard value. Deficiency of calcium can result in the weakening of stem structure of the plants. It also affects the structure and quality of the leaves of plants

(Silva and Uchida, 2000). According to Azim et al., 2011. The concentration of calcium and phosphorus in Russian olive in district Chakwal, Pakistan are 2.09% and 0.15 % respectively which is quite closer to my analyzed values.

Potassium content is supposed to be an enzyme activator and it promotes metabolism (Silva and Uchida, 2000). Generally it is said that sandy soils in high rainfall areas are susceptible to potassium deficiency (soil quality.org.au). It may be one of the reasons for the greater analyzed value of potassium in this study as Gilgit is a dry temperate area with lower rainfall and loamy soil have very good concentration of potassium in the dry foliage of Russian olive. According to Ersoy, (2013) the concentration of potassium in dry foliage of Russian olive in Turkey is 1.03 % which is very low as compare to the analyzed values. Among all the macro elements potassium content is varying due to location.

Deficiency of magnesium in plants may cause the reduction of green color in the leaves of the plant because magnesium supports the formation of chlorophyll (Sawyer, 2004). According to the analyzed value the concentration of magnesium is 0.769 % this value of magnesium is palatable as the standard value of plant magnesium concentration is 0.625%. According to the study of Ersoy, 2013 the concentration of magnesium in the dry foliage of Russian olive in Turkey is 0.0762 this value is quite similar with the analyzed value of magnesium.

CONCLUSION

It is concluded from the results that the range of N, P, K, Ca and Mg in the leaves of *E. angustifolia* was very close to sufficient range of plant N, P, K, Ca and Mg content. Therefore it may be recommended to use *E. angustifolia* as a fodder for livestock feeding.

ACKNOWLEDGEMENT

The author is grateful to Department of Environmental Sciences Karakoram International University, Gilgit, Gilgit-Baltistan Pakistan.

REFERENCES

Abid H, Ghulam A, Ashfaq A, Sayed AW, 2005. Water use efficiency of maize of affected by irrigation schedules and nitrogen rates. Journal of Agriculture. 4: 339-342.

Azim A, Ghazanfar S, Latif A, Nadeem MA, 2011. Nutritional evaluation of some top fodder tree leaves and shrubs of district Chakwal, Pakistan in relation to ruminants requirements. Pakistan Journal of Nutrition. 10 (1): 54-59.

Ayub M, Nadeem MA, Sharar MS, Mahmood N. 2002. Response of maize (*Zea mays* L.) fodder to different levels of nitrogen and phosphorus. Asian Journal of Plant Sciences. 1: 352-354.

Assuero SG, Mollier A, Pellerin S, 2004. The decrease in growth of phosphorus-deficient maize leaves is related to a lower cell production. Plant, Cell and Environment. 27: 887-895.

Borell, A. E. 1962. Russian-olive for wildlife and other conservation uses. Leaflet no. 517. U.S. Department of Agriculture, Washington, D.C.

Cakmak I, 2002. Plant nutrition research: Priorities to meet human needs for food in sustainable ways. Plant Soil. 247: 3-24.

Castro MS, Steudler PA, Melillo JM, Aber JD and Millhamet S, 1993. Exchange of N_2O and CH_4 between the atmosphere and soils in spruce-fir forests in Northeastern United States. Biogeochemistry. 18: 119-135.

Cote B, Carlson, RM and Dawson JO, 1988. Leaf photosynthetic characteristics of seedlings of actinorhizal Alnus spp. and Elaeagnus spp. Photosynthesis Research. 16: 211-218.

Creech JE, Westphal A, Ferris VR, Faghihi J, Vyn TJ, Santini JB, Johnson WG, 2008. Influence of winter annual weed management and crop rotation on soybean cyst nematode (*Heterodera glycines*) and winter annual weeds. Weed Sci. 56: 103-111.

Dawson JO, Van Sambeek JW, 1993. Inter-planting woody nurse crops promotes differential growth of black walnut saplings. In: Gillespie, A.R.; Parker, G.R.; Pope, P.E.; Rink, G., eds. Proceedings, ninth central hardwood forest conference. St. Paul, MN: U.S. Department of

Agriculture, Forest Service, North Central Forest Experiment Station. pp. 455-464.

Domenach AM, Moiroud A and Jocteur-Monrozier L, 1994. Leaf carbon and nitrogen consituents of some actinorhizal tree species. Soil Biology and Biochemistry. 26(5): 649-653.

Ersoy N, Kalyoncu IH, Elidemir AY, Tolay I, 2013. Some Physico-Chemical and Nutritional Properties of Russion Olive (*Elaeagnus angustifolia* L.) Fruit Grown in Turkey. International Journal of Biological, Biomolecular, Agricultural, Food and Biotechnological Engineering. Vol. 7(6): 427-429.

ITRC (Interstate Technology & Regulatory Council), 2012. Incremental Sampling Methodology (ISM-1): Interstate Technology & Regulatory Council, Incremental Sampling Methodology Team, Washington, D.C., www.itrcweb.org.

Katz GL and Shafroth PB, 2003. Biology, ecology and management of *Elaeagnus angustifolia* L. (Russian olive) in western North America. Wetlands. 23 (4): 763-777.

Kelly JF, Smith R, Finch DM, Moore FR, Yong W, 1999. Influence of summer biogeography on wood warbler stopover abundance. The Condor. 101: 76-85.

Knopf FL and Olson TE, 1984. Naturalization of Russian-olive: implications for Rocky Mountain wildlife. Wildlife Society Bulletin 12:289-298.

Motsara, M.R., Roy, R.N., 2008. Guide to Laboratory Establishment for Plant Nutrient Analysis. FAO-Fertilizer and Plant Nutrition Bulletin 19 (Retrieved April, 13 2014.

Muldavin E, Durkin P, Bradley M, Stuever M. Mehlhop P, 2000. Handbook of wetland vegetation communities of New Mexico. Volume I: Classification and community descriptions. Albuquerque, NM: University of New Mexico, Biology Department; New Mexico Natural Heritage Program. pp. 172.

Paterson RT, Karanja GM, Nyaata OZ, Kariuki IW and Roothaert RL, 1998. A review of tree fodder production and utilization within smallholder agroforestry systems in Kenya. Agroforestry Syst. 41: 181-199.

Sawyer J, 2004. Nutrient deficiencies and application injuries in field crops. Iowa State University, IA, USA.

Shafroth PR, Aubla GT and Scott ML, 1995. Germination and establishment of the native plains cottonwood (*Populus deltoides* Marshall subsp. *moniifera*) and the exotic Russian-olive (*Elaeagnus angustifolia* L.). Conservation Biology 9: 1169-1175.

Silva JA and Uchida RS, 2000. Essential Nutrients for Plant Growth: Nutrient Functions and Deficiency Symptoms. Plant Nutrient Management in Hawaii's Soils, Approaches for Tropical and Subtropical Agriculture. Chapter 3. pp. 31-55.

USDA, 2015. Integrated invasive plant management for the lakeview resource area revised environmental assessment. (DOI-BLM-OR-L050-2014-0021-EA), U.S. Department of the Interior Bureau of Land Management. pp. 11-371.

Zouhar and Kris, 2005. *Elaeagnus angustifolia*. In: Fire Effects Information System, U.S. Department of Agriculture, Forest Service, Rocky Mountain Research Station, Fire Sciences Laboratory (Producer). Available: http://www.fs.fed.us/database/feis/.

EFFECTS OF COBALT INDUCED STRESS ON *TRITICUM AESTIVUM* L. CROP

Banashree Sarma, Prema Devi, Nirmali Gogoi,* Yendrembam Menaka Devi
Department of Environmental Science, Tezpur University, Napaam, Assam - 784028, India

ABSTRACT

This experiment was conducted with different concentrations of cobalt (100, 200, 300, 400 and 500 ppm) on wheat (*Triticum aestivum* L.) plants for quantification of its effects on growth performance in a sand culture medium using Hoagland solution. Germination was tested in petri dishes using same treatments. Inhibitory effect of Co on germination percent was observed from 200ppm of Co while vigor index decreased with increasing Co concentration. No negative effect of Co on germination index was found up to 300 ppm while beyond that higher percent inhibition of wheat seed germination was recorded. An enhancing effect on plant height, leaf number, leaf area and dry matter production was observed up to 200ppm of Co treatment while higher concentrations showed detrimental effect of the same. Chlorophyll a/b increased and chlorophyll stability index decreased with increasing Co concentration from 300 ppm onwards. Results from analysis of above ground plant biomass showed higher accumulation of Co with increasing concentration in the medium. However, at 100 ppm of Co treatment, significantly higher grain accumulation of Co was recorded compared to 500ppm of Co treatment. From these findings, it can be summarised that at lower concentrations (up to 200ppm) Co has an enhancing effect on growth of wheat crop and it has a good phyto-extracting ability for cobalt

Keywords: Cobalt, Wheat, Morphological parameters, Chlorophyll pigment, Plant accumulation

INTRODUCTION

Cobalt is a heavy metal that occurs naturally in many different chemical forms in our environment. Though, it is an essential element for both plant and animal but at higher concentration Co is reported to be toxic (Jayakumar and Vijayarengan, 2006). The toxicity of Co is quite low compared to many other metals in soil, but its accumulation in agricultural crops is especially very important because it contributes toxicity into the human food chain. The levels of Co toxicity depend on plant species, soil type and soil chemistry (Bakkus *et al.*, 2005). More acidic the soil, the greater is the potential for Co toxicity at any concentration (Hasan *et al.,* 2011).

Co has both beneficial as well as harmful effects to plants. At lower concentration, Co has been reported to have a positive effect on plants. Being a component of cobalamin, Co at lower concentrations has positive effect on legume plants in nitrogen fixation (Palit *et al.*, 1994). Rathsooriya and Nagarajah (2003) reported that Co has beneficial effect on growth of salinized pea plants to increase leaf water potential relative to those untreated plants. At higher Co concentration detrimental effects on plant growth along with chlorosis and necrosis was reported by Caselles *et al.* (1997). It also inhibits root growth by retarding cell division hindering the uptake and translocation of nutrient and water (Jayakumar *et al.*, 2008). Accumulation of free proline in response to heavy metal exposure helps to maintain the water balance in higher plants (Costa and Morel, 1994). Co inhibits the synthesis of chlorophyll pigments by obstructing its biosynthesis pathway (Mysliva-Kurdziel *et al.,* 2004).

Wheat *(Triticum aestivum* L.*)*, one of the most important strategy crops is grown worldwide in a variety of climates. The present experiment was conducted in a sand culture medium to evaluate the effects of various levels of Co on morphological and biochemical parameters of wheat. The accumulation of Co in different above ground parts of the plant was also estimated at harvest.

MATERIALS AND METHODS

Wheat (*Triticum aestivum* L.) seeds cv. DBW-39 was used for this experiment. The seeds were collected from Regional Agricultural Research Station, Shilongoni, Nagaon, Assam, India. Hoagland solution as developed by Hoagland and Arnon (1950) was prepared and used as the nutrient solution for this experiment. For preparation of Hoagland

**Corresponding author:* e-mail: nirmali@tezu.ernet.in

solution $CaNO_3$, KNO_3, $MgSO_4$, KH_2PO_4, Fe-EDTA were used as sources of macronutrient and H_3BO_3, $ZnSO_4.7H_2O$, $CuSO_4.5H_2O$, $MnSO_4$and $NaMoO_4$ were used as micronutrients in proportion as described by Hoagland and Arnon (1950). Cobalt in the form of $CoNO_3$ was applied in different concentrations to the nutrient solution. The treatments include 100ppm (T_1), 200ppm (T_2), 300ppm (T_3), 400ppm (T_4), 500ppm (T_5) of Co and control (C) with only Hoagland solution. Treatments were applied from 30 days after sowing (DAS).

Germination test

Different concentrations (100ppm, 200ppm, 300ppm, 400ppm, 500ppm) of $CoNO_3$ solution in distilled water were prepared. Distilled water was taken as control. Germination was investigated in sterilized petri plates with whatmann No. 1 filter paper. Ten seeds were placed in each petri plates and the treatments were replicated 5 times. The plates were then kept at 25$\pm$2°C for 7 days. Proper care was taken to keep the plates wet throughout the experiment with addition of respective solutions periodically. The radicle and plumule lengths were recorded at the end of the experiment (7 days). The germination performances under different treatments were calculated as follows:

Percent germination:

(No. of seeds germinated/ Total no of seeds taken) × 100

Germination index (GI):

$$2(7x + 6x + 5x + 4x + 3x + 2x + x)$$

Where, x is the number of seeds germinated in 24 hours.

Percent inhibition of germination:

100 – (GI of treatment /GI of control ×100)

Vigour index (VI):

radicle length + plumule length × percent germination.

Preparation of sand culture medium and experimental layout

Sand was collected and repeatedly washed under running tap water and then with distilled water till the soil, mud or dust portion washed out. Then it was treated with 0.1% of HCL and kept for 2 days followed by repeated washing with distilled water to make it free from acid. To ascertain the presence of no acid, the sand was tested with litmus paper. Then the pots were filled with 1.5kg of the nutrient free sand. Total thirty six pots were taken for the experiment (6 for each treatment). 3 replications were kept intact and were used to measure the morpho-physiological and yield attributes. The remaining three replicates were used for destructive sampling for analyzing various biochemical parameters at regular interval. Wheat seeds (@5 seeds per pot) were sown in sand medium on 5th November, 2012. After germination of the wheat seeds, nutrient solution was added to the pots at 3-4 days interval till harvest of the crop.

Morphological parameters

Changes in various plant morphological parameters like plant height, leaf number, leaf area and dry biomass were recorded at 15 days interval from 30 DAS till harvest. Leaf area was recorded by using laser leaf area meter (CI-203, USA). Dry biomass of the plant was taken after oven drying the plant at 70°C for 24 hours.

Biochemical parameter

Total chlorophyll, chlorophyll-a and chlorophyll-b was estimated by using standard method described by Anderson and Boardman (1964) and chlorophyll a/b was calculated. The pigment composition was measured by a double-beam spectrophotometer (UV 1700 Pharma Spec, Japan). Chlorophyll stability index (CSI) was calculated using the following formula:

CSI= (Total chlorophyll of the stressed / Total chlorophyll of the control) × 100.

The leaf proline content was estimated using the method of Bates *et al.* (1973).

Estimation of cobalt in plant, sand medium and grain

For estimation of Co in sand, the sand was digested by following the method of Shapiro and Brannock (1962). 0.5g of sand was digested with 10 ml concentrated HF and 5ml concentrated HNO_3 with 1ml $HClO_4$ in sealed Teflon crucible at 85°C to 90°C on a hot plate for 4 hours. After 4 hours, the crucible was opened and heated until almost dry. In the second phase, 5ml concentrated HF and 10ml concentrated HNO_3 were added and heated till dryness at 90°C. In the third phase 10ml concentrated HNO_3 was added and heated at 90°C till dryness. Finally 20ml of 2N HCl was added to dry crucible and transferred to 100 ml volumetric flask and cooled and volume make up was done with ultra pure water.

At harvest, the grains and above ground plant biomass were collected, dried and ground properly and subsequently digested to analyse for Co concentration. 1gm of grind material was taken for di-acid digestion using HNO_3 and $HClO_4$ (9:4).

The digested samples (sand, plant and grains) were then analysed for Co concentration in Atomic absorption spectrophotometer (AAS).

Statistical analysis

Standard errors of means (of three replicates) were calculated for all the parameters. Data obtained were subjected to one-way analysis of variance (ANOVA) in SPSS for windows 16.0.20. Critical Differences (CD) among the means and Duncan's Multiple Range Test (DMRT) at 5% level of probability were used to test the significance of differences between treatment means.

Table 1: Effect of different Co concentrations on germination parameters of wheat seed

	% Germination	Germination Index	% Inhibition	Vigor Index
C	100 ± 0^{c}	69.67 ± 2.37^{b}	0	1032.08 ± 17.11^{c}
T_1	100 ± 0^{c}	82 ± 2.31^{d}	-18.30 ± 7.68^{a}	672.29 ± 20.21^{b}
T_2	91.67 ± 4.17^{c}	79 ± 1.0^{cd}	-13.84 ± 5.68^{ab}	658.33 ± 50.88^{b}
T_3	70.83 ± 4.17^{b}	71.67 ± 3.84^{bc}	-3.09 ± 6.27^{ab}	101.52 ± 7.23^{a}
T_4	60.32 ± 2.04^{a}	50.67 ± 1.33^{a}	27.10 ± 2.96^{b}	96.82 ± 14.64^{a}
T_5	54.17 ± 2.08^{a}	43.67 ± 2.73^{a}	36.89 ± 6.07^{c}	61.57 ± 8.97^{a}
CD	8.27	7.73	16.70	75.87
P	0.00	0.00	0.00	0.00

Different letters within each column for each parameter indicate significant differences between treatments at 5% level of significance according to DMRT.

Table 2: Effect of different Co concentrations on plant height of wheat

	30 DAS	45 DAS	60 DAS	75 DAS	90 DAS	110 DAS
C	13.467 ± 0.33^{a}	20.33 ± 0.40^{ab}	28.50 ± 0.98^{a}	34.30 ± 0.57^{b}	36.50 ± 2.75^{b}	37.30 ± 3.65^{ab}
T_1	12.70 ± 0.56^{a}	18.80 ± 0.26^{ab}	28.93 ± 2.16^{a}	36.50 ± 0.29^{b}	37.70 ± 2.55^{b}	39.00 ± 3.18^{b}
T_2	13.11 ± 0.72^{a}	19.93 ± 0.76^{ab}	30.97 ± 2.30^{a}	37.03 ± 1.47^{b}	38.97 ± 3.15^{b}	41.17 ± 1.59^{b}
T_3	13.27 ± 0.27^{a}	20.78 ± 1.97^{ab}	28.47 ± 1.30^{a}	36.67 ± 2.24^{b}	37.00 ± 2.18^{b}	37.07 ± 2.75^{ab}
T_4	13.10 ± 0.23^{a}	15.26 ± 1.75^{a}	26.47 ± 1.67^{a}	31.17 ± 2.33^{ab}	32.20 ± 2.25^{ab}	33.30 ± 3.67^{ab}
T_5	13.73 ± 0.27^{a}	20.37 ± 2.78^{b}	24.17 ± 2.95^{a}	27.77 ± 2.63^{a}	28.13 ± 1.48^{a}	28.40 ± 2.39^{a}
CD	1.34	4.95	6.18	5.62	7.56	9.13
P	0.66	0.22	0.31	0.02	0.07	0.10

All values are Mean$\pm$ SE (cm). Different letters within each column indicate significant differences between treatments at 5% level of significance according to DMRT.

Table 3: Effect of different Co concentrations on dry matter content of wheat

	30 DAS	45 DAS	60 DAS	75 DAS	90 DAS	110 DAS
C	0.232 ± 0.016^{a}	0.493 ± 0.009^{a}	0.782 ± 0.004^{bc}	2.981 ± 0.374^{a}	3.459 ± 0.037^{b}	4.593 ± 0.064^{ab}
T_1	0.270 ± 0.011^{ab}	0.511 ± 0.008^{ab}	0.952 ± 0.078^{d}	3.018 ± 0.118^{a}	4.392 ± 0.015^{d}	5.340 ± 0.279^{b}
T_2	0.317 ± 0.027^{b}	0.495 ± 0.014^{a}	0.899 ± 0.029^{cd}	4.128 ± 0.007^{b}	4.693 ± 0.041^{e}	5.541 ± 0.110^{b}
T_3	0.219 ± 0.017^{a}	0.503 ± 0.004^{a}	0.551 ± 0.036^{a}	3.163 ± 0.198^{a}	3.549 ± 0.023^{c}	4.447 ± 0.107^{ab}
T_4	0.248 ± 0.013^{a}	0.535 ± 0.005^{b}	0.712 ± 0.011^{b}	3.229 ± 0.039^{a}	3.599 ± 0.012^{c}	4.064 ± 0.172^{a}
T_5	0.242 ± 0.020^{a}	0.509 ± 0.009^{ab}	0.784 ± 0.011^{bc}	2.650 ± 0.017^{a}	3.219 ± 0.025^{a}	3.841 ± 0.372^{a}
CD	0.056	0.018	0.113	0.557	0.080	1.043
P	0.02	0.05	0.00	0.00	0.00	0.02

All values are Mean± SE (g/plant). Different letters within each column indicate significant differences between treatments at 5% level of significance according to DMRT.

Table 4: Correlation between the different parameters at harvest

	Plant Concentration	Proline	Chlorophyll a:b	CSI	Dry Matter	Leaf Area	Plant Height	Leaf Number
Plant Concentration	1							
Proline	0.982**	1						
Chlorophyll a:b	0.724	0.820*	1					
CSI	-0.885*	-0.912*	-0.802	1				
Dry Matter	-0.615	-0.714	-0.894*	0.635	1			
Leaf Area	-0.881*	-0.934**	-0.870*	0.925**	0.852*	1		
Plant Height	-0.791	-0.859*	-0.808	0.707	0.912*	0.915*	1	
Leaf Number	-0.861*	-0.926**	-0.914*	0.915*	0.884*	0.995**	0.916*	1

**. Correlation is significant at the 0.01 level (2-tailed)
*. Correlation is significant at the 0.05 level (2-tailed)

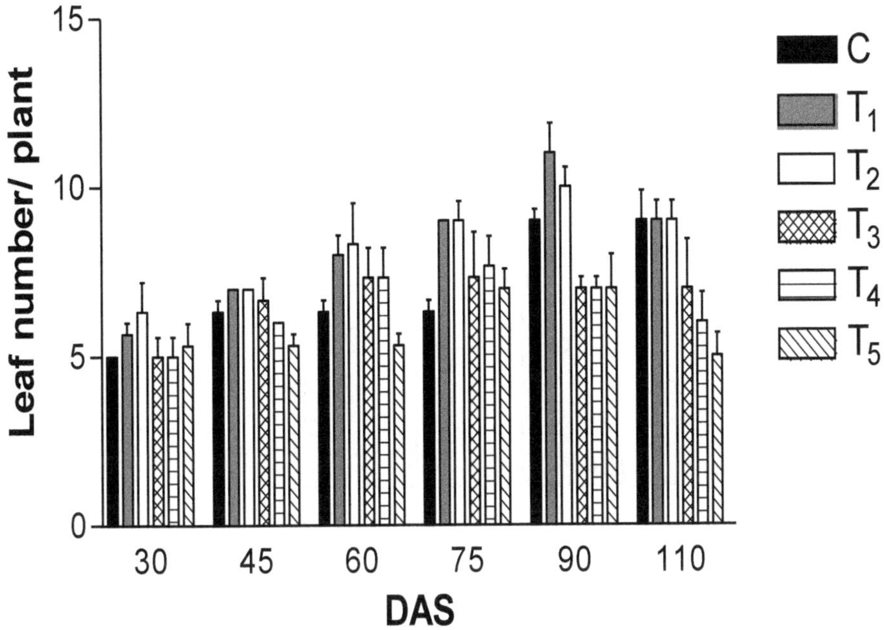

Figure 1. Effect of different Co concentrations on leaf number of wheat

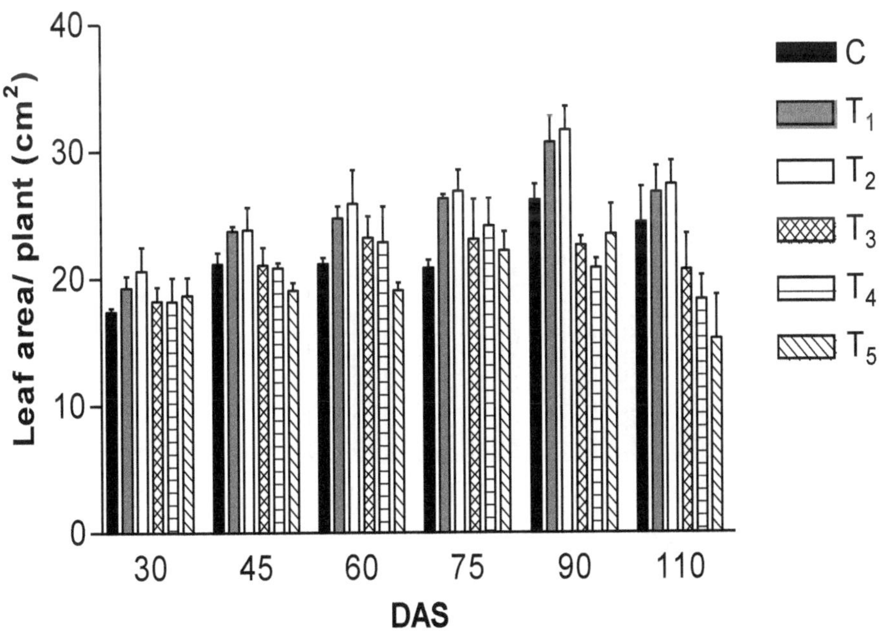

Figure 2. Effect of different Co concentrations on leaf area of wheat

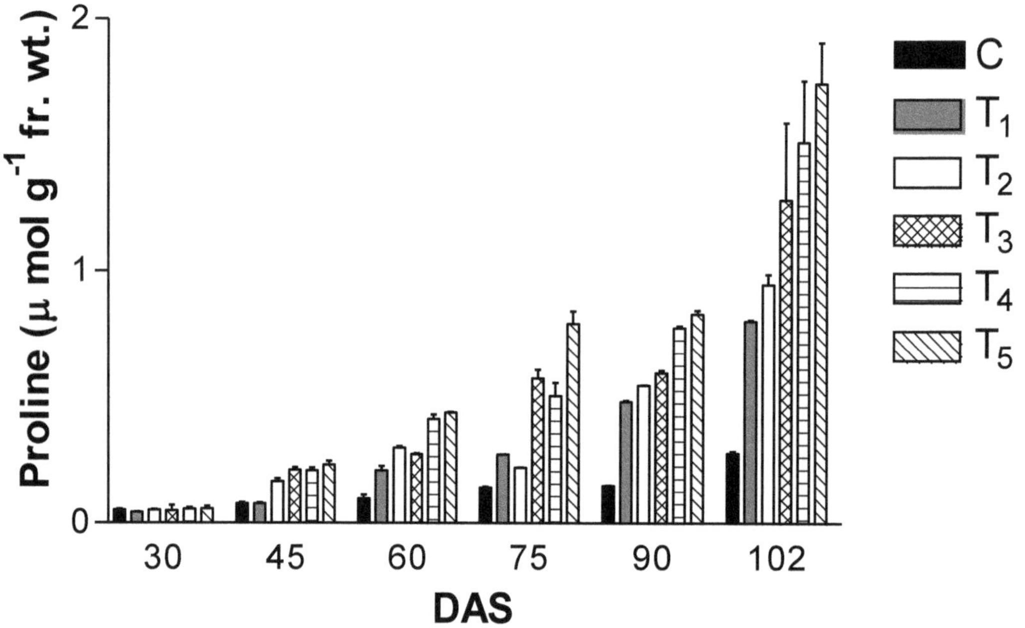

Figure 3. Effect of different Co concentrations on leaf proline content of wheat

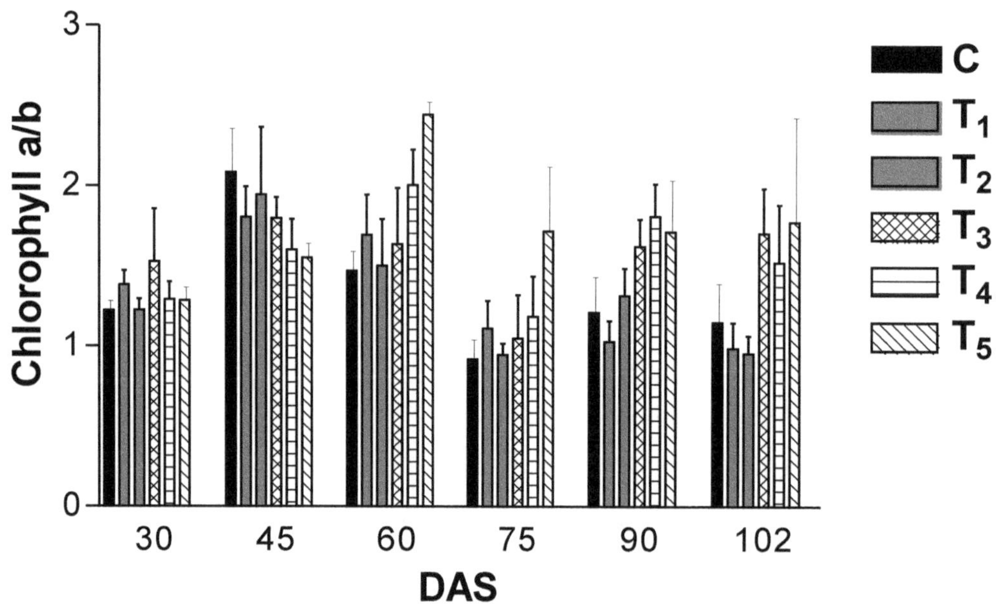

Figure 4. Effect of different Co concentrations on chlorophyll a/b of wheat

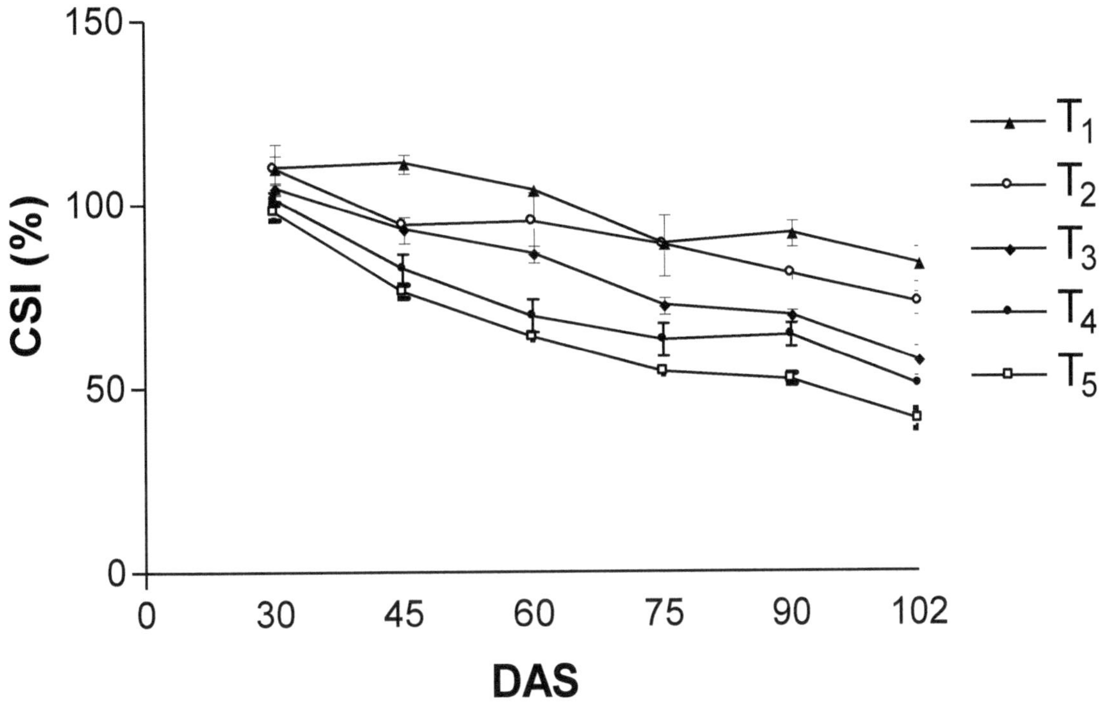

Figure 5. Effect of different Co concentrations on chlorophyll stability index of wheat

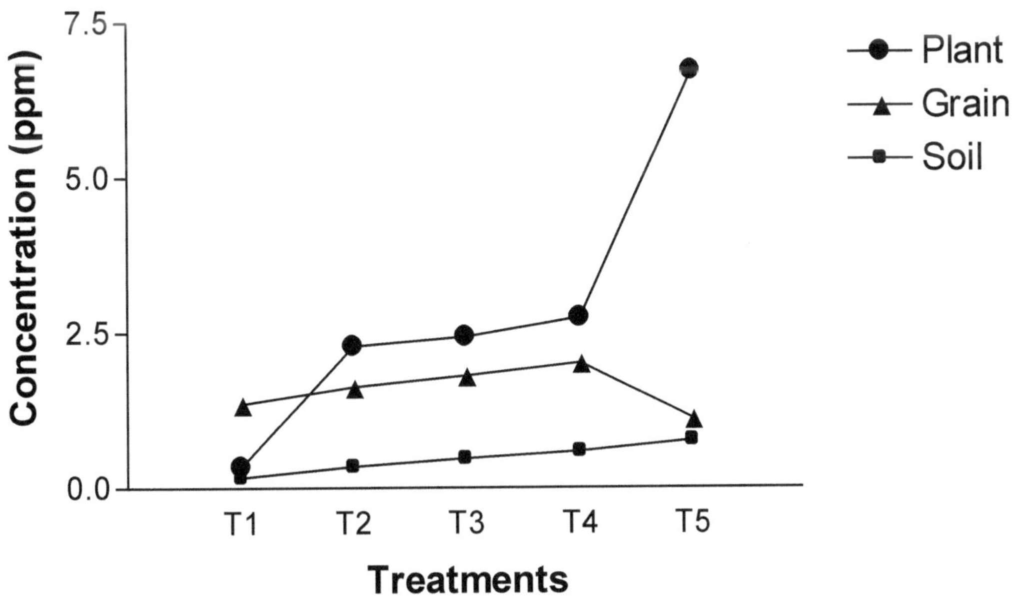

Figure 6. Co concentrations in plant, grain and sand after harvest of wheat

RESULT AND DISCUSSION

Germination responses of wheat to different Co concentration

Cobalt exposure significantly reduced the germination of wheat seed with increasing Co concentration (table 1). No negative effect on percent germination was recorded up to 100ppm of Co. The inhibitory effect of Co on germination of wheat seed was noted from 200ppm (T_2) onwards. At the highest concentration of Co (T_5), the reduction in percent germination was 50% of the control. Ionic toxicity induced by higher concentrations of Co might be the cause of reduced seed germination or it could be due to osmotic effect (Shaukat *et al.*, 1999). Similar results of reduced germination under Co exposure had been reported in various crops such as ragi (Jayakumar *et al.,* 2008), paddy (Jayakumar and Vijayarengan, 2006) and *Vigna mungo* (Munzuroglu and Geckil, 2002). Highest germination index was obtained in T_1 with increasing values up to 300ppm of Co (table 1). This may be attributed to the triggering effect of Co on germination of wheat seed as Co enhances the activity of enzyme α-amylase, one of the important enzymes involved in the process of seed germination (Zeid, 2001). The percent inhibition of germination was recorded from 400 ppm of Co onwards with highest value at 500 ppm (table 1). Cobalt had a stimulatory effect up to 300ppm. The recorded vigor index was concentration dependent and the control showed the highest vigor index of 1032.08. These results of vigor index indicate that addition of Co even at lowest concentration did not have any positive impact on elongation of plumule and radicle. This might be due to the effect of Co in the process of cell division of these growing parts.

Morphological responses of wheat to different Co concentration

Plant height was recorded throughout the experimental period and enhancement in the linear growth of shoot was observed up to 200ppm of Co at all the stages of growth when compared with control plants (table 2). While significant reduction of shoot growth rate was observed after 60 DAS with increasing (from 300ppm onwards) concentration of Co. T_3 maintained the equal plant height with that of control plants. However, at lower concentrations (up to 200 ppm) Co has an enhancing effect on all these physiological parameters as been previously reported by Jaleel *et al.* (2009). Enhancing effect of lower doses of Co on plant height and leaf area has previously been reported by Kadhim (2011).

Throughout the growth period more leaf number and leaf area were recorded in treatment T_1 and T_2 (100 and 200 ppm of Co) compared to control and other treatments. Leaf number and area increased till 90DAS and then started to decrease in all the treatments (figure 1 and 2). Highest leaf senescence was observed in T_4 and T_5 treatments. The decreasing growth rate of these morphological parameters under higher concentration of Co (300ppm and above) is the direct inhibitory effect of Co in cell division or cell elongation or combination of both (Jayakumar *et al.*, 2007). Plant accumulation of Co at harvest had showed a significant negative correlation with leaf area (r=0.881) and leaf number (r=0.861) (table 4).

Plant dry matter increased in all the treatments with the growth of the crop, but the rate of increase was lesser in T_3, T_4 and T_5. Enhanced production of plant biomass was seen up to 200ppm of Co concentration while same was retarded from 300ppm onwards (table 3). Due to panicle initiation, higher dry plant biomass was recorded at 75 DAS compared to 45 and 60 DAS irrespective of treatments. These results showed an enhancing effect of Co on wheat plant biomass production at lower concentrations (up to 200ppm). Similar results of higher dry matter production at lower Co concentration have been reported by Aery and Jagetiya (2000). At higher doses, Co ion might inhibit cell division resulting in stunted growth and lower dry matter production of wheat plants.

Biochemical responses of wheat to different Co concentration

Figure 3 clearly indicate that leaf proline concentration increased with increasing Co concentration as well as with growth stages of wheat. From 45DAS significant (P=0.00) increase in proline content was recorded with increasing levels of Co. Highly significant increase in proline content was obtained at 102 DAS in all the Co treatments. Significant positive correlation of proline (r = 0.982) was observed with increased accumulation of Co in plants at harvest (table 4). Treatment T_5 maintained the highest proline concentration throughout the experiment while control plants were found to accumulate the least amount of proline. Increased proline content under Co stress might be due to Co induced physiological drought. As osmo-protectant, proline accumulation up to 200 ppm of Co, can induce resistance in wheat plants and nullify the Co induced stress on wheat crop (Zengin and Munzuroglu, 2005). Proline accumulation in response to Co stress is reported to increase tolerance by osmoregulation, resisting enzyme denaturation and stabilization of protein synthesis (Kuznetsov and Shevyakova, 1997).

At 45 DAS, increase in chlorophyll a/b was found in all the treatments. However, at 60 DAS significant increase in chlorophyll a/b ratio was obtained in T_4 and T_5 (figure 4). However, from 90 DAS a/b ratio were found to increase with increasing Co concentration except for T_1. At 102 DAS up to 200 ppm of Co chlorophyll a/b ratio was at par to that of control. Also, addition of 100ppm of Co (T_1) maintained highest CSI value throughout the growing period whereas, significantly higher reduction in CSI value were observed from 300ppm of Co (T_3) treatments from 45 DAS till end of the crop growth period (figure 5). Co up to 200 ppm had anabolic effect on chlorophyll content of wheat crop due to its positive role in hormonal synthesis and metabolic activities. Anabolic effect of lower Co concentration (up to 200 ppm) on chlorophyll biosynthesis resulted higher a/b ratio throughout the experiment through enhanced uptake of Mg (Zeid, 2001). Heavy metals were found to affect the chlorophyll biosynthesis/degradation by enzymatic modulation. Co enhances the activity of chlorophyll degrading enzyme chlorophyllase (Reddy and Vora, 1986) and inhibits the activity of chlorophyll synthesizing enzyme 5-aminolevulinic acid and protoporphyrin (Mysliva-Kurdziel *et al.*, 2004). At higher concentration Co showed prominent catabolic effect on chlorophyll synthesis. However, the negative effect of Co on chlorophyll b was lesser than on chlorophyll a as has previously been described by Zengin and Munzuroglu (2005).

Plant, grain and sand accumulation of Co

From figure 6, it was observed that when wheat plants were exposed to higher Co concentration, the plant accumulation was relatively higher compared to grain. But under lower (100ppm) soil Co exposure, wheat plants accumulate more amount of Co in grain compared to other plant parts. This is due to the tendency to accumulate in the sink compared to other plant parts under low Co exposure. In control (plants, grain and sand) Co concentration was beyond the detectable limit. Plant accumulation of Co was found to be lowest in 100ppm (T_1) of Co exposure while with increasing concentration, Co content in plant parts was found to be higher. 500 ppm (T_5) exposure of Co recorded the highest Co accumulation in above ground plant biomass. Accumulation of Co is always higher in the above ground plant parts of wheat than that of grains. This is attributed to the fact that Co has a lesser mobility in the leaf tissues compared to the vascular system of wheat (Collins *et al.*, 2010). Significant increase in wheat grain Co concentration was found up to 400 ppm (T_4) of Co exposure. T_4 recorded the highest (2.013ppm) grain accumulation while interestingly T_5 showed the lowest (1.117ppm) grain Co concentration. This might be due to stimulatory effect of Co in phloem translocation pathway which might be disrupted under higher Co concentration. Lesser accumulation of Co in grains of wheat has already been confirmed by Zeller and Feller, 1998.

Co content in the sand medium also showed an increasing trend with increasing dose of Co application. T_1 (0.167ppm) recorded the lowest while T_5 (0.757ppm) recorded the highest concentration of Co in sand medium.

SUMMARY

Results of germination index and percent inhibition of germination showed that Co up to 300 ppm has an enhancing effect on germination of wheat seed while it has a negative impact on enlargement of plumule and radicle. Exposure of wheat crop to lower concentration of Co (up to 200 ppm) is useful for growth and development of morpho-physiological attributes but higher had detrimental effects of the same. Results on various biochemical (CSI, chlorophyll a/b, proline) analyses also support this. Wheat is found to be a good phyto accumulator for cobalt.

REFERENCES

Aery NC and Jagetiya BL, 2000. Effect of Cobalt Treatments on Dry Matter Production of Wheat and DTPA Extractable Cobalt Content in Soils. Commun. Soil Sci. Plant Anal.31(9&10):1275-1286.

Anderson JM and Boardman NK, 1964. Studies on greening of dark grown bean plants. Aust. J. Biol. Sci.17:93-101.

Bakkaus E, Gouget B, Gallien JP, Khodja H, Carrot F, Morel JL and Collins RN, 2005. Concentration and distribution of Co in higher plants: the use of micro-PIXE spectroscopy. Nucl. Instr. Meth. Phy. Res. B., 231(1):350–356.

Bates LS, Waldren RP and Teare ID, 1973. Rapid determination of free proline for water-stress studies. Plant Soil.39:205-207.

Caselles JM, Pérez-Espinosa A, Pérez-Murcia MD, Moral R and Gómez I, 1997. Effect of Increased Cobalt Treatments on Cobalt Concentration and Growth of Tomato Plants.J. Plant Nutr. 20(7&8):805-811.

Collins RN, Bakkaus E, Carriere M, Khodja H, Proux O, Morel JL and Gouget B, 2010. Uptake, Localization, and Speciation of Cobalt in *Triticum aestivum* L.(Wheat) and *Lycopersicon esculentum* M. (Tomato). Environ. Sci. Technol.44(8):2904-2910.

Hasan SA, Hayat S, Wani AS and Ahmad A, 2011. Establishment of sensitive and resistant variety of tomato on the basis of photosynthesis and antioxidative enzymes in the presence of cobalt applied as shotgun approach. Braz. J. Plant Physiol. 23(3):175-185.

Hoagland DR and Arnon DI, 1950. The water-culture method for growing plants without soil. California Agricultural Experiment Station Circular 347:1-32.

Jaleel CA, Jayakumar K, Xing ZC and Iqbal M, 2009. Low Concentration of Cobalt Increases Growth, Biochemical Constituents, Mineral Status and Yield in *Zea Mays*. J. Sci. Res. 1(1):128-137.

Jayakumar K and Vijayarengan P, 2006. Influence of zinc on seed germination and seedling growth of *Vigna mungo*(L.) Hepper. Plant Arch.6:681-682.

Jayakumar K, Jaleel CA and Vijayarengan P, 2007. Changes in growth, biochemical constituents and antioxidant potentials in radish (*Raphanus sativus* L.) under cobalt stress. Turkish J. Biol. 31:127-136.

Jayakumar K, Jaleel CA and Azooz MM, 2008.Impact of Cobalt on Germination and Seedling Growth of *Eleusine coracana* L.and *Oryza sativa* L. under Hydroponic Culture. Global J. Mol. Sci. 3(1):18-20.

Kadhim RE, 2011. Effect of Pb, Ni and Co in growth parameters and metabolism of *Phaseolus aureus* Roxb. Euphrates J. of Agri. Sci. 3(3):10-14.

Kuznetsov VV and Shevyakova NI, 1997. Stress responses of tobacco cells to high temperature and salinity. Proline accumulation and phosphorilation of polypeptides. Physiol. Plant. 100:320-326.

Munzuroglu O and Geckil H, 2002. Effects of Metals on Seed Germination, Root Elongation, and Coleoptile and Hypocotyl Growth in *Triticuma estivum* and *Cucumis sativus*.Arch. Environ. Contam.Toxicol.43:203–213.

Mysliva-Kurdziel B, Prasad MNV and Strzalka K, 2004. Photosynthesis in metal stressed plants,pp. 146-181. In: M.N.V. Prasad (eds.), Heavy metal stress in plants: from biomolecule to ecosystems, Springer-Verlag, Berlin.

Palit S, Sharma A and Talukder G, 1994. The effect of cobalt on plants. Bot.Rev.60:149–181.

Rathsooriya, GB and Nagarajah S, 2003. Studies effect of salinity with cobalt on tomato plants. Plant Soil,44: 158.

Reddy MP and Vora AB, 1986. Changes in pigment composition, hill reaction activity and saccharide metabolism in bajra (*Pennisetum typhoides* S&H) leaves under NaCl salinity. Photosynthetica. 20:50-55.

Shapiro L and Brannock WW, 1962. Rapid analysis of silicate, carbonate, and phosphate rocks. U.S. Geol. Sur. Bull.1144A:56.

Shaukat SS, Mushtaq M and Siddiqui ZS, 1999. Effects of cadmium, chromium and lead on seed germination, early seedling growth and phenolic contents of *Parkinsonia aculeatt* L. and *Pennisetum americanum*(L.). *Schumann*. Pak. J. Biol. Sci. 2:1307-1313.

Zeller S and Feller U, 1998. Redistribution of cobalt and nickel in detached wheat shoots: effects of stream-grinding and of cobalt and nickel supply. Biol. Plant. 41(3):427-434.

Zeid JM, 2001. Responses of *Phaseolus vulgaris* to chromium and cobalt treatments. Biol. Plant. 44:111-115.

Zengin FK and Munzuroglu O, 2005.Effects of some heavy metals on content of Chlorophyll, Proline and some Antioxidant chemical in bean (*Phaseolus vulgaris* L.) seedlings. Acta Biol. Cracov. Bot. 47(2):157-164.

GENETIC VARIABILITY AND HERITABILITY STUDIES IN F_2 POPULATION OF SPRING WHEAT

Asghar, K.[1], Z.A Soomro[1*], S.N. Mari[1], T.A. Baloch[2], A.A. Rajper[1], U.A. Kasi[1] and F. Asghar[3]
[1]*Department of Plant Breeding and Genetics, Sindh Agriculture University Tandojam-70060, Pakistan*
[2]*Department of Agriculture Research, Government of Balochistan, Pakistan*
[3]*Department of Entomology, Sindh Agriculture University Tandojam-70060, Pakistan*

ABSTRACT

A research experiment was conducted to investigate the variability and heritability in F_2 population of spring wheat at the experimental field of Sothern Wheat Research Station, Tandojam during 2014-15.The experimental material consisted of 8F_2 hybrid *viz:* Sassui x Anmol-91, WL-711 x Anmol-91, Sassui x Johar, Abadgar x SKD-1, Sassui x Abadgar-93, Marvi x Noori, Moomal- 2002 x Anmol-91, Marvi x Sassui and their parents were sown in Randomized Complete Block Design (RCBD) with 3 replications. Genotype were highly significant at 0.01 level of probability for days to 75 % heading, days to 75% maturity, plant height, spike length and spikelets spike^{-1}. The highest heritability (95.39%) regarding spike length was displayed by cross WL-711 x Anmol-91. Spike length is the main contributors to the grain yield and may be kept as selection criteria for developing high yielding varieties. The F_2 population Sassui x Anmol-91 and WL-711 x Anmol-91 could be selected for higher yield in advanced generations.

Keywords: Heritability, genetic advance, genetic variability, segregating population

INTRODUCTION

Wheat is an important cereal crop and the distribution of the world wheat and grains and believed to progenitors of the cultivated wheat supports the belief that wheat was originated in South Asia (Ansari *et al.*, 2005). Some species were cultivated in Greece Persia, Turkey and Egypt in pre-historic times, while the cultivation of other species may be more recent origin. In Pakistan evidence from Mohen-jo-Daro excavation, indicates that wheat was cultivated and used as staple food some 7000 years ago (Bharat *et al.*, 2013).

The breeders are interested in population with high mean performance and concerned with genetic variation within the population so that further progress can be made. The genetic variance and heritability provide useful guide lines in the plant breeding programme. The heritability variance is the total variance among the phenotypes when grown over the range of environments. Heritability of a character helps to predict the performance of subsequent invention for superior results in terms of future generation. However, reliable comparison of genetic variance or heritability of the population requires large sample of progeny from each population. (Thebo *et al.*, 2005). Keeping in view the above facts the present research was undertaken to assess the genetic variability and heritability in F_2 population of spring wheat.

*Corresponding author: e-mail: zasoomro_cap@hotmail.com

MATERIALS AND METHODS

The research experiment was conducted to study the genetic variability and heritability studies in F_2 population of spring wheat in the experimental field of Southern Wheat Research Station Tando Jam during 2014-15.The experimental material consisted 8 F_2 population Sassui x Anmol-91, WL-711 x Anmol-91, Sassui x Johar, Abadgar x SKD-1, Sassui x Abadgar-93, Marvi x Noori, Moomal- 2002 x Anmol-91, Marvi x Sassui and their parents were sown in Randomized Complete Block Design (RCBD) with 3 replications. Each genotype was planted in four rows in each replication with plant to plant and row to row spacing of 15 and 30 cm, respectively. The data were recorded on days to 75% heading, days to 75% maturity, plant height, spike length and spikelets spike^{-1}.

Statistical analysis

The collected data were subjected to analysis of variance as described by Gomez and Gomez (1984), the comparison of means was done through least significant difference at 0.5% and heritability was worked out following Burtain (1951).

RESULTS

The analysis of variance for the traits studied are presented in table 1, whereas the mean performance are presented in table 2.The heritability and the genetic advance are presented in table 3.

Table 1 showed that genotype are highly significant at 0.01 for days to 75 % heading , days to 75% maturity, plant height, spike length and spikelets spike^{-1}. The mean performance of yield and its component are presented in table 2, which further exhibited that cross Sassui x Abadgar-93 took maximum days to 75% heading (84.67),while the variety Anmol-91 took 62.00 days to 75% heading, whereas the cross Abadgar -93 x SKD-1 take maximum days (139.7) to 75% maturity while cross Sassui x Anmol-91 take minimum 121.71 days to 75% maturity .For the character plant height Johar take maximum (93.00) while the cross Sassui x Anmol-91 (66.40 cm) displayed short stature. In case of spike length the large spike (13.43cm) displayed by cross Sassui x Anmol-91 followed by cross Moomal-2002 x Anmol-91 (11.83cm) whereas WL-711 x Anmol-91 (22.53) had maximum spikelets spike^{-1}, followed by Moomal-2002 x Anmol-91(22.17).

Heritability:

During the course of breeding for improvements of any particular trait must be practical for effective change. The success of selection is governed by the degree to which the derived character is transmitted to the off-springs of the selected parents which is called heritability. The heritability and genetic advance of important traits are presented in table 3.

Days to 75% heading:

Table 3 revealed that crosses showed wide range of phenotypic variance from (21.98 to 83.33) and showed low, moderate and high heritability for this trait. The highest heritability (83.33%) exhibited by cross Marvi x Noori followed by cross Sassui x Anmol-91 (57.14%). The lowest heritability (17.21%) showed by cross Marvi x Sassui. The cross showed highest heritability showed high genetic advance (1.717). Whereas low heritability cross showed the lowest genetic advance (0.355).

Days to 75% maturity:

Regarding days to 75% maturity, this trait showed the wide range of genotypic, environmental and phenotypic variances (0.23 to 39.73). The heritability observed from high, moderate and low. The highest heritability (98.70%) exhibited by cross Sassui x Johar followed heritability (93.77%) for the cross Abadgar-93 x SKD-1.The high genetic advance showed by the cross. Sassui x Johar (2.033). The lowest heritability showed by cross Marvi x Noori which is (0.82 %) with the lowest genetic advance (0.017) respectively.

Plant height (cm):

The genotypic, environmental and phenotypic variance ranged from 7.73 to 67.725. The highest heritability 81.06% exhibited by cross Sassui x Anmol-91 coupled with high genetic advance (1.670) followed by the cross WL-711 x Anmol-91 (69.27%). The lowest heritability 17.59% showed by cross Abadgar-93 x SKD-1 alongside with lowest genetic advance (0.362).

Spike length (cm)

Spike length showed high, moderate and low genotypic, environmental and phenotypic variance. The highest heritability (95.39%) shown by cross WL-711x Anmol-91 along with high genetic advance (1.965) followed by cross Marvi x Noori (79.70%). Lowest heritability 8.14% was shown by cross Moomal-2002 x Anmol-91 together with the minimum genetic advance (0.168).

Spikelets spike^{-1}

High, moderate and low heritability observed for the trait spikelets spike^{-1}. The genotypic, environmental and phenotypic variance ranges from (0.9 to 8.805). The high heritability (53.35%) shown by cross Moomal-2002 x Anmol-91 coupled with moderate genetic advance (1.099).Wheras the minimum heritability 16.37 % was shown by cross Abadgar-93 x SKD-1 with low genetic advance (0.337).

Table 1: Mean Squares from analysis of variance for grain yield and its components of wheat

Source of variation	DF	Days to 75% heading	Days to 75% maturity	Plant height	Spike length	Spikelets per spike
Replication	2	3.784	0.255	3.153	0.034	0.512
Genotype	16	125.294**	88.230**	165.603**	6.478**	15.853**
Parents	8	126.620**	52.750**	147.628**	3.691**	3.426*
Crosses	7	94.190**	141.214**	195.138**	4.343**	7.315**
Error	32	6.680	0.734	4.483	0.517	1.303
Total	65					

**= Significant at 0.01 level of probability

Table 2: Mean performance of nine wheat varieties and their eight F_2 population for grain yield and its contributing characters

Genotypes	Days to 75% heading	Days to 75% maturity	Plant height (cm)	Spike length (cm)	Spikelets per spike
Sassui	72.67 fg	129.0 g	79.03 f	10.30 c d	17.60 fg
Anmol-91	62.00 j	122.3 hi	80.37 ef	8.233 h	16.60 gh
Wl-711	75.00 ef	131.0 e	74.10 h	8.833 fgh	16.40 h
Johar	77.00 de	130.7 ef	93.00 a	8.700 gh	17.40 fgh
Abadgar-93	81.67 b	138.7 b	85.00 c d	9.733 de	18.87 de
Skd-1	62.33 j	130.7 ef	68.47 i	8.233 h	16.37 h
Marvi	74.67 ef	132.7 d	74.73 gh	11.57 b	17.90 ef
Noori	71.33 gh	131.3 e	81.67 e	9.067 efg	17.73 fg
Moomal-2002	69.00 hi	130.7 ef	80.87 ef	8.500 gh	15.27 i
Sassui X Anmol-91	73.33 fg	121.7 i	66.40 j	13.43 a	21.77 abc
WL-711 X Anmol-91	69.00 hi	128.3 g	80.33 ef	10.83 c	22.53 a
Sassui X Johar	78.00 cd	137.0 c	90.70 b	10.63 c	21.13 bc
Abadgar-93 X SKD-1	80.33 bc	139.7 a	86.13 c	10.23 cd	19.53 d
Sassui X Abadgar-93	84.67 a	138.3 b	86.37 c	10.70 c	21.17 bc
Marvi X Noori	67.33 i	130.0 f	76.50 g	9.467 ef	17.73 fg
Moomal-2002 X Anmol-91	67.33 i	122.7 h	83.83 d	11.83 b	22.17 ab
Marvi X Sassui	77.67 d	131.0 e	90.13 b	11.63 b	20.67 c

The values with similar alphabets are statistically similar

Table 3: Heritability estimates and genetic advance of eight F_2 population for important traits

Genotypes	Genotypic Variance	Environmental Variance	Phenotypic Variance	Heritability %	Genetic advance
Days to 75% heading					
Sassui x Anmol- 91	0.32	0.24	0.56	57.14	1.177
WL 711 x Anmol-91	0.4	1.42	1.82	21.97	0.453
Sassui x Johar	0.23	0.47	0.7	32.85	0.677
Abadgar- 93 x SKD-1	0.23	0.255	0.485	47.42	0.977
Sassui x Abadgar-93	0.32	0.96	1.28	25	0.515
Marvi x Noori	3.6	0.72	4.32	83.33	1.717
Moomal- 2002 x Anmol-91	0.27	0.89	1.16	23.27	0.479
Marvi x Sassui	0.26	1.25	1.51	17.21	0.355
Days to 75% maturity					
Sassui x Anmol- 91	0.54	1.045	1.585	34.06	0.702
WL 711 x Anmol-91	0.62	0.47	1.11	55.85	1.151
Sassui x Johar	29.66	0.395	30.05	98.7	2.033
Abadgar-93 x SKD-1	2.71	0.18	2.89	93.77	1.932
Sassui x Abadgar- 93	0.23	0.36	0.59	38.98	0.803
Marvi x Noori	0.32	38.99	38.98	0.82	0.017
Moomal-2002 x Anmol- 91	0.54	0.485	1.025	52.68	1.085
Marvi x Sassui	0.76	38.97	39.73	1.91	0.039
Plant height (cm)					
Sassui x Anmol- 91	54.9	12.825	67.725	81.06	1.67
WL 711 x Anmol-91	19.78	8.775	28.555	69.27	1.427
Sassui x Johar	29.37	16.825	46.195	63.57	1.31
Abadgar -93 x SKD-1	7.73	36.215	43.945	17.59	0.362
Sassui x Abadgar-93	11.78	33.935	45.715	25.76	0.531
Marvi x Noori	45.34	21.935	67.275	67.39	1.388
Moomal-2002 x Anmol- 91	10.71	13.625	24.335	44.01	0.907
Marvi x Sassui	20.98	25.52	46.5	45.11	0.929
Spike length (cm)					
Sassui x Anmol- 91	2.04	2.33	4.37	46.68	0.962
WL 711 x Anmol-91	19.78	0.955	20.735	95.39	1.965
Sassui x Johar	0.26	2.485	2.745	9.47	0.195
Abadgar 93 x SKD-1	0.48	1.02	1.5	32	0.659
Sassui x Abadgar-93	0.3	2.51	2.81	10.67	0.22
Marvi x Noori	5.95	1.515	7.465	79.7	1.642
Moomal-2002 x Anmol-91	0.48	5.415	5.895	8.14	0.168
Marvi x Sassui	1.38	3.485	4.865	28.36	0.584
Spikelets per spike					
Sassui x Anmol-91	3.6	3.285	6.885	52.28	1.077
WL 711 x Anmol-91	2.5	3.67	6.17	40.51	0.835
Sassui x Johar	1.87	2.66	4.53	41.28	0.85
Abadgar- 93 x SKD-1	1.15	5.875	7.025	16.37	0.337
Sassui x Abadgar-93	0.9	3.58	4.48	20.08	0.414
Marvi x Noori	2.05	5.05	7.1	28.87	0.595
Moomal-2002 x Anmol -91	2.9	2.535	5.435	53.35	1.099
Marvi x Sassui	3.56	5.245	8.805	40.43	0.833

DISCUSSION

Genetic improvement of any crop depends upon the trend and frequency of the genetic heritability and the degree of relationship of heritable and non-heritable variation between yield and its contributing traits. The analysis of variance displayed that genotypes are highly significant at 0.01 levels and displayed highest variation in F_2 population and their parents for the traits studied. The results regarding the heritability in broad sense for the character days to 75% maturity exhibited that cross WL-711 x Anmol-91 showed highest, heritability for days to 75 % heading whereas Sassui x Anmol-91 displayed highest heritability for days to 75% maturity and the highest heritability for plant height, displayed by Sassui x Anmol-91 .It showed that higher heritability for heading and particular maturity produced the shorter plant statures. The previous workers like Khilwat *et al.* (2014) concluded in their research that higher heritability for heading, maturity and plant height produced early matured and dwarf statured plants.

The highest heritability regarding spike length, showed by cross WL-711 x Anmol-91 which displayed moderate heritability for this character. As the spike length increases the biological yield plant^{-1} increases. Whereas Sultana and Malik, (2005) stated that spike length, biological yield and 1000-grain weight are the main contributes to the grain yield and may be kept as selection criteria for higher yield plant^{-1}.

For the character spikelets spike^{-1} cross Moomal-2002 x Anmol-91 showed moderate heritability. The other workers like Saifullah *et al.* (2009) got the similar results. Grain yield plant^{-1} showed low high heritability in the F_2 population, the highest heritability (91.94%) displayed by the cross Marvi x Abadgar-93 observed great variability in their material and suggest that improvement achieved through selection.

The cross Sassui x Anmol-91 displayed minimum days to maturity, shorter plant height and larger spike length compared to the rest of the genotypes the previous works like Shoukat *et al.* (2007) and Hussain *et al.* (2013) got highly significant differences among the genotypes which is similar results as obtained from this research, which may be due to similar type of experimental material and environmental.

The cross Moomal-2002 x Anmol-91 displayed minimum days to 75% heading and higher in grain yield spike^{-1} whereas crosses WL-711 x Anmol-91 and Sassui x Abadgar-93 showed higher spikelets spike^{-1}.

CONCLUSION

Cross Sassui x Anmol-91 perform early in days to 75 % maturity and produced larger spike length. Cross Moomal-2002 x Anmol-91 perform higher heritability coupled with genetic advance for days to 75 % heading and Spikilets spike^{-1} whereas cross WL-711 x Anmol-91 displayed higher heritability for spike length. Selection of the genotypes could be made on the longer spike length and crosses Sassui x Anmol-91, WL-711 x Anmol-91and Marvi x Sassui could be selected for further evaluation in future successive generations.

REFERENCES:

Ansari K. A., Soomro Z. A., B.A. Ansari and M. H.Leghari 2005 Genetic variability and heritability studies of some quantitative traits in bread wheat (*Triticum aestivum* L.). Pakistan J. Agri. Agril. Engg. Vet. Sci., 21(1):18-24.

Bharat B., S. S. Gaurav, Kumar, Pal, Panday, Kumar, Bharti, S. S. Nagar and VP Rahu.2013. Genetic variability, heritability and genetic advance in bread Wheat (*Triticum aestivum* L.). Environment and Ecology, 31 (2):405-407.

Hussain. M. A., H. S. Askandar and Z. A. Hassan.2013. Selecting high yielding wheat hybrids from a restricted factorial mating design. Sarhad J. Agric., 29(2): 173-179.

Khilwat A. Ahmad, Ishaq, I. A. Khalil, I. Shah, Saeed and N.Ahamd. 2014. Genetic potential and variability for morpho-yield traits in durum wheat (*Triticum turgidum* L. var. durum). Int. J. of Farming and Allied Sci., 3(12):1206-1212.

Saifullah A., N. Zakir and Mujahid.2009. Estimation of genetic parameters and character association in wheat. J. Agric. Biol. Sci., 1(1):15-18.

Shoukat A. M., M. A. Sial, B. A. Ansari and M. A. Arain. 2007. Study of genetic parameters in segregating populations of spring wheat. Pak. J. Bot., 39(7): 2407-2413.

Soshma J., M. Fida M. and F. U. Khan. 2015. Genetic potential and heritability estimates of yield traits in F_3 segregating populations of bread wheat. Int. J. Envir., 4(4):106-115.

Sultana, R. and S.K. Malik. 2005. Genetic variability and character association between yield and yield attributing traits in bread wheat (*Triticum aestivum* L.em. Thell). Annals of Agricultural Res., 26 (1):118-125.

Thebo S. K., Ansari B. A., L. A., Bhutto B., Munaiza and B. Sheereen 2005 Estimation of genetic variability in F3 generation of spring wheat. Indus J. Pl. Sci., 4(4):570-575.

Yousaf A. B. M. Atta, J. Akhter, P. Monneveux and Z. Lateef. 2008. Genetic variability, association and diversity studies in wheat (*Triticum aestivum* L.) germplasm. Pak. J. Bot., 40(5): 2087-2097.

EFFECT OF MICRONUTRIENT (ZINC) ON GROWTH AND YIELD OF MUSTARD VARIETIES

Hakim Ali Sahito*, Abdul Waheed Solangi, Abdul Ghani Lanjar, Abdul Hafeez Solangi and Sajjad Ali Khuhro
Department of Entomology, Sindh Agriculture University, Tandojam- Sindh.

ABSTRACT

Two varieties (Early Mustard and S-9) were evaluated against six Zn levels (0, 2, 4, 6, 8 and 10 kg Zn ha^{-1}). Significant improvements in the plant growth, seed yield and oil contents increased with increasing Zn levels. The results revealed that there was a significant improvement in the growth, seed yield and oil content with increasing Zn levels, irrespectively. The highest Zn level of 10 kg ha^{-1} resulted 216 cm plant height, 10.86 branches plant^{-1}, took 55.66 days to initiate flowering, 574.50 pods plant^{-1}, 17.61 g weight of seeds plant^{-1}, 3.63 g seed index, 2037.20 seed yield kg ha^{-1} and 36.80 percent oil as the highest output. In case of varieties, S-9 ranked 1st with 216.50 cm plant height, 10.84 branches plant^{-1}, took 56.33 days to initiate flowering, 581.11 pods plant^{-1}, 17.82 g weight of seeds plant^{-1}, 3.66 g seed index, 1960.30 seed yield kg ha^{-1} and 36.80 percent oil content; while variety Early Mustard resulted 186.56 cm plant height, 9.25 branches plant^{-1}, took 52.72 days to initiate flowering, 484.67 pods plant^{-1}, 14.50 g weight of seeds plant^{-1}, 2.90 g seed index, 1677.90 seed yield kg ha^{-1} and 35.13 percent oil content. It is suggested that for achieving economically higher seed yields in mustard, the Zn application to mustard may be done at the rate of 8 kg ha^{-1}. Moreover, variety S-9 may preferably grow for obtaining higher seed and oil contact yields.

Keywords: Zinc; growth; yield and mustard

INTRODUCTION

Mustard, *Brassica juncea* L., is an important member of family Brassicaceae (Bayer, 2010). Oilseeds contribute the second highest share in the economy of Pakistan after grain crops. There are number of species producing edible oil mainly the species and sub-species in the Brassicaceae family. The most popular species are *B. juncea, B. napus, B. oleracea* and *B. rapa* (*B. campestris*). The scientists have sequenced the entire genome of rapeseed/canola (*B. napus*) and its constituent genomes present in *Brassica rapa* and *Brassica oleracea* in 2009. This also represents the A genome component of the amphidiploid crop species *B. napus* and *B. juncea* (Bayer, 2010). Mustard has primary center of its origin in central Asia (northwest India), with secondary centers in central and western China, eastern India, Burma and through Iran to Near East cultivated for centuries in many parts of Eurasia. However, the principal growing countries are Bangladesh, Central Africa, China, India, Japan, Nepal, and Pakistan, as well as southern Russia in north of the Caspian Sea (Perry, 1999).

The total availability of edible oil in 2009-10 was 2.9 million tons. Local production of edible oil was 662 thousand tons which accounted for 23 percent of total availability in the country, while the remaining 77 percent availability was ensured through imports. During the year 2010-11 (July-March), a quantity of 1.7 million tons edible oil/oilseeds worth US$ 1.65 billion has been imported. The local production in 2010-11 is provisionally estimated at 696 thousand tons. Total availability from all sources is thus reduced to 2.35 million tons so far. The area under rapeseed and mustard during 2009-10 was 178,000 hectares with production of 151,000 tons, while during 2010-11 the area increased to 194,000 hectares and production to 168,000 tons showing an increase of 11.3 percent. During 2010-11, the oil from the rapeseed and mustard was 50,000 tons (GoP, 2011). Similarly, the area under mustard cultivation in Sindh province during the year 2008-2009 was 45,412 hectares with production of 48,478 tons, showing seed yield of 1.067 tons ha^{-1}. Mustard is often used at the table as a condiment on meat (Sawyer, 1990). Due to continuous cropping and non-practice of green manuring

**Corresponding author:* e-mail: hakimsahito@gmail.com

or non-use of farm yard manure, the soils have become deficient in various macro and micronutrients. Therefore, the major soil fertility problems in our soils are deficiencies of macro and some micronutrients (Rashid *et al.*, 1993).

Zinc is one of the first micronutrients recognized as essential for plants that transported to plant root surface through diffusion (Maqsood *et al.*, 2009). Zn is a micronutrient and in case of its severe deficiency the symptoms may last throughout the entire crop season (Asad and Rafique, 2000). Zn deficient plant also appears to be stunted (Torun *et al.*, 2001) as a result approximately 2 billion people suffer from Zn deficiency all over the world (Asad and Rafique, 2002). The grain yield can be improved by addition of Zn fertilization (Maqsood *et al.*, 2009). Bora and Hazarika, (1997) reported highest stover yield (2770 kg ha^{-1}) with Zn and almost the same trend of seed yield. The seed yield can be improved by addition of Zn fertilization. Chen and Aviad, (1990) found that application of Zn alongwith other micronutrients improved soil organic matter and resulted in increasing mustard yields. Kutuk *et al.*, (2000) also suggested that the application of Zn has become necessary for improved crop yields. Mandal and Sinha, (2004) recommended application of $ZnSO_4$ at the rate of 20 kg ha-1 for oilseeds including mustard. Moniruzzaman *et al.*, (2008) applied zinc at the concentrations of 0, 2.5, 5.0 and 7.5 kg ha^{-1} and suggested 8 kg Zn ha^{-1} for brassica species. In view of the significance of zinc in crop production process, the experiment was conducted to investigate the effect of micronutrient (zinc) on growth and yield of mustard varieties at Tandojam.

MATERIALS AND METHODS

The study was carried out to investigate the effect of micronutrient (zinc) on growth and yield of mustard varieties at the experimental fields of physiology Section, Agriculture Research Institute, Tandojam in a three replicated Randomized Complete Block Design (Factorial) having plot size of 3 x 5m ($15m^2$). The land was prepared by giving two dry plowings, followed by land leveling. After soaking dose and when land came in condition, the land was plowed with crosswise cultivator, followed rotavator and planking for a good seed bed preparation. The sowing of the seed was done by hand drilling method. The treatments details are as under:

Treatments: A Zinc levels = 6
Z_1=0 kg ha^{-1}, Z_2=2 kg ha^{-1}, Z_3=4 kg ha^{-1}, Z_4=6 kg ha^{-1}, Z_5=8 kg ha^{-1}, Z_6=10 kg ha^{-1}

Treatments: B Varieties = 2
V1= Early Mustard and V2= S-9.

The zinc was applied in the form of $ZnSO_4$ after sowing at first irrigation with the use of N (@ 100 kg ha^{-1}) and P (@ 50 kg ha^{-1}). All P along with 1/3rd of N was applied at the time of sowing and remaining two N splits were applied at flowering and pod formation stages, respectively. For recording observations on various agronomical traits, five plants in each plot were selected at random and labeled. After completion of observations on growth parameters, and when crop matured, the labeled plants were harvested manually and tied in small bundles, and were shifted to threshing yard. Threshing was performed manually to record observations on the following traits:

Observations recorded
1. Soil analysis (pH, EC, Organic Matter), **2.** Plant height (cm), **3.** Number of branches $plant^{-1}$, **3.** Days to flowering, **4.** Number of pods $plant^{-1}$, **5.** Weight of seeds $plant^{-1}$ (g), **6.** Seed index (1000 seed weight, g), **7.** Seed yield (kg ha^{-1}) and **8.** Oil content (%).

Methodology for recording observations
Electrical conductivity: 1:5 soil water extract by using conductivity digital meter model Hi-8333 as dSm^{-1}.

Soil reaction (pH): 1:5 soil water extract by using Suntex pH meter (Model SP-34).

Organic matter: By Walkely-Black method "Soil and Chemical Analysis" By M.L. Jackson (1959).

Plant height (cm): The plant height (cm) was measured at maturity of the crop by measuring tape from bottom to tip of plant in labelled plants in each treatment.

Number of branches $plant^{-1}$: The number of branches $plant^{-1}$ was counted at maturity of the crop in labelled plants in each treatment and mean was worked out.

Days to flowering: The number of days from sowing to initiation of flowering was counted in each treatment and mean was worked out.

Number of pods plant^{-1}: The number of pods plant^{-1} was counted at maturity of the crop in labelled plants in each treatment.

Weight of seed plant^{-1}: The total seeds of labelled plants were weighed and divided with total number of plants examined to obtain weight of seeds plant^{-1} in each treatment.

Seed index (1000 seeds, g): Seed index value was observed on the basis of 100 seeds manually separated by counting in each treatment and was weighed to record seed index in grams.

Seed yield (kg ha^{-1}): The seed yield (kg ha^{-1}) was worked out by using the following formula:

$$\text{Seed yield (kgha}^{-1}\text{)} = \frac{\text{Yield plot}^{-1}\text{ of given treatment (kg)}}{\text{Plot area (m}^2\text{)}} \text{ X } 10000$$

Oil content (%): The extracted oil was weighed from a certain quantity of seed and percentage of the total weight was calculated.

STATISTICAL ANALYSIS

The collected data were subjected to analysis of variance using the statistical software Statix-8.1. The difference among the treatment means were compared by the least significant difference (LSD) test at P=0.05 level (Steel and Torrie, 1984).

RESULTS

Soil analysis

The soil analysis (Table-1) indicates that before sowing of mustard the soil pH was 7.8 which decreased to 7.2 when soil samples were determined after harvesting of the crop. The EC before sowing of mustard was 4.52 dSm^{-1} which slightly decreased to 4.34 dSm^{-1} while, the soil organic matter before sowing of the experimental crop was 0.63 percent which increased to 0.84 percent and soil samples were determined after harvest of the crop. This indicates that the application of Zn reduced soil pH and soil EC, while considerably improved the soil organic matter.

Table 1. Soil analysis before sowing and after harvesting

Soil properties	Before sowing	After harvest
pH	7.8	7.2
EC (dSm^{-1})	4.52	4.34
Organic matter (%)	0.63	0.84

Plant height (cm)

The results presented in Table- 2 about the analysis of variance indicated that the plant height of mustard was significantly (P<0.05) affected by Zn levels and varieties, while the effect of their interaction on plant height was non-significant (P>0.05). The application of Zn at the rate of 10 kg ha^{-1} resulted in maximum plant height of 216 cm, followed by Zn application at the rates of 8 kg and 6 kg ha^{-1}, resulting average plant height of 212.50 cm and 206.33 cm, respectively. The plant height was significantly declined to 197.33 and 191.50 cm when the Zn levels were reduced to 4 kg and 2 kg ha^{-1}, respectively. The lowest plant height (185.50 cm) was recorded in control plots (without Zn). This indicates that application of Zn had marked effect on plant height of mustard. However, the increase in plant height was not statistically positive when Zn was applied beyond 6 kg ha^{-1}, because the differences in plant height under Zn levels of 6, 8 and 10 kg ha^{-1} were non-significant (P>0.05). In case of varieties, the plants of S-9 were found to be genetically tall growing (216.50 cm) more than those for variety Early Mustard (186.56 cm). The interactive effect of Zn levels and varieties indicates that mustard variety S-9 when given 10 kg Zn ha^{-1} resulted in significantly maximum plant height (232 cm), while the variety Early Mustard resulted in minimum plant height (171.67 cm) when given no Zn (control). Both the mustard varieties were found to be responsive to Zn levels, but S-9 showed more promising results than Early Mustard for plant height.

Table 2. Effect of zinc (Zn) on the plant height (cm) on different varieties of mustard

Treatments (Zinc levels)	Varieties		Mean for Zn levels
	Early Mustard	S-9	
Zn @ 0 kg ha^{-1} (control)	171.67	199.33	185.50 c
Zn @ 2 kg ha^{-1}	177.33	205.67	191.50 b
Zn @ 4 kg ha^{-1}	182.67	212.00	197.33 b
Zn @ 6 kg ha^{-1}	191.00	221.67	206.33 a
Zn @ 8 kg ha^{-1}	196.67	228.33	212.50 a
Zn @ 10 kg ha^{-1}	200.00	232.00	216.00 a
Mean for varieties	186.56 b	216.50 a	-

Number of branches $plant^{-1}$

The data in relation to the number of branches $plant^{-1}$ of two mustard varieties as affected by different Zinc (Zn) levels are shown in Table-3. The analysis of variance suggested that the number of branches $plant^{-1}$ of mustard was significantly (P<0.05) influenced by Zn levels and varieties, while the effect of their interaction was non-significant (P>0.05). The results indicated that the application of Zn at the rate of 10 kg ha^{-1} resulted in maximum number of branches (10.86) $plant^{-1}$, followed by Zn application at the rates of 8 kg and 6 kg ha^{-1}, resulting 10.62 and 10.31 average number of branches $plant^{-1}$, respectively. The number of branches diminished to 9.86 and 9.62 $plant^{-1}$ when Zn was applied at the lower rates of 4 kg and 2 kg ha^{-1}, respectively. However, the minimum number of branches (9.00) $plant^{-1}$ was observed in control plots, where Zn was not applied. Among varieties, the number of branches $plant^{-1}$ was significantly higher (10.84) in variety S-9 as compared to variety Early Mustard (9.25). The interactive effect of Zn levels and varieties showed that mustard variety S-9 when fertilized with 10 kg Zn ha^{-1} resulted in significantly maximum branches (11.60) $plant^{-1}$, while the variety Early Mustard resulted in minimum number of branches (8.04) $plant^{-1}$ under control (no Zn). The application of Zn had positive impact on the number of branches $plant^{-1}$ of mustard; but Zn application beyond 6 kg ha^{-1} did not show economical impact on the number of branches $plant^{-1}$, because the differences in the number of branches under 6, 8 and 10 kg ha^{-1} Zn levels were non-significant (P>0.05).

Table- 3. Effect of zinc (Zn) on the number of branches $plant^{-1}$

Treatments (Zinc levels)	Varieties		Mean for Zn levels
	Early Mustard	S-9	
Zn @ 0 kg ha^{-1} (control)	8.04	9.95	9.00 c
Zn @ 2 kg ha^{-1}	8.86	10.98	9.62 b
Zn @ 4 kg ha^{-1}	9.13	10.59	9.86 b
Zn @ 6 kg ha^{-1}	9.55	11.07	10.31 a
Zn @ 8 kg ha^{-1}	9.83	11.40	10.62 a
Zn @ 10 kg ha^{-1}	10.11	11.60	10.86 a
Mean for varieties	9.25 b	10.84 a	-

Days to flowering

The results pertaining to days to flowering of mustard varieties as influenced by different Zinc (Zn) levels are presented in Table- 4. The analysis of variance suggested that the days to flowering of mustard was significantly (P<0.05) influenced by Zn levels and varieties, while the effect of their interaction was non-significant (P>0.05) on this parameter. It is evident from the results that the crop supplied with Zn at the rates of 10 kg ha^{-1} and 8 kg ha^{-1} took equally 55.66 days to flowering, followed by Zn application at the rates of 6 kg and 6 kg ha^{-1}, where the crop took 55.50 days to flowering. The mustard crop initiated flowering little earlier i.e. in 54.16 and 53.50 days when the crop supplied with Zn at the lower rates of 4 kg and 2 kg ha^{-1}, respectively. However, the minimum number of days (52.66) to flowering was noted in plots given no Zn application (control). In varieties, variety S-9 took significantly greater number of days (56.33) to flowering as compared to variety Early Mustard (52.72 days). The interactive effect of Zn levels and varieties showed that mustard variety S-9 given 10 kg or 8 kg Zn ha^{-1} initiated flowering equally in maximum number of days (57.33), while the variety Early Mustard initiated flowering earliest in 51 days when sown without Zn application. This indicates that irrespective of varieties, the application of Zn delayed the flowering, but this delay in flowering was Zn dose specific. The LSD test suggested that the differences in the days to flowering of mustard was non-significant (P>0.05) between Zn levels of 6, 8 and 10 kg ha^{-1}. However, Early Mustard was found to be genetically early flowering initiation when compared with variety S-9.

Table- 4. Effect of zinc (Zn) on the number of days to flowering

Treatments (Zinc levels)	Varieties		Mean for Zn levels
	Early Mustard	S-9	
Zn @ 0 kg ha^{-1} (control)	51.00	54.33	52.66 c
Zn @ 2 kg ha^{-1}	51.66	55.33	53.50 b
Zn @ 4 kg ha^{-1}	52.33	56.00	54.16 b
Zn @ 6 kg ha^{-1}	53.33	57.66	55.50 a
Zn @ 8 kg ha^{-1}	54.00	57.33	55.66 a
Zn @ 10 kg ha^{-1}	54.00	57.33	55.66 a
Mean for varieties	52.42 b	56.33 a	-

Number of pods plant^{-1}

The data in regards to number of pods plant^{-1} of mustard varieties as affected by different Zinc (Zn) levels are given in Table-5. The analysis of variance demonstrated that the number of pods plant^{-1} of mustard was significantly (P<0.05) affected by Zn levels and varieties, while the effect of their interaction was non-significant (P>0.05). It is apparent from the data (Table-5) that the application of Zn at the rate of 10 kg ha^{-1} produced maximum number of pods (574.50) plant^{-1}, followed by Zn application at the rates of 8 kg and 6 kg ha^{-1}, producing 569.67 and 548.67 average number of pods plant^{-1}, respectively. The number of pods reduced to 516.67 and 510.67 plant^{-1} when Zn was applied at the lower rates of 4 kg and 2 kg ha^{-1}, respectively. However, the lowest number of pods (477.17) plant^{-1} was noted in control, where no Zn was applied. Among varieties, the number of pods plant^{-1} was significantly higher (581.11) in variety S-9 as compared to variety Early Mustard (484.67). The interactive effect of Zn levels and varieties indicated that highest number of pods (621.33) plant^{-1} was recorded under interaction of S-9 x 10 kg Zn ha^{-1}, while the lowest number of pods (419.33) plant^{-1} under interaction of Early Mustard x 0 kg Zn ha^{-1}. The application of Zn had positive impact on the number of pods plant^{-1}; but Zn application beyond 6 kg ha^{-1} did not show economical impact on this trait, because the differences in pods under 6, 8 and 10 kg ha^{-1} Zn levels were non-significant (P>0.05). The pod bearing capacity was found to be genetically higher in variety S-9 as compared to variety Early Mustard.

Table- 5. Effect of zinc (Zn) on the number of pods plant^{-1}

Treatments (Zinc levels)	Varieties		Mean for Zn levels
	Early Mustard	S-9	
Zn @ 0 kg ha^{-1} (control)	419.33	535.00	477.17 c
Zn @ 2 kg ha^{-1}	454.00	567.33	510.67 b
Zn @ 4 kg ha^{-1}	470.33	563.00	516.67 b
Zn @ 6 kg ha^{-1}	503.33	594.00	548.67 a
Zn @ 8 kg ha^{-1}	533.33	606.00	569.67 a
Zn @ 10 kg ha^{-1}	527.67	621.33	574.50 a
Mean for varieties	484.67 b	581.11 a	-

Weight of seeds plant^{-1}

The results in relation to weight of seeds plant^{-1} of mustard varieties as affected by different Zinc (Zn) levels are presented in Table-6. The analysis of variance suggested that the weight of seeds plant^{-1} of mustard was significantly (P<0.05) affected by Zn levels and varieties, while the effect of their interaction was non-significant (P>0.05). The application of Zn at the highest rate of 10 kg ha^{-1} produced maximum weight of seeds (17.61 g) plant^{-1}, closely followed by Zn application at the rates of 8 kg and 6 kg ha^{-1}, producing average seed weight of 17.46 g and 16.84 g plant^{-1}, respectively. The weight of seeds decreased markedly to 15.70 g and 15.52 g plant^{-1} when Zn was applied at the lower rates of 4 kg and 2 kg ha^{1}, respectively. However, the minimum weight of seeds (13.82 g) plant^{-1} was obtained in control, where no Zn was applied. Among varieties, the weight of seeds plant^{-1} was significantly higher (17.82 g) in variety S-9 as compared to variety Early Mustard (14.50 g). The interactive effect of Zn levels and varieties showed that highest weight of seeds (19.02 g) plant^{-1} was recorded under interaction of S-9 x 10 kg Zn ha^{-1} while the lowest weight of seeds (11.38 g) plant^{-1} under interaction of Early Mustard x 0 kg Zn ha^{-1}. The increasing Zn application resulted in a positive effect on the weight of seeds plant^{-1}; but Zn application beyond 6 kg ha^{-1} did not show economical impact on this parameter, because the differences in weight of seeds plant^{-1} in plots supplied with 6, 8 and 10 kg ha^{-1} Zn were non-significant (P>0.05). Moreover, the seeds were found to be heavier in variety S-9 as compared to Early Mustard, because regardless the Zn levels, S-9 showed its superiority in seed weight plant^{-1}.

Table- 6. Effect of (Zn) on the weight of seeds (g) plant^{-1}

Treatments (Zinc levels)	Varieties		Mean for Zn levels
	Early Mustard	S-9	
Zn @ 0 kg ha^{-1} (control)	11.38	16.26	13.82 c
Zn @ 2 kg ha^{-1}	13.80	17.25	15.52 b
Zn @ 4 kg ha^{-1}	14.28	17.11	15.70 b
Zn @ 6 kg ha^{-1}	15.30	19.39	16.84 a
Zn @ 8 kg ha^{-1}	16.03	18.89	17.46 a
Zn @ 10 kg ha^{-1}	16.21	19.02	17.61 a
Mean for varieties	14.50 b	17.82 a	-

Seed index

The data regarding the seed index (1000 seeds weight) of mustard varieties as affected by various Zinc (Zn) levels are shown in Table-7. The analysis of variance indicated that the seed index of mustard was significantly (P<0.05) influenced by Zn levels and varieties, while the influence of their interaction was non-significant (P>0.05) on seed index. The data indicates that the application of highest Zn level of 10 kg ha^{-1} resulted in maximum seed index (3.63 g), closely followed by Zn levels of 8 kg and 6 kg ha^{-1}, resulting average seed index of 3.56 g and 3.49 g, respectively. The seed index value declined to 3.14 g and 3.10 g when Zn was applied at the lower rates of 4 kg and 2 kg ha^{-1}, respectively. However, the lowest seed index value (2.76 g) was noted in control, where no Zn was applied. Among varieties, the seed index was significantly higher (3.66 g) in variety S-9 as compared to variety Early Mustard (2.90 g). The interactive effect of Zn levels and varieties indicated that highest seed index value (4.02 g) was recorded under interaction of S-9 x 10 kg Zn ha^{-1} while the lowest seed index (2.28 g) under interaction of Early Mustard x 0 kg Zn ha^{-1}. The increasing Zn application resulted in an increased seed index in mustard; but Zn application beyond 6 kg ha^{-1} remained uneconomical for this trait, because the differences in seed index in plots supplied with 6, 8 and 10 kg ha^{-1} Zn were non-significant ($P>0.05$) statistically. Moreover, the seed index represents the seed quality; thus variety S-9 supposed to have better quality seed as compared to Early Mustard on the basis of their seed index values.

Table- 7. Effect of zinc (Zn) on the seed index (1000 seeds weight g)

Treatments (Zinc levels)	Varieties		Mean for Zn levels
	Early Mustard	S-9	
Zn @ 0 kg ha^{-1} (control)	2.28	3.25	2.76 c
Zn @ 2 kg ha^{-1}	2.76	3.45	3.10 b
Zn @ 4 kg ha^{-1}	2.86	3.42	3.14 b
Zn @ 6 kg ha^{-1}	3.06	3.93	3.49 a
Zn @ 8 kg ha^{-1}	3.20	3.92	3.56 a
Zn @ 10 kg ha^{-1}	3.24	4.02	3.63 a
Mean for varieties	2.90 b	3.66 a	-

Seed yield (kg ha^{-1})

The results pertaining to seed yield kg ha^{-1} of mustard varieties as affected by different Zinc (Zn) levels are reported in Table-8. The analysis of variance indicated that the seed yield kg ha^{-1} of mustard was significantly ($P<0.05$) affected by Zn levels and varieties, while the effect of their interaction was non-significant ($P>0.05$) on seed yield kg ha^{-1}. It can be seen from the results (Table-8) that Zn application at the highest level of 10 kg ha^{-1} produced maximum seed yield (2037.20 kg ha^{-1}), closely followed by Zn levels of 8 kg and 6 kg ha^{-1}, producing average seed yields of 1936.50 kg and 1883.70 kg ha^{-1}, respectively. The seed yield declined to 1746 kg and 1701.30 kg ha^{-1} when Zn was applied at the lower rates of 4 kg and 2 kg ha^{-1}, respectively. However, the lowest seed yield kg (1610.20 kg ha^{-1}) was recorded in control, where no Zn was applied. In case of varieties, the seed yield kg ha^{-1} was significantly higher (1960.30 kg) in variety S-9 as compared to variety Early Mustard (1677.90 kg ha^{-1}). The interactive effect of Zn levels and varieties indicated that highest seed yield (2092.30 kg ha^{-1}) was achieved under interaction of S-9 x 10 kg Zn ha^{-1} while the lowest seed yield (1431.70 kg ha^{-1}) under interaction of Early Mustard x 0 kg Zn ha^{-1}. There was a simultaneous increase in seed yield ha^{-1} with increasing Zn levels, but the seed yield did not improve significantly when Zn was applied beyond 8 kg ha^{-1}, because the differences in seed yield ha^{-1} were statistically non-significant ($P>0.05$) when Zn levels of 8 kg and 10 kg ha^{-1} were compared. It was also observed that genetically, variety S-9 showed its superiority over Early Mustard irrespective of the Zn levels.

Table- 8. Effect of zinc (Zn) on the seed yield (kg ha^{-1})

Treatments (Zinc levels)	Varieties		Mean for Zn levels
	Early Mustard	S-9	
Zn @ 0 kg ha^{-1} (control)	1431.70	1788.70	1610.20 d
Zn @ 2 kg ha^{-1}	1505.30	1897.30	1701.30 c
Zn @ 4 kg ha^{-1}	1609.30	1882.70	1746.00 c
Zn @ 6 kg ha^{-1}	1744.30	2023.00	1883.70 b
Zn @ 8 kg ha^{-1}	1795.00	2078.00	1936.50 a`
Zn @ 10 kg ha^{-1}	1982.00	2092.30	2037.20 a
Mean for varieties	1677.90 b	1960.30 a	-

Oil content (%)

The data in regards to oil content of mustard varieties as influenced by different Zinc (Zn) levels are shown in Table-9. The analysis of variance suggested that the oil content of mustard was significantly ($P<0.05$) affected by Zn levels and varieties, while the effect of interaction between Zn levels and mustard varieties was non-significant ($P>0.05$) on oil content. It is evident from the results (Table-9) that maximum oil content (36.80%) in mustard was achieved from the plots supplied with highest Zn level of 10 kg ha^{-1}, followed by Zn levels of 8 kg and 6 kg ha^{-1} where the average oil content was 35.74 and 35.85 percent, respectively. The oil content was 35.31 and 35.44 percent when the mustard was supplied with Zn levels of 4 kg and 2 kg ha^{-1}, respectively. The lowest oil content (34.97%) was recorded in control plots (without Zn). This indicates that application of Zn had positive impact on oil content of mustard. In case of varieties, the oil content was significantly higher (36.24%) in seeds of mustard variety S-9 as compared to variety Early Mustard (35.13 %). The interactive effect of Zn levels and varieties indicates that mustard variety S-9 when given 10 kg Zn ha^{-1} resulted in significantly maximum oil content (37.48%), while the variety Early Mustard receiving zero Zn (control) resulted in minimum oil content (34.45 %). Both the mustard varieties were found to be responsive to Zn levels for oil content, but S-9 showed more promising results than Early Mustard for this character.

Table- 9. Effect of zinc (Zn) on the oil content (%)

Treatments (Zinc levels)	Varieties		Mean for Zn levels
	Early Mustard	S-9	
Zn @ 0 kg ha^{-1} (control)	34.45	35.49	34.94 c
Zn @ 2 kg ha^{-1}	34.91	35.97	35.44 b
Zn @ 4 kg ha^{-1}	34.81	35.84	35.31 b
Zn @ 6 kg ha^{-1}	35.32	36.38	35.85 b
Zn @ 8 kg ha^{-1}	35.21	36.27	35.74 b
Zn @ 10 kg ha^{-1}	36.12	37.48	36.80 a
Mean for varieties	35.13 b	36.24 a	-

DISCUSSION

The study showed that Zn level resulted to increase the plant height, branches per plant, that took less days to reach at flowering stage, pods per plant, weight of seeds per plant, seed index, seed yield and oil percent level. The results further are in agreed with the soils have become deficient of micronutrients, particularly Zinc deficiency has resulted significant reduction in crop yields (Rashid *et al.*, 1993; Grewal *et al.*, 1998). These results are partially supported by Maqsood *et al.*, (2009) who reported that the seed yield can be improved by addition of Zn fertilization; and Chen and Aviad (1990) found that application of Zn alongwith other micronutrients improved soil organic matter and resulted in increasing mustard yields. Kutuk *et al.*, (2000) also suggested that the application of Zn has become necessary for improved crop yields. Moniruzzaman *et al.*, (2008) applied zinc at the concentrations of 0, 2.5, 5.0 and 7.5 kg ha^{-1} and suggested 8 kg Zn ha^{-1} for brassica species. Mandal and Sinha, (2004) recommended application of $ZnSO_4$ at the rate of 20 kg ha-1 for oilseeds including mustard.

Moreover, the variety S-9 may preferably grow for obtaining higher seed and oil yields in mustard so far. These results are categorically supported by many researchers, who worked on the similar aspects of rape seed and mustard in different parts of the world. Chen and Aviad, (1990), Rashid *et al.*, (1994), Asad *et al.*, (1997) examined the effect of foliar applied Boron at concentrations from 0.04 to.0.3 percent on *B. napus* (L) found the higher seed and oil yield from mustard. Bora and Hazarika, (1997) carried out studies to investigate the effect of micronutrients on the growth, seed yield and stover yield of rape seed. The results showed that the highest stover yield (2770 kg ha^{-1}) was obtained with Zn + B + Mo and followed almost the same trend of seed yield. In a similar investigation, the study of Mandal and Sinha, (2004) examined the effect of Zn and B application in addition to NPK fertilizers on production and yield of Indian mustard. Plant height, number of branches per plant, number of siliqua per plant, number of seeds per siliquae, 1000-seed weight with 20 kg ha^{-1} $ZnSO_4$. Yang *et al.,* (2009) reported that the combined application of B with Zn resulted in higher seed yield than the application of B or Zn alone, and the seed yield of the B+Zn treatment was the highest in all treatments, 68.1% above the control. The above discussion on the comparative research results obtained in the present study and findings reviewed from various studies carried out in different parts of the world clearly suggested that addition of zinc along with recommended dose of NPK fertilizers have resulted improved mustard yields. The results indicated that the seed yield kg ha^{-1} was highest under 10 kg Zn ha^{-1}, but the differences in seed yield kg ha^{-1} were non-significant (P>0.05) when compared with those obtained under 8 kg Zn ha^{-1}. Hence, it was concluded that for achieving economically higher seed yields in mustard, the Zn application to mustard may be done at the rate of 8 kg ha^{-1}. Moreover, variety S-9 may preferably grow for obtaining higher seed and oil yields in mustard.

REFERENCES

Asad A, Rafique R, 2000. Effect of Zinc, Copper, Iron, Manganese and Boron on the yield and yield components of wheat crop in Tehsil Peshawar. Pakistan J. Biol. Sci., 3 (10):1815-1820.

Asad A, Rafique R, 2002. Identification of micronutrient deficiency of wheat in the Peshawar valley, Pakistan. Communications in Soil Science and Plant Analysis, 33 (3/4): 349-364.

Asad AR, Bell W, Dell B, Huang L, 1997. External Boron Requirements for canola (*brassica Napusl*) in Boron Buffered Solution Culture. Annals of Botany, 80 (5): 65-73.

Bayer, 2010. Bayer Sequence Genome of Canola. Bayer Crop Science: The Bio-energy Site, Retrieved 8 November, 2010.

Bora PC, Hazarika V, 1997. (*Brassica campestrissub* sp. oleifera var. toria). Indian J. Agron., 42: 361-364.

Chen Y, Aviad T, 1990. Use of humic acid for crop production. J. Am Soc. of Agronomy, 12 (3): 86-90.

GoP, 2011. Agriculture (Oilseeds): Economic survey of Pakistan, Ministry of Food and Agriculture, Federal Bureau of Statistics, Government of Pakistan, Islamabad, Pp. 21.

Grewal HS, Grahamb RD, Stangoulisb J, 1998. Zinc-boron interaction effects in oilseed rape. J. Plant Nutrition, 21 (10): 2231-2243.

Kutuk C, Gökhan C, Abdullah B, Başkan O, 2000. Effect of Humic Acid on some soil properties. Bildiri Özetleri, Soil Science Department, Agricultural Faculty, Ankara University, 06110- Ankara Turkey.

Mandal, Sinha AC, 2004. Nutrient Management Effects on Light Interception, Photosynthesis, Growth, Dry-matter Production and Yield of Indian Mustard (*Brassica juncea*). J. Agronomy and Crop Sci., 55 (4): 287-291.

Maqsood MA, Rahmatullah S, Kanwal T, Aziz, Ashraf M, 2009. Evaluation of Zn distribution among grain and straw of twelve indigenous wheat, *Triticum aestivum* L. genotypes. Pak. J. Bot., 41 (1): 225-231.

Moniruzzaman M, Mozumder SN, Islam MR, 2008. Effects of Sulfur, Boron, Zinc and Molybdenum on Yield and Profitability of Broccoli (*Brassica oleracea* L. Var. italica). J Agric Rural Dev., 6 (1&2): 55-61.

Perry LM, 1999. Medicinal Plants of East and Southeast Asia. MIT Press, Cambridge.

Rashid A, 1996. Monitoring of soil fertility in Pakistan. National Book of Soil Sci., 2 (1): 262-270.

Rashid AE, Rafique, Bughio N, 1994. Diagnosing boron deficiency in rapessed and mustard by plant analysis and soil testing communication in soil and plant analysis. 25 (17-18): 2883-2897.

Sawyer S, 1990. Gourmet mustards: how to make and cook with them. Culinary arts Ltd., Pp. 24.

Torun A, Gültekin I, Kalayci M, Yilmaz A, Eker S, 2001. Effects of zinc fertilization on grain yield and shoot concentrations of zinc, boron, and phosphorus of 25 wheat cultivars grown on a zinc-deficient and boron-toxic soil. Journal of plant nutrition, 24 (11): 1817-1829.

Yang M, Shi L, Xu FS, Lu JW, Wang YH, 2009. Effects of B, Mo, Zn, and their interactions on seed yield of rapeseed (*Brassica napus* L.). Pedosphere. 19 (1): 53-59.

PHYTOTOXICITY OF DIFFERENT AQUEOUS EXTRACT CONCENTRATIONS OF MESQUITE (*PROSOPIS JULIFLORA*) ON GERMINATION AND SEEDLING GROWTH OF WHEAT

Rafaqat Hussain Shah[1*], Mohammad Safdar Baloch[1], Muhammad Zubair[2], Ejaz Ahmad Khan[1], Muhammad Ahmad[3] and Ghulam Abbas[4]

[1] *Department of Agronomy, Faculty of Agriculture, Gomal University, Dera Ismail Khan, Pakistan*

[2] *The Department of Forestry and Range Management, Faculty of Agricultural Sciences and Technology, Bahauddin Zakariya University, Multan, Pakistan*

[3]*Agrivcultural Training Institute, Karor Lal Eason, Layyah, Pakistan*

[4]*Adaptive Research Farm, Karor Lal Eason, Layyah, Pakistan*

ABSTRACT

The research trial was conducted under laboratory conditions at Department of Agronomy, Faculty of Agriculture, Gomal University, Dera Ismail Khan, Pakistan in order to study phytotoxic effects of mesquite (*Prosopis juliflora*) aqueous extract taken from its leaves, stem and root on germination and seedling growth of wheat. There were five treatments *viz.* 100g, 200g, 300g, and 400g aqueous leaves, stem and root extract of mesquite including tap water as control. It was evident from the results that by using different concentrations of aqueous extract of mesquite, seed germination and seedling growth of wheat were inhibited. The maximum reduction in germination and seedling growth was noted by applying higher concentration (400g) of aqueous extracts of mesquite. Hence, it is concluded that seed germination and seedling growth of wheat is inhibited due to presence of water soluble allelochemicals in aqueous leaves, stem and root extracts of mesquite at 400 g for 72 hours. These allelochemicals have negative effects on growth and development of wheat. It is, therefore, recommended not to cultivate wheat crop near the mesquite.

Keywords: Phytotoxicity, *Prosopis juliflora*, extract, germination, seedling vigor, wheat

INTRODUCTION

Wheat (*Triticum aestivum* L.) is the leading food grain of Pakistan, which contributes 10.3% value added in agriculture and 2.2% to GDP. It is annually cultivated on an area of 9.03 million hectares with total production of 25.9 million tons in the country (Anonymous, 2015).

The yield of wheat is low in Pakistan due to number of abiotic and biotic factors (Qureshi and Bhatti, 2001; Ullah *et al*., 2013). Weeds are the most serious but less noticeable yield-deteriorating factor that also poses allelopathic effects against crops (Khan *et al*., 2004) due to presence of organic compounds, which release phytotoxic substance. These phytotoxic substances release chemicals into the ecosystem (Tanveer, 2008) in the form of secondary metabolites which may leach out from various parts of plants to the surrounding rhizosphere either as exudates or rain-residues which impact on germination and growth processes of other plants (Sajjad *et al*., 2007; Iqbal *et al*., 2010). In allelopathy, plant growth is negatively affected due to competition of the two organisms when using the same resources (Burhan and Shaukat, 2000). It has been reported that the plants which release phytotoxic substances, can have both positive and negative effects with the interaction of weeds and crops (Rebaz *et al.*, 2001; Shaukat *et al*., 2002). Mehar (2011) reported that allelopathic plants have certain compounds effecting the environment is totally different from competition. Competition is to remove and reduce all available resources of neighboring plants.

Phytotoxic plants discharge toxic chemicals or secondary plant metabolites, which affect seed germination, cell division, cell elongation, membrane permeability and ion uptake of the nearby plants (Dongre *et al*., 2007). Oweyegha-Afunaduula, (2008) has reported that interaction of trees with crop plants may have positive or negative effects on both crops and forest plants. Therefore, before introducing to an agro-forestry system it is necessary to check the phytotoxic compatibility of plants with the trees. Mesquite plant is considered to be both

**Corresponding author:* e-mail:

invasive and phytotoxic, because it can delay the growth of some nearby plants species by discharging allelochemicals in the environment (Mehar, 2011).

Mesquite (*Prosopis juliflora*) is generally known as Kabuli kikar, a well adopted shrub to harsh environment conditions of arid zones. It is large inhabitant shrub of the Northern South America, West Indies, Central America and Mexico. It was introduced in Pakistan in 1950s. It has been noted that the fallen leaves, leachates or root exudates of mesquite release phytotoxic substances which affect all ground vegetation under its canopy (Siddiqui *et al.*, 2009). Therefore, discharging of toxic chemicals into the soil delayed seed germination and establishment of field crops and vegetation. Due to its weedy nature, mesquite is a threat to indigenous biodiversity in Ethiopia and in the Middle Awash. The reported area under mesquite invasion is about 30,000 hectare in the Middle Awash (Mehari, 2008), however in Pakistan its area under cultivation is still under study.

Mesquite not only decreases grass availability for the livestock but also restricts many physiological and biochemical processes of the crop plants due to release of phytotoxic compounds. (Getachew *et al.*, 2012). Since the allelopaths restrict plant growth at a certain concentration, therefore, their proper screening tests should be conducted to ensure their effects on economic crop (Tanveer, 2008). The present research was therefore designed to investigate the phytotoxic potential of different concentrations of aqueous leaves, stem and root extract of mesquite on germination and early seedling growth of wheat.

MATERIALS AND METHODS

Phytotoxic effect of mesquite (*Prosopis juliflora*) was assessed under laboratory conditions at Department of Agronomy, Faculty of Agriculture, Gomal University, Dera Ismail Khan, Khyber Pakhtunkhwa (KPK), Pakistan, in a completely randomized design with 4 replications for 28 days. Fresh Vegetative growing tissues of the weed species were collected from fields nearby the experimental site. For the weed specie, plants were separated into three leaves, stem and root parts, crushed and grinded. Tap water was used as control treatment. The ground plant material of mesquite was soaked in distilled water in different ratios, *viz-a-viz*, T_1 = 100g leaves/ liter of water for 72 h, T_2 = 200g leaves/liter of water for 72 h, T_3 = 300g leaves/liter of water for 72 hours for 72 hours, T_4 = 400g leaves/liter of water for 72 h, T_5 = 100g stem/liter of water for 72 h, T_6 = 200g stem/liter of water for 72 h, T_7 = 300g stem/liter of water for 72 h, T_8 = 400g stem/liter of water for 72 h, T_9 = 100g roots/liter of water for 72 h, T_{10} = 200g roots/liter of water for 72 h, T_{11} = 300g roots/liter of water for 72 h and T_{12} = 400g roots/liter of water for 72 h of *P.juliflora*. The aqueous extract was collected in bottles and tagged by filtering through 10 and 60 mesh sieves. Thirteen trays, replicated 4 times, filled with sand, silt and clay (1:1:1) were taken for sowing 100 wheat seeds in each tray viz. Tray-1: T_0: Tap water (check) and Tray-2 to 13: Treatments T_1–T_{12} (*P.juliflora*). All treatments were applied 5 and 10 days after sowing (DAS). Water was applied on daily/alternate basis to keep the soil moist/field capacity. The parameters recorded during the course of experimentation were speed of germination, mean germination time, germination rate, germination (%), germination energy (%), root and shoot length (cm), fresh and dry root weight (g), fresh and dry shoot weight (g), root: shoot ratio and chlorophyll content (μ g cm^{-2}).

Data for the individual parameters were subjected to statistical analysis by using analysis of variance technique (Steel *et al.*, 1997). The difference among the treatment means was computed by using least significant difference (LSD) test at 1% probability level using MSTATC computer software (MSTATC, 1991).

RESULTS

Speed of germination, mean germination time, germination rate

The results showed that aqueous extract from leaves, stems and root of mesquite negatively affected the speed of germination, mean germination time and germination rate (Table -- 1). All the treatments showed significant reduction at 1 percent level of probability in the speed of germination, germination rate and also delay in mean germination time as compared to control (Tap water). The maximum speed of germination and germination rate was noted in control where no extract was used. Among different concentrations applied, aqueous leaves extract (400g) showed maximum reduction in speed of germination and germination rate (7.81 and 57.25) respectively, by taking more mean germination time (13.73). Similar trend was also noted with increasing concentration of aqueous stems and root extract which significantly reduced the speed of germination and germination rate and by delaying mean germination time.

Germination percentage and germination energy

Germination percentage and germination energy (recorded 7 and 14 DAS) was significantly_reduced at 1 per cent level of probability with the increasing concentration of aqueous leaf, stem and root extract of mesquite when compared with control (Table - 2). Significantly lowest germination percentage was recorded (2.25 and 59.50%) with the higher concentration (400g) of aqueous leaf extract over control (Tap water) which showed higher germination percentage (13.50 and 100.0%) 7 and 14 DAS respectively. Similarly, the higher concentration (400g) of aqueous stems and roots extract recorded the minimum germination percentage (4.50, 69.50% and 4.50, 71.75%) on both 7 and 14 DAS respectively. Similar trend was also noted for germination energy. In control, maximum germination energy was recorded (13.50 and 100.00%) 7 and 14 DAS respectively, whereas minimum germination energy (2.09 and 59.51%, 4.51 and 69.51% and 4.51and 71.74%) was recorded 7 and 14 DAS with higher concentration (400g) of aqueous extract of leaves, stems and roots of mesquite, respectively.

Shoot and root length

Data pertaining to shoot and root length are presented in Table - 3, which indicated that aqueous leaf, stems and root extracts significantly at 1 percent level of probability inhibited shoot and root length as compared to tap water. Maximum shoot length (18.70, 25.61 and 29.35cm) was recorded in control 14, 21 and 28 DAS. Among different concentrations, the maximum concentration (400g) of aqueous extract of all three parts of mesquite produced minimum shoot length of wheat 14, 21 and 28 DAS. Similarly, maximum root length was measured in control (3.75, 5.50 and 6.20cm) 14, 21 and 28 DAS respectively. The root length was subsequently reduced with increase in extract concentration. The highest concentration (400g) of aqueous leaf, stem and root extracts produced minimum root length compared to untreated plot.

Fresh shoot and root biomass

Fresh shoot and root biomass was reduced by increasing concentration of aqueous leaf, stem and root extracts as compared to control (Table - 4). Maximum shoot weight was noted in control (1.32, 2.37 and 2.60g) 14, 21 and 28 DAS, respectively. The highest concentration (400g) of aqueous leaf extract considerably reduced the shoot weight (0.84, 1.24 and 1.66g) 14, 21 and 28 DAS, respectively. Similar reduction in fresh shoot weight was also noted with the application of higher concentration (400g) of aqueous stem and root extract. Similarly, all three aqueous leaves, stem and roots extracts of plant significantly reduced fresh root biomass as compared with tap water (0.61,0.68 and 0.79g) 14, 21 and 28 DAS, respectively.

Dry shoot and root weight

Data given in Table - 5 revealed significant reduction in dry shoot and root weight by applying different concentration of aqueous extract of leave, stem and root than control (Tap water). The minimum dry shoot weight (0.079, 0.110 and 0.137g) was noted by applying maximum concentration (400g) of aqueous leaves extract over control, which had maximum dry shoot weight (0.106, 0.145 and 0.176g) 14, 21 and 28 DAS respectively. Similarly, aqueous stem and root extracts also reduced dry shoot weight as compared to untreated control. Similar results were noted in case of dry root weight in which the increasing concentrations of aqueous leave, stem and root extracts of mesquite correspondingly decreased dry root weight as compared to control that had maximum dry root weight (0.073, 0.087 and 0.11g) 14, 21 and 28 DAS, respectively.

Root: Shoot ratio, Chlorophyll Content

Applications of aqueous leave, stem and root extract of mesquite significantly at 1 percent level of probability reduced the root: shoot ratio and chlorophyll content as shown in Table - 6. By increasing concentration of aqueous extract of all three plant parts correspondingly decreased the root: shoot ratio as well as chlorophyll content compared with control. The maximum root: shoot ratio was noted in control (39.67, 40.17 and 41.0) 14, 21 and 28 DAS, respectively. Application of highest concentration (400g) of all three plant parts (leaves, stem and root) correspondingly reduced the root: shoot ratio. Similarly, the control treatment showed maximum chlorophyll content (27.05, 33.48 and 26.23) while it decreased considerably by increasing the concentration of aqueous extract. Minimum chlorophyll content (21.79, 25.74 and 19.47) was recorded 14, 21 and 28 DAS, respectively, by applying the highest concentration (400g) of aqueous leave extract of mesquite.

Table – 1: Phytotoxic effect of mesquite (*Prosopis juliflora*) on speed of germination, mean germination time and germination rate of wheat

Treatments	Speed of germination	Mean germination time	Germination rate
T_0 = Tap water	10.61 a	9.62 h	87.50 a
T_1 = 100g leaves	9.10 def	11.81 cd	68.50 de
T_2 = 200g leaves	8.61 gh	12.54 b	62.25 fg
T_3 = 300g leaves	8.06 i	13.37 a	59.00 gh
T_4 = 400g leaves	7.81 i	13.73 a	57.25 h
T_5 = 100g stem	9.45 cd	11.28 ef	76.00 bc
T_6 = 200g stem	9.21 de	11.64 de	71.50 cd
T_7 = 300g stem	8.82 fgh	12.12 bc	70.25 d
T_8 = 400g stem	8.53 h	12.57 b	65.00 ef
T_9 = 100g roots	9.93 b	10.76 g	84.00 a
T_{10} = 200g roots	9.60 bc	11.07 fg	78.50 b
T_{11} = 300g roots	8.92 efg	12.15 bc	71.50 cd
T_{12} = 400g roots	8.57 gh	12.45 b	67.25 de
$LSD_{0.01}$	0.376	0.460	4.691

Means followed by different letter(s) in a column are statistically significant at 1% level of probability.

Table – 2: Phytotoxic effect of mesquite (*Prosopis juliflora*) on germination percentage and germination energy of wheat

Treatments	Germination (%)		Germination energy (%)	
	7 DAS	14 DAS	7 DAS	14 DAS
T_0 = Tap water	13.50 a	100.0 a	13.50 a	100.00 a
T_1 = 100g leaves	8.00 cd	76.50 c	8.08 e	76.51 g
T_2 = 200g leaves	5.00 def	67.25 e	5.09 g	67.23 k
T_3 = 300g leaves	2.50 f	61.50 f	2.25 i	61.23 l
T_4 = 400g leaves	2.25 f	59.50 f	2.09 i	59.51 m
T_5 = 100g stem	9.00 bc	85.00 b	9.09 d	85.10 d
T_6 = 200g stem	8.00 cd	79.50 c	8.09 e	79.53 e
T_7 = 300g stem	5.75 de	76.00 cd	5.76 f	76.10 h
T_8 = 400g stem	4.50 ef	69.50 e	4.51 h	69.51 j
T_9 = 100g roots	11.50 ab	95.50 a	11.51 c	95.55 b
T_{10} = 200g roots	9.25 bc	87.75 b	9.25 d	87.74 c
T_{11} = 300g roots	5.75 de	77.25 c	11.76 b	77.24 f
T_{12} = 400g roots	4.50 ef	71.75 de	4.51 h	71.74 i
$LSD_{0.01}$	3.008	4.531	0.226	0.065

Means followed by different letter(s) in a column are statistically significant at 1% level of probability.

Table – 3: Phytotoxic effect of mesquite (*Prosopis juliflora*) on shoot length and root length of wheat

Treatments	Shoot length(cm)			Root length(cm)		
	14 DAS	21DAS	28 DAS	14 DAS	21 DAS	28 DAS
T_0 = Tap water	18.70 a	25.61 a	29.35 a	3.75 a	5.50 a	6.20 a
T_1 = 100g leaves	12.50 d	19.52 c	22.22 c	2.72 def	3.80 bc	4.80 c
T_2 = 200g leaves	11.75 e	18.38 d	21.42 d	2.57 d-g	3.62 cd	4.57 de
T_3 = 300g leaves	10.55 f	17.58 e	20.25 e	2.47fg	3.47 de	4.35 fg
T_4 = 400g leaves	9.61 g	16.27 f	19.29 f	2.22 g	3.10 f	4.15 h
T_5 = 100g stem	13.58 b	20.51 b	23.34 b	3.37 b	3.75 bc	4.77 c
T_6 = 200g stem	12.54 d	19.50 c	22.44 c	2.92 cd	3.65 cd	4.57 de
T_7 = 300g stem	11.46 e	18.62 d	21.48 d	2.77 c-f	3.50 d	4.42 ef
T_8 = 400g stem	10.42 f	17.57 e	20.49 e	2.52 efg	3.27 ef	4.25 gh
T_9 = 100g roots	13.15 c	20.52 b	23.36 b	3.12 bc	3.90 b	5.00 b
T_{10} = 200g roots	12.53 d	19.45 c	22.37 c	2.87cde	3.72 bc	4.67 cd
T_{11} = 300g roots	11.64 e	18.48 d	21.38 d	2.65 def	3.47 de	4.57 de
T_{12} = 400g roots	10.43 f	17.56 e	20.58 e	2.50 fg	3.25 f	4.40 fg
$LSD_{0.01}$	0.406	0.370	0.443	0.361	0.211	0.170

Means followed by different letter(s) in a column are statistically significant at 1% level of probability.

Table – 4: Phytotoxic effect of mesquite (*Prosopis juliflora*) on fresh shoot weight and fresh root weight of wheat

Treatments	Fresh shoot weight (g)			Fresh root weight (g)		
	14 DAS	21 DAS	28 DAS	14 DAS	21 DAS	28 DAS
T_0 = Tap water	1.32 a	2.37 a	2.60 a	0.61 a	0.68 a	0.79 a
T_1 = 100g leaves	0.95 b	1.34 bcd	1.76 de	0.37 b	0.46 bc	0.59 bc
T_2 = 200g leaves	0.92 bcd	1.31 cde	1.73 de	0.36 b	0.44 bcd	0.57 bcd
T_3 = 300g leaves	0.86 cde	1.29 de	1.70 de	0.35 b	0.43 cd	0.55 def
T_4 = 400g leaves	0.84 e	1.24 e	1.66 e	0.33 b	0.41 d	0.53 f
T_5 = 100g stem	0.96 b	1.38 bc	1.95 b	0.38 b	0.47 b	0.60 b
T_6 = 200g stem	0.95 b	1.37 bc	1.92 b	0.37 b	0.44 bcd	0.58 bcd
T_7 = 300g stem	0.92 bc	1.35 bcd	1.87 bc	0.34 b	0.44 bcd	0.56 cde
T_8 = 400g stem	0.87 cde	1.31 cde	1.79 cd	0.34 b	0.42 d	0.55 def
T_9 = 100g roots	0.96 b	1.39 b	1.94 b	0.38 b	0.46 bc	0.59 bc
T_{10} = 200g roots	0.92 bc	1.33 bcd	1.92 b	0.36 b	0.44 bcd	0.58 bcd
T_{11} = 300g roots	0.87 cde	1.31 cde	1.90 b	0.35 b	0.42 d	0.56 def
T_{12} = 400g roots	0.85 de	1.28 de	1.86 bc	0.34 b	0.42 d	0.54 ef
$LSD_{0.01}$	0.065	0.075	0.103	0.053	0.034	0.034

Means followed by different letter(s) in a column are statistically significant at 1% level of probability.

Table – 5: Phytotoxic effect of mesquite (*Prosopis juliflora*) on dry shoot weight and dry root weight of wheat

Treatments	Dry shoot weight (g)			Dry root weight (g)		
	14 DAS	21 DAS	28 DAS	14 DAS	21 DAS	28 DAS
T_0 = Tap water	0.106 a	0.145 a	0.176 a	0.073 a	0.087 a	0.11 a
T_1 = 100g leaves	0.085 b	0.117 bc	0.143 b	0.052 b	0.071 bc	0.089 b
T_2 = 200g leaves	0.083 bc	0.115 cd	0.140 cd	0.051 b	0.070 bc	0.087 bcd
T_3 = 300g leaves	0.081 cd	0.113 de	0.138 de	0.050 b	0.069 bc	0.085 cde
T_4 = 400g leaves	0.079 d	0.110 e	0.137 e	0.048 b	0.066 c	0.081 f
T_5 = 100g stem	0.085 b	0.119 b	0.144 b	0.053 b	0.072 b	0.088 bc
T_6 = 200g stem	0.084 bc	0.116 bc	0.142 bc	0.051 b	0.070 bc	0.086 bcd
T_7 = 300g stem	0.083 bc	0.114 cd	0.140 cd	0.050 b	0.069 bc	0.085 cde
T_8 = 400g stem	0.081 cd	0.113 de	0.138 de	0.050 b	0.068 bc	0.082 f
T_9 = 100g roots	0.086 b	0.119 b	0.144 b	0.054 b	0.071 bc	0.088 bc
T_{10} = 200g roots	0.084 bc	0.116 bc	0.141 bc	0.052 b	0.070 bc	0.086 bcd
T_{11} = 300g roots	0.083 bc	0.114 cd	0.138 de	0.050 b	0.069 bc	0.084 def
T_{12} = 400g roots	0.085 b	0.113 de	0.136 e	0.050 b	0.069 bc	0.082 ef
$LSD_{0.01}$	3.390	3.274	2.765	2.821	4.835	3.241

Means followed by different letter(s) in a column are statistically significant at 1% level of probability.

Table – 6: Phytotoxic effect of mesquite (*Prosopis juliflora*) on root:shoot ratio and chlorophyll content of wheat

Treatments	Root shoot ratio			Chlorophyll content (μ g cm^{-2})		
	14 DAS	21 DAS	28 DAS	14 DAS	21 DAS	28 DAS
T_0 = Tap water	39.67 a	40.17a	41.01 a	27.05 a	33.48 a	26.23 a
T_1 = 100g leaves	37.81 b	37.66 b	38.27 bc	24.14 def	27.81 bc	22.29 bcd
T_2 = 200g leaves	37.91 b	37.84 b	38.45 b	23.34 g	27.56 bc	21.37 ef
T_3 = 300g leaves	38.02 ab	37.91 b	38.25 bc	22.71 h	26.54 de	20.26 g
T_4 = 400g leaves	37.90 b	37.56 b	37.33 d	21.79 i	25.74 f	19.47 h
T_5 = 100g stem	38.49 ab	37.69 b	38.06 bcd	24.38 cd	28.19 b	22.59 b
T_6 = 200g stem	37.86 b	37.53 b	37.85 bcd	24.32 de	27.38 c	21.57 e
T_7 = 300g stem	37.44 b	37.60 b	37.91 bcd	23.96 ef	27.29 c	20.89 fg
T_8 = 400g stem	38.01 ab	37.74 b	37.44 cd	23.77 f	26.32 ef	20.41 g
T_9 = 100g roots	38.43 ab	37.37 b	38.06 bcd	24.85 b	28.10 b	22.85 b
T_{10} = 200g roots	38.09 ab	37.53 b	37.79 bcd	24.74 bc	27.88 bc	22.44 bc
T_{11} = 300g roots	37.68 b	37.59 b	37.88 bcd	24.34 cde	27.52 bc	21.91 cde
T_{12} = 400g roots	36.91 b	37.94 b	37.75 bcd	24.02d ef	27.21 cd	21.77 de
$LSD_{0.01}$	1.683	1.611	0.862	0.407	0.698	0.667

Means followed by different letter(s) in a column are statistically significant at 1% level of probability.

DISCUSSION

Some plant species or their residues selectively delayed growth of other specific plants (Al-Zahrani and Al-Robai, 2007). This sensitivity was observed in field and laboratory experiments with extracts and allelopathic substances. In this study, the aqueous extract of leaves, stems and roots of mesquite has resulted in reduced germination and seedling growth of wheat. The maximum reduction in germination was recorded at the highest concentration of aqueous extract of leaves, stems and root of mesquite. Similarly, the highest concentration (400g) of all the three parts significantly reduced the speed of germination and germination rate and also delayed the mean germination time as compared to control (Tap water). This was due to some phytotoxic compound in mesquite which includes tannins flavonoides, wax, alkaloids and phenolic acids, those negatively affected the seed germination and radical length of wheat (Pragnesh *et al*., 2013). The percent seed germination is reported to decrease with increasing aqueous leaf extract concentration of mesquite (Siddiqui *et al*., 2009). Germination percentage and germination energy was correspondingly reduced in the present study by using different concentration of aqueous leaves, stem and root extracts as compared to control. This was due to presence of plant growth inhibitory alkaloids which were extracted from the mesquite leaves (Nakano *et al*., 2004). Phytochemical analysis of mesquite showed that mesquite contains phenolics, tannins, steroids, flavonoids, alkaloids and terpenoids in leaf extracts. Stem contains, steroids, phenolics, flavonoids and terpenes in minimum concentrations, while root has saponin, alkaloids, phenolics, steroids, flavonoids, tannins and terpenes (Singh, 2012).

In the present study, aqueous extracts of mesquite were also tested for root and shoot length, fresh root and shoot weight and dry root and shoot weight which showed a corresponding reduction in all these parameter as compared to control where no extract was used. Several reports revealed the phytotoxic effects of various plant extracts e.g. *E.camaldulensis*, *P.juliflora* and *A.nilotica* which significantly affected seed germination and seed seedling growth of several crops and weed species (Khan *et al*., 2004). Their inhibitory effect on seed germination, root length and other growth parameters is well established (Rafique *et al*., 2003). The mesquite extract is reported to cause maximum reduction in root length (Siddiqui *et al*., 2009). Similarly, aqueous extract application of mesquite significantly reduced the germination and seedling growth of a number of crop plants (Khan, 2005). An especially maximum degree of inhibition occurred with aqueous extract of leaves at the maximum concentrations in wheat in comparison with aqueous extracts of stem and root.

REFERENCES

Anonymous, 2015. Agricultural Statistics of Pakistan. Ministry of Food and Agriculture, Govt. of Pakistan, Islamabad.

Al-Zahrani HS and Al-Robai SA, 2007. Allelopathic effect of Calotropis procera leaves extract on seed germination of Some Plants. JKAU: Sci. 19: 115-126.

Burhan N and Shaukat SS, 2000. Effects of atrazine and phenolic compounds on germination and seedling growth of some crop plants. Pak. J. Biol. Sci. 3: 269-274.

Cheema ZA, Khaliq A and Saeed S, 2004. Weed control in maize (Zea mays L.) through sorghum allelopathy. J. Sustain. Agric. 23: 73-86.

Dongre PN and Singh AK, 2007. Inhibition effects of weeds on growth of wheat seedlings. Allelopathy J. 20(2): 387-394.

Getachew S, Demissew S and Woldemariam T, 2012. Allelopathic effects of the invasive Prosopis juliflora (Sw.) DC. on selected native plant species in Middle Awash, Southern Afar Rift of Ethiopia. Manag. Biol. Invasions. 3(2): 105–114.

Iqbal J, Karim F and Hussain S, 2010. Response of the wheat crop (Triticum aestivum L.) and its weeds to allelopathic crop water extracts in combination with reduced herbicides rates. Pak. J. Agric. Sci. 47: 309-316.

Khan I, Hassan G, Khan MI and Khan IA, 2004. Efficacy of some new herbicidal molecules on grassy and broadleaf weeds in wheat-II. Pak. J. Weed Sci. Res. 10(1-2): 33-38.

Khan AZ, Marwet KB, Hussan G and Hussain Z, 2005. Bio-herbicidal effects of tree extracts on seed germination and growth of crops and weeds. Pak. J. Weed Sci. Res. 11(3-4): 89 – 94.

Khan MA, Marwat KB, and Hassan Z, 2004. Allelopathic potential of some multipurpose trees species (MPTS) on the wheat and some of its associate's weeds. Intl. J. Biol. And Biotechnol. 1(3): 275-278.

Kato-Noguchi H and Tanaka Y, 2006. Abscisic acid-glucose ester as an allelopathy agent from citrus fruit. Acta Physiolgiae Plantatrum. 28: 635-639.

Mehari Z, 2008. Invasion of Prosopis juliflora (SW.) DC and rural livelihoods. The case of Afar Pastoralists at Middle Awash area of Ethiopia. M.Sc. Thesis, Norwegian University of Life Science, Oslo, 3 pp.

Mehar KS, 2011. Assessment of effect of Prosopis juliflora litter extract on seed germination and growth of rice. Food Sci. and Quality Mang. 2: 9-19.

MSTATC, 1991. MSTATC Package, Version 1. Michigan State Univ. USA.

Nakano H, Nakajima E, Hiradate S, Fujii Y, Shigemori H and Hasegawa K, 2004. Growth inhibitory alkaloids from mesquite (Prosopis juliflora (Sw.) DC.) leaves. Pytochemistry. 65(5): 587-591.

Oweyegha-Afunaduula FC, 2008. Agro-forestry strategy of food production: Principles, practices, and recommendations. Zoology Department, Makerere University, Kampala, Uganda.

Pragnesh N, Dave and Bhandari J, 2013. Prosopis juliflora: A review. Intl. J. Chem. Studies, 1(3): 181-196.

Qureshi R and Batti GR, 2001. Determination of weed communities in wheat (Triticum aestivum L.) fields of district Sakkhur. Pak. J. Bot. 33(1): 109-115.

Rafique HATM, Ahmed R, Uddin MB and Hossain MK, 2003. Allelopathic effect of different concentration of water extract of Acacia auriculiformis leaf on some initial growth parameters of five common agricultural crops. Pak. J. Agron. 2(2): 92-100.

Rebaz Z, Shaukat SS and Siddiqui IA, 2001. Allelopathic potential of Anagallis arvensis: A cosmopolitan weed. Pak. J. Biol. Sci. 4: 446-450.

Shaukat SS, Siddiqui IA, Khan GH and Zaki MJ, 2002. Nematacidal and allelopathic potential of Aregemon mexicana, a tropical weed. Plant and Soil. 245: 239-247.

Sajjad H, Sadar S, Khalid S, Jamal A, Qayyum A and Ahmad Z, 2007. Allelopathic potential of senna (Cassia angustifolia Vahl.) on germination and seedling characters of some major cereal crops and their associated grassy weeds. Pak. J. Bot. 39(4): 1145-1153.

Siddiqui S, Bhardwaj S, Khan SS and Meghvanshi MK, 2009. Allelopathic effect of different concentration of water extract of leaf of Prosopis juliflora on seed germination and radicle length of wheat (Triticum aestivum Var-Lok-1). Am-Eur. J. Sci. Res., 4 (2): 81-84.

Steel RGD, Torrie JH and Dicky DA, 1997. Principles and Procedures of Statistics. A Biometrical Approach. 3rd Ed. McGraw Hill, Inc. Book Co. N.Y. USA. pp. 352-358.

Singh S, 2012. Phytochemical analysis of different parts of Prosopis juliflora. Intl. J. Curr. Pharm. Res. 4(3): 59-61.

Tanveer A, 2008. Biology and Ecology of Weeds. Higher Education Commission, Pakistan. p.109.

Ullah A, Khan EA, Baloch MS, Nadim MA, Sadiq M and Noor K, 2013. Allelopathic effects of herbaceous and woody plant species on seed germination and seedling growth of wheat. Pak. J. Weed Sci. Res. 19(3): 357-375.

EFFECT OF DIFFERENT BRASSICA VEGETABLES ON BIOLOGY AND DEMOGRAPHIC PARAMETERS OF *BREVICORYNE BRASSICAE* (HOMOPTERA: APHIDIDAE) UNDER LABORATORY CONDITIONS

Muhammad Asif Aziz[1], Muhammad Irfan[1], Imran Bodlah[1] and Muhammad Hanif[2]
[1]Department of Entomology, [2]Department of Mathematics and Statistics, Pir Mehr Ali Shah, Arid Agriculture University, Rawalpindi

ABSTRACT

Biological and demographic parameters of cabbage aphid, *Brevicoryne brassicae* (L.) were studied on four different Brassica hosts; Cabbage, Broccoli, China cabbage, and Knol khol under laboratory conditions. Age-stage, two-sex life table studies showed significantly higher intrinsic rate of increase (r_m) and finite rate of increase (λ) on China cabbage followed by Broccoli and Cabbage. Performance of cabbage aphids remained comparatively low on Knol khol in terms of intrinsic rate of increase (r_m), finite rate of increase (λ) and net reproductive rate. Total duration of immature stages was significantly longer on Knol khol than Cabbage and China cabbage, however, adult pre-reproductive period showed non-significant variation. Reproductive period, adult longevity and female fecundity upon all the studied plant species were significantly different. The minimum fecundity and adult longevity were observed on Knol khol. The probability of surviving nymph to adult female was 0.25, 0.35, 0.50 and 0.60 on Cabbage, Knol khol, Broccoli and China cabbage, respectively. The gross reproductive rate, life expectancy and age stage reproductive values were also minimal on Knol khol. On the basis of present life table studies, China cabbage remained the most susceptible host followed by both Broccoli and Cabbage while Knol khol proved to be the most unsuitable host. Present studies confirm Knol khol as comparatively more resistant against cabbage aphid and suitable to be cultivated as important vegetable crop in Pothwar region of Punjab.

Keywords: *Brevicoryne brassicae*, Brassica, life table, development, fecundity, age-stage two-sex

INTRODUCTION

Cabbage aphid, *Brevicoryne brassicae* (L.) is one of the most serious pests of cruciferous crops like cabbage, broccoli, cauliflower, kale and knol khol (Bashir et al., 2013). Being highly host specific insect, it feeds exclusively on the phloem sap of brassica and closely related plants (Cole, 1996). Feeding sap from phloem causes stress in host plants, (Will et al., 2007) curling and chlorosis of leaves and transmission of different disease causing viruses (Raybould et al., 1999).

Different morphological and bio-chemical characteristics of host plants affect the survival rate, fecundity and mean generation time of herbivore insects (Awmack and Leather, 2002). Shiny and smooth surface of cabbage and cauliflower accessions exert antibiosis effects on cabbage aphid in field conditions (Ellis, 1996). High lectin contents in brassica sap can cause high mortality in *B. brassicae* due to binding with its stylets and foregut (Cole, 1994).The profile and concentration of glucosinlates and free amino acids in wild brassica species have significant effect on intrinsic rate of increase of cabbage aphid (Cole, 1996). Phenylalanine contents of brassica plants have negative effect on the aphid development (GaBrys, 2008).

To assess the impact of different morphological, physiological and biochemical characteristics of host plant on the biology and life history of insects, life table studies are too much necessary. Life table is a powerful tool to determine survival rate, net reproductive rate, mean generation time and intrinsic rate of increase of any insect to elucidate the existence of antibiosis in host plants (Chi and Su, 2006). Intrinsic rate of increase (rate of increase of females/female/day) is most useful life table parameter to compare the fitness of populations of a specific species across different food resources (Southwood, 1966). Present studies were conducted to explore the effect of different brassica vegetables on biological and demographic parameters of cabbage aphid and to find out the overlaps between different stages of insect development regarding age-stage specific survival rate, life expectancy and age-stage reproductive value on the basis of age-stage, two-sex life table theory. Because

**Corresponding author:* e-mail: asifaziz572@gmail.com

the cabbage aphid produces only females through parthenogenesis during active season, therefore only one sex was used in the analysis of life table. This study will be helpful for the researchers to devise push and pull strategy for ecological management of cabbage aphid on brassica vegetables depending on the host suitability through intercultural practices.

MARERIALS AND METHODS

Host plants

Different varieties of brassica viz., Cabbage (*B. oleracea* var. capitata), Broccoli (*B. oleracea* var. italica), China cabbage (*B. rapa* var. pekinensis) and Knol khol (*B. oleracea* var. gongylodes) were sown in pots containing equal proportion of soil, sand and rotten farmyard manure to ensure the availability of the leaves for the experiment. The pots were kept in green house of Department of Horticulture, PMAS-Arid Agriculture University Rawalpindi during cropping season 2012-13 to provide suitable condition for development of plants.

Aphid Culture

Cabbage aphid population was reared and maintained in Biocontrol laboratory, Department of Entomology, PMAS-Arid Agriculture University Rawalpindi for one month on the leaves of respective host plants to be tested in the experiment. For this purpose, leaves of uniform age (four-to-six leaf stage) were detached from the plants and placed in petri dishes and 20-30 adults of *B. brassicae* were placed on leaves for 24 hours. Newly emerged nymphs were separated on the next day using camel hair brush and placed on leaf discs in plastic petri dishes (10 cm diameter) with lids having small hole in the centre. The petri dishes were lined with filter paper. Forty aphids (one aphid per petri dish) were used for each host plant. The leaves were changed daily and nymphs were again placed on them in natural position (Madahi and Sahragard, 2012). The culture was maintained under controlled laboratory conditions at 22±1°C, 75±5% relative humidity and a photoperiod of 16:8 (L: D).

Development time, survival, adult longevity and reproduction

Forty newly emerged nymphs of *B. brassicae* were transferred in petri dishes (one aphid per petri dish) and data were recorded daily for survival and molting up to 4th instar. After reaching the adult stage, the data for each female were recorded regarding total pre- and post-reproductive period, adult longevity and fecundity until it died.

Life table construction and statistical analysis

We constructed age-stage, two-sex life table following to the method described previously by Chi (1988). Both adult females and individuals dying at immature stages were considered in life table construction. Raw data of all individual aphids were analyzed in accordance with age-stage, two-sex life table theory (Chi and Liu, 1985) by using a program package TWOSEX-MS Chart (Chi, 2012) to calculate life table parameters like, age-stage specific survival rate (S_{xj}) (where x=age in days and j= stage and reflects the probability that a newly emerged nymph will survive to age x while in stage j), age-stage specific fecundity (f_{xj}), age-specific survival rate (l_x), age-specific fecundity (m_x), age-stage specific life expectancy (e_{xj}) (expected life time that an individual of age x and stage j may be expected to live.), age-stage specific reproductive value (V_{xj}) (the contribution of age 'x' stage 'j' to the further population) and the population dynamics parameters (r_m, the intrinsic rate of increase; λ, the finite rate of increase; Ro, the net reproductive rate; T, the mean generation time).

Differences in the development time, longevity, reproduction and population dynamic parameters among *B. brassicae* on four Brassica species were analyzed using one way ANOVA followed by multiple comparison with Tukey-Kramer test ($P<0.05$). Moreover Bootstrap techniques was used to estimate the means/ and standard errors of population paramaters (Huang and Chi, 2013).

RESULTS

Development and survival of cabbage aphid

Maximum immature duration of cabbage aphid was observed on Knol khol (7.45 days) and minimum on both cabbage (6.64 days) and China cabbage (6.33 days) (Table 1).The survival curves illustrated in detail the survivorship and stage differentiation and depicted clear overlaps within different instars of immature stage and adult stage of cabbage aphid. The probability that a newly emerged nymph will survive to adult stage varied highly as 0.25, 0.35, 0.50, and 0.60 on Cabbage, Knol khol, Broccoli, and China cabbage, respectively (Fig 1).

Reproduction of cabbage aphid on different *Brassica* vegetables

There were non-significant differences in adult pre-reproductive period of *B. brassicae* against all brassica vegetables species (Table 1). The maximum pre-reproductive period was observed on Knol khol (9.3 days) followed by Broccoli (8.11days), while the minimum on Cabbage (7.00 days) and China cabbage (7.42 days). The maximum adult longevity was observed on Cabbage which was significantly different from that on Knol khol. Total fecundity per female was significantly higher on Cabbage (37.8 nymphs/female) followed by Broccoli (22.0 nymphs/female) and China cabbage (20.0 nymphs/female) with the lowest on Knol khol (8.6 nymphs/female). Significantly lower age-specific survival rate (l_x) was observed on Cabbage than other hosts on 5th day, with l_x values of 0.35, 0.6, 0.7 and 0.6 for Cabbage, Broccoli , China Cabbage and Knol khol, respectively. First nymph was laid at the age of 6, 7, 6 and 8 days on Cabbage, Broccoli, China cabbage and Knol khol (Fig.2). The fecundity curve showed two peaks both in Cabbage and China cabbage which were significantly higher than other hosts, while shortest peak was observed in case of Knol khol (Fig.2).

Life expectancy and Reproductive value

The life expectancy of newly emerged nymphs (e_{01}) was 6.4, 7.95, 8.45 and 8.05 days on Cabbage, Broccoli, China cabbage and Knol khol, respectively (Fig 3). Age stage specific reproductive value (V_{xj}) reflects the contribution of age '*x*' stage '*j*' to the further population (Fig 4). The age-stage reproductive value of newly emerged nymph (v_{01}) followed the pattern of finite rate of increase (λ) (Fig 4). . In Cabbage, two major peaks were observed for the reproductive values of females; first at the age of 11 days and second at the age 15 days ($V_{11,5}$=5.0 and $V_{15,5}$ =5.0). In other host species, the peak started from the age of 9 days and remained up to 11 days as $V_{9,5}$ =2.0, $V_{9,5}$ =5.36, $V_{10,5}$ =4.87, $V_{11,5}$ =4.1 days for China cabbage, Broccoli and Knol khol, respectively. Both the number of peaks and individuals at peak reproduction contributed significantly to the population much more than any other age.

Population Dynamics Parameters

Significantly higher intrinsic rate of increase (r_m) was observed on China cabbage followed by Broccoli and Cabbage, whereas lowest value was observed in Knol khol (Table 2). Finite rate of increase (λ) which reflects number of individuals added to the population per female per day reflected the same trend. Net reproductive rate (R_o) expresses the number of female offsprings that replace each female of previous generation. Significantly higher R_o was observed on China cabbage (12.654) followed by both Broccoli (10.211) and Cabbage (9.461) and lowest on Knol khol (3.261). Significantly higher mean generation time (T) was observed for Knol khol followed by Broccoli, Cabbage and China cabbage, respectively.

Table – 1: Developmental and Reproductive characters of *B. brassicae* on five Brassica vegetables

Statistics	Cabbage	Broccoli	China Cabbage	Knol Khol
1st Instar	1.40±0.078ab	1.5±0.095ab	1.35±0.092b	1.7±0.073a
2nd Instar	1.18±0.070 c	2.32±0.204ab	1.7±0.144bc	2.5±0.163a
3rd Instar	2.08±0.123b	1.63±0.1021b	1.73±0.144b	1.65±0.150b
4th Instar	1.66±0.141a	1.5±0.104a	1.67±0.153a	1.72±0.162a
Total immature duration	6.33±0.225b	7.17±0.115ab	6.64±0.236b	7.45±0.157a
Adult pre reproductive period	0.8±0.25a	0.78±0.18a	0.91±0.20a	1.3±0.41a
Total pr reproductive period	7.0±0.21c	8.11±0.14b	7.42±0.18c	9.3±0.30a
Female Fecundity	37.80±4.16 a	22.00±3.17 b	20.00±2.55 b	8.60±1.66 c
Adult longevity	8.80±0.82 a	6.1±0.46 b	5.58±0.44 bc	4.44±0.345 cd

The data in the table are values ±SE; means within a row followed by different letters indicate significant differences at 5% level by using Tukey-Kramer test.

Table – 2: Population dynamics parameter of *B. brassicae* on different Brassica species at 22±1 °C

	Parameters	Cabbage	Broccoli	China cabbage	Knol khol	LSD	F Value
r	**Original**	0.209	0.214	0.249	0.104		
	Bootstrap	0.206±0.025b	0.208±0.02 b	0.247±0.02 a	0.101±0.02 c	0.01	$F_{159,3}$=299.56; F≤0.00
λ	**Original**	1.232	1.239	1.283	1.109		
	Bootstrap	1.229± 0.031b	1.231±0.03b	1.280±0.03 a	1.106±0.03 c	0.01	$F_{159,3}$=291.29; F≤0.00
R_o	**Original**	9.45	10.8	12.52	3.25		
	Bootstrap	9.461±2.46 b	10.211±2.31 b	12.654±2.29 a	3.261±0.82 c	0.92	$F_{159,3}$=147.891 ;F≤0.00
T	**Original**	10.76	11.11	10.14	11.39		
	Bootstrap	10.74±0.47 ab	11.06±0.18 b	10.22±0.18 a	11.43±0.31b	0.14	$F_{159,3}$=110.66; F≤0.00

Values (means ±SE) within a row followed by different letters indicate significant differences at 5% level by using Tukey-Kramer test.

Where, r = Intrinsic rate of increase λ= Finite rate of increase

R_o = Net reproduction rate T = Mean length of a generation

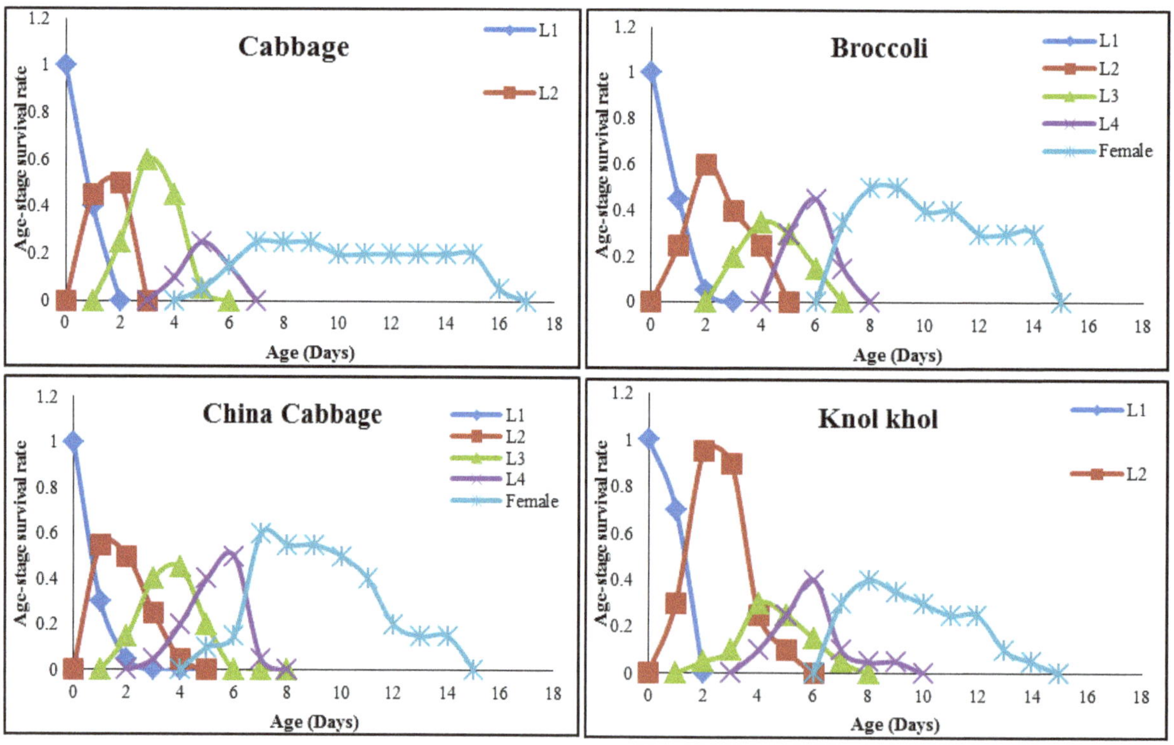

Fig – 1: Age-stage survival rate (days) of B. brassicae reared on different Brassica species at 22±1°C, 70±5% relative humidity, photoperiod 16 : 8 (L : D).

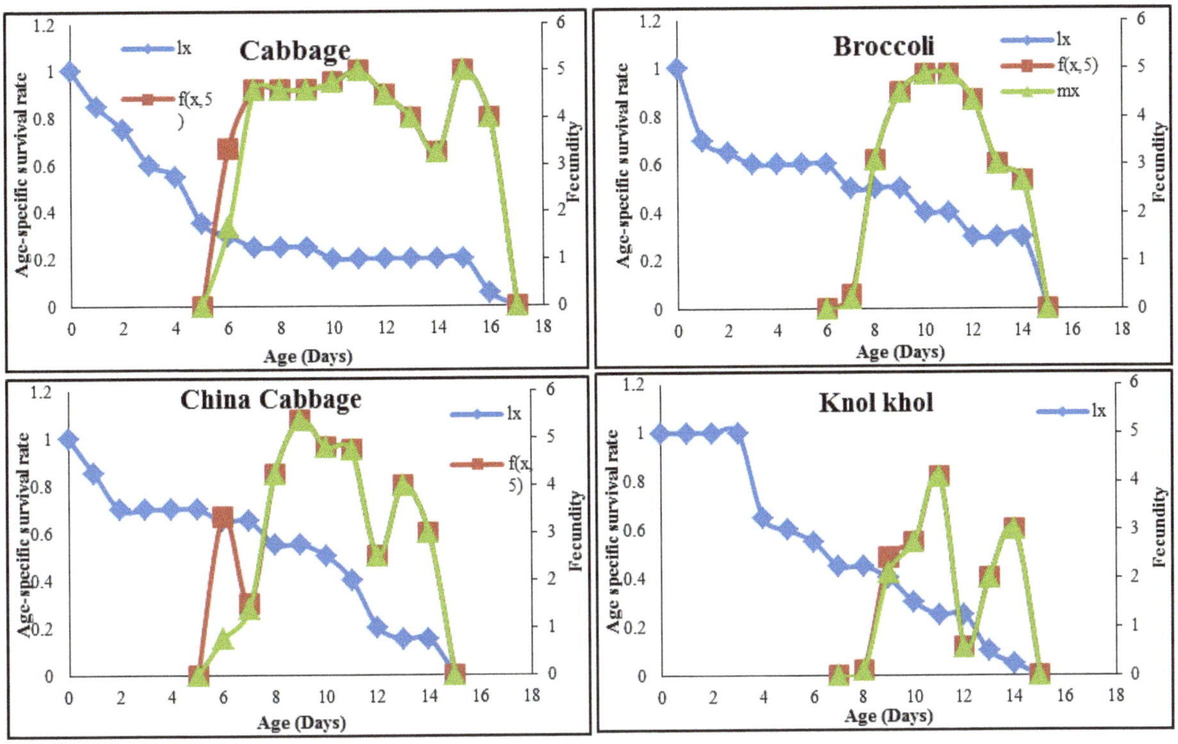

Fig – 2: Gross Reproductive rate of *B. brassicae* reared on different Brassica species at 22±1°C, 70±5% relative humidity, photoperiod 16 : 8 (L : D).

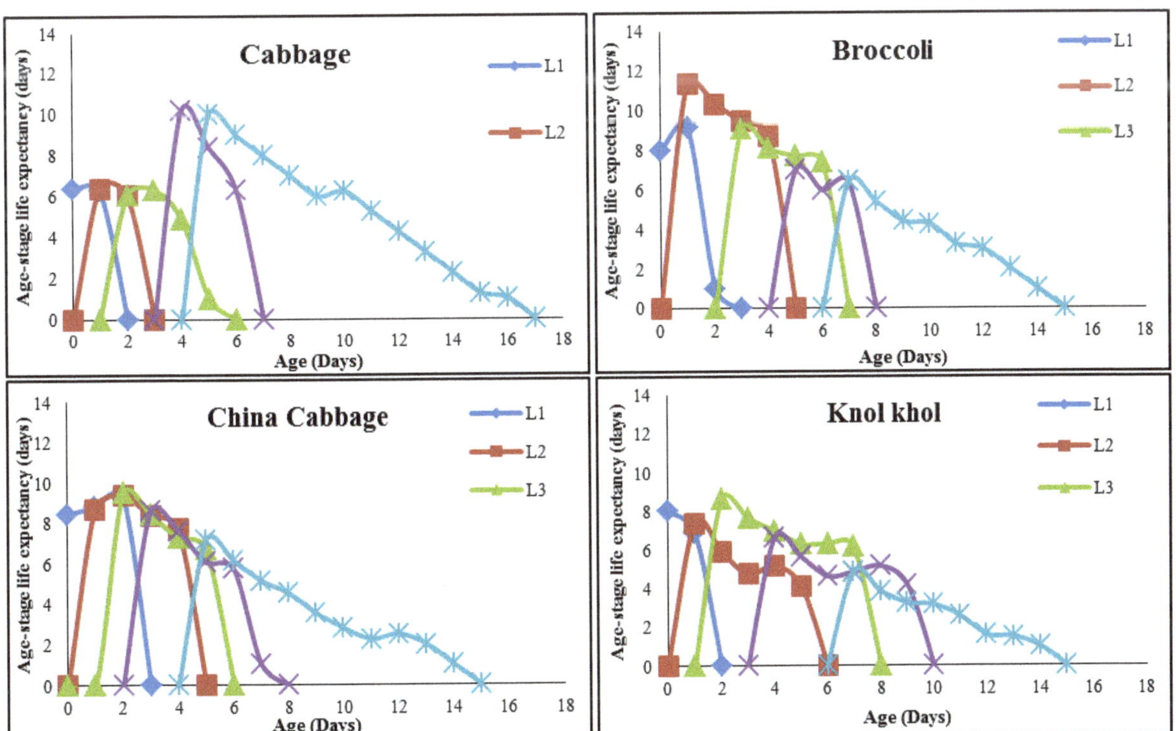

Fig 3 Life expectancy of *B. brassicae* reared on different Brassica species at 22±1°C, 70±5% relative humidity, photoperiod 16 : 8 (L : D).

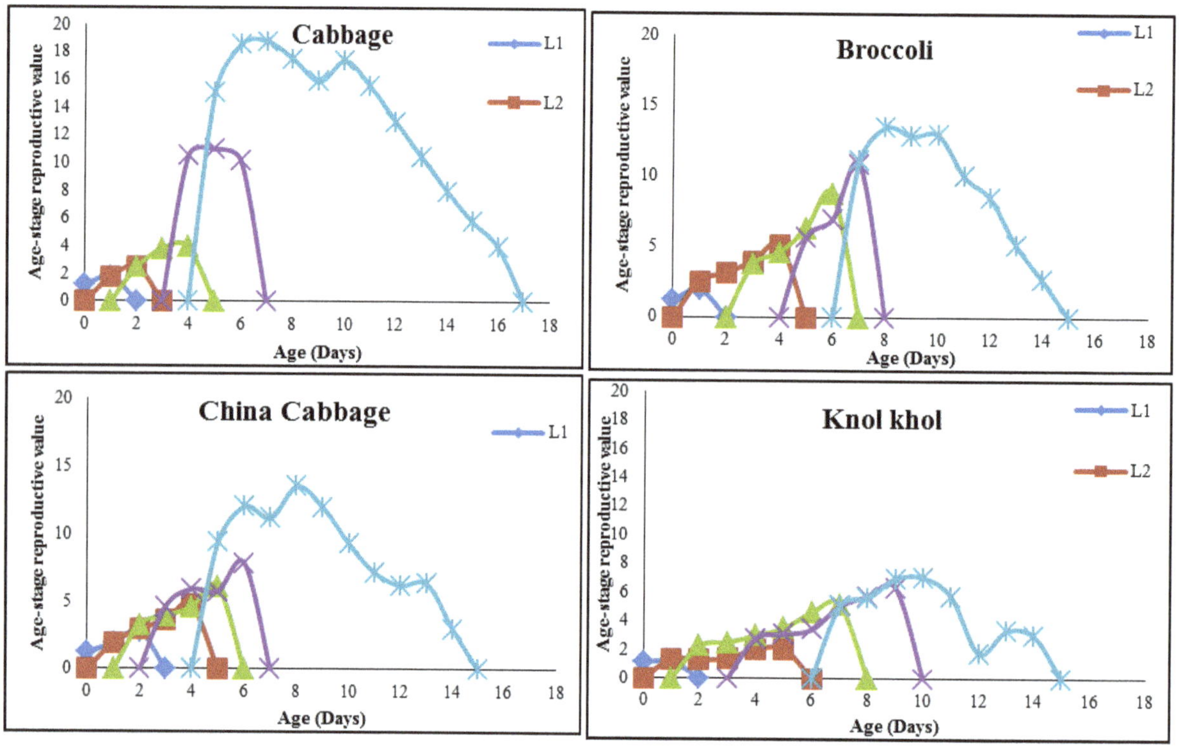

Fig 4 Age-Stage reproductive value of *B. brassicae* reared on different Brassica species at 22±1°C, 70±5% relative humidity, photoperiod 16 : 8 (L : D).

DISCUSSION

A number of variations in biological and demographic parameters of cabbage aphids were observed on different brassica host crops in the present study. For example, a significantly longer duration of immature stages was recorded on Knol khol as compared to that on Cabbage and China cabbage. Moreover, total pre-reproductive period was recorded as 9.3, 8.11, 7.42 and 7.00 days on Knol khol, Broccoli, China cabbage and Cabbage, respectively. These findings are in agreement with the previous studies reflecting effect of host plants on development time of immature stages as Ulusoy and Bayhan (2006) reported shortest development time of immature stage of *B. brassicae* on cauliflower (8.9 days) and longest on cabbage (10.4 days). Recently, Jahen et al. (2014) observed immature duration of 6.7 days on Smilla and 8.10 days on White Cloud (different cauliflower cultivars). Anzabi et al. (2014) found the minimum immature duration of cabbage aphid on Geronimo (8.38 days) and the maximum on Okapi (9.16 days) genotype of canola in Iran. These variation in the rate of development may occur due to specific nutritional qualities (Zarghami et al., 2010), morphological structures (Ulusoy and Bayhan, 2006) and biochemical characters (Cole, 1997) of different host plants offered to the insects.

Mean number of nymphs of cabbage aphid/female varied from 37.80 (Cabbage) to 8.60 (Knol khol), which were less than those (52.43 to 60.8) on oilseed rape (Mirmohammadi et al., 2009). On cauliflower cultivars, mean number of nymphs laid per female ranged from 58.6 to 30.9 (Jahen et al., 2014). On canola, Aslam et al. (2011) reported 30.79 nymphs per cabbage aphid while Ulusoy and Bayhan (2006) recorded 52.91 nymph per female on cauliflower and 16.20 nymphs per female on turnip. This difference of reproductive rate on different cultivars reflects that some host plants are less suitable for *B. brassicae* than others to produce offsprings. Low level of fecundity on Knol khol in the present studies may be presumed to be the result of this factor.

Age-stage specific survival curves indicated clear overlaps between different stages of cabbage aphid i.e., at ages 4 and 7 days, we observed 2nd, 3rd, 4th instars and adult females at the same time in China cabbage (Fig 1). These overlaps document clear variations in stage differentiation and provide detailed information regarding life history of an insect. This is why age-stage, two-sex life table theory is more useful than traditional female age-specific life tables (Birch, 1948) in which variable developmental rate among individuals is ignored (Yu et al. 2005).

We observed maximum net reproductive rate on China Cabbage and minimum on Knol khol. Jahan et al. (2014) reported comparatively higher values of R_o (30.83–56.550) for cabbage aphid on cauliflower cultivars. Fathipour et al. (2005) reported 15.92 females/female/day on the Green Cornet cultivar for cabbage aphid. Comparatively lower net reproductive rate on Knol khol (3.25 females/female/day) refers that it is more resistant to the cabbage aphid than China Cabbage (12.52 females/female/day).

The minimum generation time was observed in case of China cabbage (10.22 d) and the maximum on Knol khol (11.43 d). Singh and Parihar (1988) reported that faster developmental time on a particular host may allow a shorter life cycle, high reproductive productivity, more rapid population growth and reduce generation time.

In our studies, significantly higher value of r_m was observed in case of China cabbage (2.47) followed by Broccoli (0.218) and Cabbage (0.209), while the minimum on Knol khol (0.104). Ulusoy and Bayhan (2006) found intrinsic rates of 0.2345, 0.2009 and 0.1976 for natural increase of cabbage aphid on Cauliflower, Cabbage, and Broccoli, respectively. Mirmohammadi et al (2009) reported r_m values from 0.316 to 0.341 on Zarfam and Hyola 401 cultivars of canola. Maremela *et al.* (2013) also reported r_m values (0.37-0.42) of cabbage aphid on rape, kale, Ethopian mustard and cabbage. Jahan et al. (2014) recorded the intrinsic rate of natural increase of cabbage aphid between 0.27 to 0.35 on Smilla and White Cloud cultivars of cauliflower.

Life table parameters provide deep insight to assess the fitness of insects against their particular hosts and provide more reliable information to predict the population growth potential of insects. Knol khol may provide a source of resistance for cabbage aphid in brassica vegetables. At the same time these crops may be used by the researchers to devise intercultural techniques in the vegetable cropping scheme to reduce the pest pressure on cabbage and manage it with less application of chemical pesticides.

ACKNOWLEDGEMENTS

Authors are highly grateful to Department of Entomology, PMAS-Arid Agriculture University Rawalpindi for providing facilities to conduct the research.

REFERENCES

Anzabi SHM, Eivazi A, ZargaranMR and Gasemi-Kahrizeh A, 2014. Effect of seven canola genotypes on cabbage aphid *Brevicoryne brassicae* growth parameters. Romanian Agric Res. 31:75-80.

Awmack CS and Leather SR, 2002. Host plant quality and fecundity in herbivorous insects. Ann. Rev. Entomol., 47: 817- 503.

Bashir F, Azim MN, Akhter N and Muzaffar G, 2013. Effect of texture/morphology of host plants on the biology of *Brevicoryne brassicae* (L.) (Homoptera: Aphididae). Intl. J. Curr. Res., 5: 178-180.

Birch LC, 1948. The intrinsic rate of natural increase of an insect population. J. Anim. Ecol., 17: 15- 26.

Chi H. and Liu H, 1985. Two new methods for the study of insect population ecology. Bull Inst Zoology Acad. Sin., 24: 225-240.

Chi H and Su HY, 2006. Age-stage two sex life table of *Aphidius gifuensis* (Ashmead) (Hymenoptera: Braconidae) and its host *Myzus persicae* (Sulzer) (Homoptera: Aphididae) with mathematical proof of the relationship between female fecundity and the net reproductive rate. Environ. Entomol., 35: 10- 21.

Chi H, 1988. Life-table analysis incorporating both sexes and variable development rates among individuals. Environ. Entomol., 17:26-34.

Chi, H., 2012. TWOSEX-MSChart: a computer program for the age-stage, two-sex life table analysis. http:/140.120.197.173/Ecology.

Cole RA, 1994. Isolation of a chitin-binding lectin, with insecticidal activity in chemically-defined synthetic diets, from two wild brassica species with resistance to cabbage aphid *Brevicoryne brassicae*. Entomol. Exp. Appl., 72 (2): 181–187.

Cole R, 1996. Abiotic induction of changes to glucosinolate profiles in Brassica species and increased resistance to the specialist aphid *Brevicoryne brassicae*. Entomol. Exp. Appl., 80: 228–230.

Cole RA, 1997. The relative importance of glucosinolates and amino acids to the development of two aphid pests *Brevicoryne brassicae* and *Myzus persicae* on wild and cultivated brassicae species. Entomol. Exp. Appl., 85(2): 121- 133.

Ellis PR, Singh R, Pink DAC, Lynn JR and Saw PL, 1996. Resistance to *Brevicoryne brassicae* in horticultural brassicas. Euphytica. 88:85-96.

Fathipour Y, Hosseini A, Talebi AA, Moharramipour, S and Asgari S, 2005. Effects of different temperatures on biological parameters of cabbage aphid, *Brevicoryne brassicae* (Hom.: Aphididae). J. Sci. Technol. Agric. Nat. Resour.Water Soil Sci. 9: 185–194 [In Persian].

GaBrys B, 2008. Cabbage aphid, *Brevicoryne brassicae* (L.) (Hemiptera: Aphididae). In: *Encyclopedia of Entomology*, Springer 2nd Ed. 4: 685-687.

Huang YB and Chi H, 2013. Life tables of *Bactrocera cucurbitae* (Diptera: Tephritidae): with an invalidation of the jackknife technique. J. Appl. Entomol. 137(5):327-39.

Jahan F, Abbasipour H, Askarianzadeh A, Hassanshahi G, Saeedizadeh A, 2014. Biology and life table parameters of *Brevicoryne brassicae* (Hemiptera: Aphididae) on cauliflower cultivars. J Insect Sci. J. Insect Sci. 14(284): DOI: 10.1093/jisesa/ieu146.

Madahi K, and Sahragard A, 2012. Comparative life table of *Aphis pomi* (Hemiptera: Aphididae) on two host plants *Maluspumila* L. and *Chaenomeles japonica* under laboratory conditions. J. Crop. Prot., 4: 321-330.

Maremela M, Tiroesele B, Obopile M and Tshegofatso AB, 2013. Effect of Brassica cultivar on population growth and life parameters of the cabbage aphid *Brevicoryne brassicae* L. (Hemiptera : Aphididae). *J. Ent. Res.,* 37: 95- 100.

Mirmohammadi S. Allahyari H, Nematollahi MR and Saboori A, 2009. Effect of host plant on biology and life table parameters of *Brevicoryne brassicae* (Hemiptera: Aphididae). Ann. Entomol. Soc. Am., 102: 450-455.

Raybould AF, Maskell LC, Edwards ML Cooper JI and Gray AJ, 1999. The prevalence and spatial distribution of viruses in natural populations of *Brassica oleracea*. *New Phytologist.,* 141: 265–275.

Singh OP and Parihar SBS, 1988. Effect of different hosts on the development of *Heliothis armigera* Hub. Bulletin of Entomology, 29: 168-172.

Southwood, T. R. E., 1966. *Ecological methods*. Methuen and Co., London.

Ulusoy MR and Bayhan SO, 2006. Effect of certain *Brassica* plants on biology of the cabbage aphid *Brevicoryne brassicae* under laboratory conditions. Phytoparasitica., 34:133-138.

Will T, Tjallingii WF and Van Bel AJE, 2007. Molecular sabotage of plant defense by aphid saliva. Proc. Natl. Acad. Sci. *U. S. A.,* 104: 10536-10541.

Yu JZ, Chi H and Chen BH, 2005. Life table and predation of *Lemnia biplagiata* (Coleoptera: Coccinelidae) fed on *Aphis gossypii* (Homoptera: Aphididae) with a proof on relationship among gross reproduction rate, net reproductive rate and preadult survivorship. *Ann. Entomol. Soc. Am.,* 98: 475- 482.

Zarghami S, 2010. Effect of nitrogen fertilization on life table parameters and population growth of *Brevicoryne brassicae*. B. Insectol., 63: 39- 43.

IMPROVING MAIZE GROWTH AND DEVELOPMENT IN RELATION TO SOIL APPLIED ELEMENTAL SULFUR

Mohsin Tanveer[1*], Ehsanullah[1], Shakeel Ahmad Anjum[1, 2], Ali Ahsan Bajwa[1], Haseeb Zahid[1]
[1]*Department of Agronomy, University of Agriculture, Faisalabad, Pakistan*
[2]*College of Agronomy and Biotechnology, Southwest University, Chongqing, China*

ABSTRACT

A field experiment was conducted to ascertain the effect of varying soil applied elemental sulfur (S) levels viz; 0, 10, 20, 30, 40 and 50 kg ha^{-1}, on maize growth and development. Experiment was conducted in randomized complete block design (RCBD) with three replicates. Growth and development parameters were computed; included leaf area per plant LAI, LAD, CGR, NAR and TDM and experimental results revealed that increasing S levels were pragmatic in improving maize performance. However, soil applied elemental S @ 30 kg ha^{-1} proved to be beneficial involved in growth and yield improvement. Maximum CGR (23.44 g m^{-2} d^{-1}), NAR (6.23 g m^{-2} d^{-1}) and LAI (4.85) was in plots where S was applied at 30 kg ha^{-1} while least LAI (4.40), LAD (193.45 days), TDM (1178.13 kg ha^{-1}) was in control treatment. Therefore, elemental S @ 30 kg ha^{-1} was noted to be most suitable for sustainable maize production amongst all other treatments.

Keywords: Elemental Sulfur, Growth, Maize, NAR, TDM

INTRODUCTION

Maize (*Zea mays* L.) is an annual, cross pollinated summer season crop that belongs to family Poaceae. Its grain is a rich source of many important nutrients. Its flour contains moisture (9.6%), carbohydrates (70.4%), protein (10.7%), oil (3-18%), crude fiber (2.2%), ash (1.7%) ether extracts (5.4%) and several important vitamins and minerals (Bressani *et al.,* 1990). It is regarded as an important profitable crop of higher agro-economic worth due to its extensive use in the agro-industries. In recent years, increased quantities of corn have been used in the manufacturing of soaps, varnishes, paints and other similar products (Craig *et al.,* 2004). It accounts for approximately 5% of the total cropped area in Pakistan and an estimated area under maize is 1083 thousand hectares with annual production of 4271 thousand tons, change in the production over last year was 15.2% (GOP, 2012). Potential yield of maize is higher than that of either wheat or rice and we can expect maize to play a proportionally larger and more important role in world food security (Fischer and Palmer, 1984) whereas the yield of maize is getting lower due to poor land management practices, low organic matter, low soil productivity, conventional cropping system, imbalanced fertilization and inappropriate sowing methods.

Struggles are being made to enhance the production and diminish the space among the demand and supply for food. Land area under the farming of pulses, cereals and oilseeds is decreasing with increasing population. Now agricultural researchers are giving more attentions to maximize grain yield through correct nutrition of the crops, developing high yielding varieties and adopting the latest agronomic practices.

Sulfur (S) is emerging as a major plant nutrient for crops grown in the Indo-Gangetic plains spread over 13 million hectares in Pakistan, India and Bangladesh. The extent of Sulfur deficiency in soil in the region is continuously increasing with the adoption of high yielding cultivars of rice, wheat, maize, oil seeds and pulses and because of the increased use of fertilizers lacking Sulfur. Most of the alluvial soils of the indo-Gangetic plains were found deficient with respect to plant available Sulfur. Sulfur is considered fundamental nutrient for plant growth and development. Its demand for plants has become significant in Pakistan. However, the doses of S fertilizer should be recommended on the basis of available soil S and crop demand to attain the maximum crop yields. Plants need S equal to the amounts of phosphorus needed. Moreover, S has particular role in growth, enzymatic reactions and metabolism (Mengal and Kirkby, 1987). It is involved in the formation of amino acid like

**Corresponding author:* e-mail: mohsin_1728@hotmail.com

cystine, cysteine and methionine. Sulfur is also component of S-glycosides, coenzyme-A, vitamins i.e. thiamine and biotine (Mengel and Kirkby, 2001).

Sulfur is associated with synthesis of oil especially in oil seed crops. A dearth of S causes plants to be consistently chlorotic, under developed, week stemmed and etiolated. Intensive cropping systems require substantial amount of N, P, K and S for proper development. Reviewing the response of different crops i.e. wheat, potato, groundnut, gram, lentil, mung bean and mash bean have revealed that application of S not only increased the yield of the crops but also improved the quality of the produce (Aulakh *et al.,* 1977). S is the major element required for profitable crop production and its uptake from the soil is of the equal magnitude as that of phosphorus (Rehman and Ghani, 1989). In a study by Schonohf *et al.,* 2007, it was substantiated the growth promoter effects of S and indicated a significant increase in plant height upon S fertilization. S considerably enhanced crop growth and development by increasing net photosynthetic rate, net assimilation rate and S use efficiency, as revealed by Khan *et al.* (2005), S fertilization increased relative growth rate, net CO_2 assimilation and S use efficiency of mustard plants. Symptoms of S deficiency in plants are characterized by reduced plant growth and occurrence of uniform chlorosis on younger leaves (Havlin *et al.,* 2005). A study was therefore conducted to examine the nature and role of elemental S in improving maize growth and development under subtropical conditions of Faisalabad, Pakistan.

MATERIALS AND METHODS

A field study was carried out in randomized complete block design (RCBD) with three replications to evaluate the efficacy of different doses of elemental Sulfur (S^0) in respect of growth and development of maize crop at the Agronomic Research Area, University Of Agriculture, Faisalabad, Pakistan. The station is located between longitude 73°-74° East and latitude 30°-31.5° North, with an elevation of 184 meters above sea level. The maize hybrid Pioneer-32B33 was sown on 20th February 2012 on ridges with hand placement maintaining plant to plant distance 20 cm using seed rate of 25 kg ha^{-1}. Treatments under study (Table 1) were aimed to examine the effect of elemental S (S^0) on maize growth and development. Sulfur bentonite was employed to supply S^0 by broadcasting. NPK fertilization was done at the rate of 250, 125 and 125 kg ha^{-1} respectively. All of phosphorous, potash and half of the nitrogen were applied at the time of sowing in the form of DAP (Diammonium Phosphate), SOP (Sulphate of potash) and Urea. Remaining half nitrogen was applied in two splits, one at five leaf stage and second at tasseling stage and irrigation requirement. Five irrigations were applied to the crop according to the crop requirements. First irrigation was applied at 15 days after sowing whereas, subsequent irrigations were applied according to the need.The crop was harvested manually after its maturity on 10th of June 2012. Data were analyzed by using Fisher's analysis of variance technique. Least significant difference test at 5% probability level was applied to compare the treatment means (Steel *et al.,* 1997).

Observations regarding growth and development were computed via measuring leaf area per plant, leaf area index, leaf area duration, crop growth rate, net assimilation rate and final total dry matter according to following formulas:

Leaf area index (LAI) was calculated as the ratio of leaf area to land area (Watson, 1952).

$$LAI = \text{Leaf area} / \text{Land area}$$

Leaf area duration (LAD) for each sampling date was estimated according to Hunt (1978).

$$LAD = (LAI_1+LAI_2) \times (T_2 - T_1) / 2$$

Crop growth rate (CGR) was calculated as proposed by Hunt (1978) at each sampling date.

$$CGR = (W_2-W_1) / (T_2- T_1)$$

The mean net assimilation rate (NAR) was estimated by using the formula of Hunt (1978).

$$NAR = TDM / LAD$$

RESULTS AND DISCUSSION

Leaf area is a major factor determining canopy photosynthesis and ultimately crop yield. Fig. 1 showed that leaf area steadily increased in all the treatments and reached at maximum value at 70 days after sowing (DAS); thereafter leaf area declined until harvest. In the beginning, differences in leaf area among treatments were less visible, but with time these differences became progressively more visible and leaf area was maximum at 70 DAS (end of grand growth period) and declined thereafter as the crop progressed towards physiological maturity. S^0 had significant effect on leaf area per plant. Maximum leaf area per plant (8654.48 cm^2) was observed in S_3 treatment, while minimum leaf area per plant (7464.58 cm^2) was noted in S_1. The progressive increase in leaf area per plant was due to S^0application which leads to increase nutrient uptake which enhanced the rate of photosynthesis. Moreover, S has particular role in growth, enzymatic reactions and metabolism (Mengal and Kirkby, 2001). Findings were supported by Daniela *et al.* (2008) who reported that leaf area and LAI was significantly affected by the S application and showed the highest value. In another study, Khan *et al.* (2005) also corroborated that improvement in leaf area might be ascribed to S fertilization during its initial growth stages.

Leaf Area Index

LAI is a major factor which determines radiation interception, canopy photosynthesis and crop yield. S^0 application significantly affected LAI of maize crop during spring season. Data related to LAI of maize are presented in Fig.2 which showed that LAI was increased with the increase in crop growth and attained maximum value at 70 days after sowing and again started declining when the crop reached to it maturity. Maximum LAI (4.85) was scored where S^0 was applied at the rate of 30 kg ha^{-1}.The minimum LAI (4.40) was observed in control treatment. The progressive increase in LAI was due to increase S^0 application which leads to increase the rate of photosynthesis which resulted in more LAI. Findings were supported by Daniela *et al.* (2008), they reported that LAI was significantly affected by the S application and showed highest value. As more LAI ensured the higher photosynthesis rate which further facilitates high dry matter accumulation, consequently more LAI could be attributed to significant development in leaf expansion. Moreover, the greater leaf expansion might be attributed to high rate of cell division and cell enlargement. As evidenced by Khan *et al.* (2005), reported that LAI is improved significantly due to S supplementation.

Leaf area duration

The effect of S^0 treatments on LAD is presented in Fig 3. The S^0 effect on LAD was positive for all treatments. Maximum LAD (212.86 days) was noted where S^0 was applied @ 30 kg ha^{-1}. The minimum LAD (193.45 days) was noted where S^0 was not applied. The progressive increase in LAD was due to increase in S^0 application which leads to increase the rate of photosynthesis, which resulted more LAI and LAD. The variation in TDM in response to agronomic treatment or S^0 rate may or may not be explained by variation in their maximum LAI. However, the differences in yield among treatments might be assessed based on their LADs. High LAD depicts that plant developed their leaves for long time, associated with delayed leaf senescence. The substantial increase in LAD at 30 Kg ha^{-1} might be ascribed to growth promontory effect of S^0, beyond this concentration both above or below, plant may exposed to S^0 toxicity and deficiency respectively. These results are in line with Khan *et al.* (2005) who reported S^0 fertilization enhanced the LAD. Furthermore, the relationship among TDM and LADs are strongly dependent on agro-climatic conditions that exist in particular conditions and environment in crop is sown (Monteith, 1981)

Total dry matter

TDM production increased steadily after crop establishment until maturity in all the treatments Fig 4. S^0 application significantly affected TDM of maize during growth. Maximum dry matter (1322.5 kg ha^{-1}) was noted where S^0 was applied @ 30 kg ha^{-1}.The minimum dry matter (1178.13 kg ha^{-1}) was noted in treatment where S^0 was not applied. The increase in TDM with application of S was due to better crop growth which gave maximum plant height, LAI and ultimately produced more biological yield. Findings are quite similar with Poonia (2000) and Daniela *et al.* (2008). Poonia (2000) reported that significant increase in dry matter of sunflower was observed when S was applied @ 25 kg ha^{-1} and according to Daniela *et al.* (2008), TDM was significantly affected by the S application and showed the highest value.

Crop growth rate

CGR is an important parameter which indicates how efficiently crop is using input resources and produces photosynthates which are used by plant for production of economic yield. Data related to CGR of maize are presented in Fig. 5 S^0 application significantly affected CGR of maize crop during spring season. CGR was increased with the increase in crop growth and attained maximum value at 70 days after sowing and again started declining when the crop reached to it maturity. Maximum CGR (23.44 g m^{-2} d^{-1}) was noted in the treatment where S^0 was applied @ 30 kg ha^{-1}. The minimum CGR (21.59 g m^{-2} d^{-1}) was noted in the treatment where S^0 was not applied. The increase in CGR was due to increase nutrients uptake which promoted better crop growth, gave maximum plant height, LAI and TDM with S^0 application. Findings are quite similar with Daniela *et al.* (2008), they reported that CGR (CGR) and RGR (relative growth rate) were significantly affected by the S application and showed the highest value. A higher CGR during anthesis period may be prerequisite to gain higher ultimately result in good crop production.

Net assimilation rate

The average NAR of a crop represents the net photosynthetic production per unit LAD (Hunt, 1978). Data related to CGR of maize are presented in Fig. 6. Maximum NAR (6.23 g $m^{-2}$$d^{-1}$) was noted where S^0 was applied @ 30 kg ha^{-1}. The minimum NAR (6.09 g m^{-2} d^{-1}) was noted in treatment where no S^0 was applied. The increase in NAR with application of S^0 was due to better crop growth which gave the maximum plant height, LAI, TDM and LAD. The improvement in NAR may be attributed to more vegetative growth due to increasing rate of N fertilizer. TDM accumulation during growth especially earlier than flowering is considered to be very essential for determination TDM as sink capacity (Andrade, 1995). Therefore, the formation of large sink size may be a requirement to higher TDM production during growth, are prerequisite for higher yield and finally higher economic return.

Table 1: Treatments and application rate of S°

Treatment no.	Application rate
S_o	Control (no S^0 application)
S_1	Soil application at sowing @ 10 kg ha^{-1}
S_2	Soil application at sowing @ 20 kg ha^{-1}
S_3	Soil application at sowing @ 30 kg ha^{-1}
S_4	Soil application at sowing @ 40 kg ha^{-1}
S_5	Soil application at sowing @ 50 kg ha^{-1}

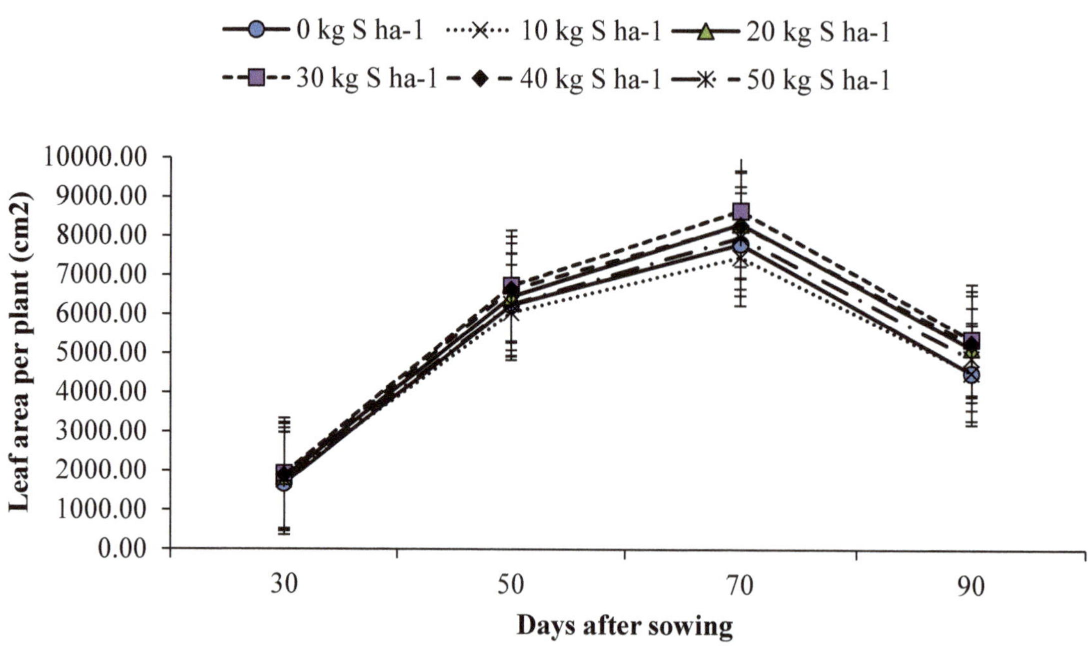

Fig. 1: Change in leaf area per plant in response to elemental sulfur

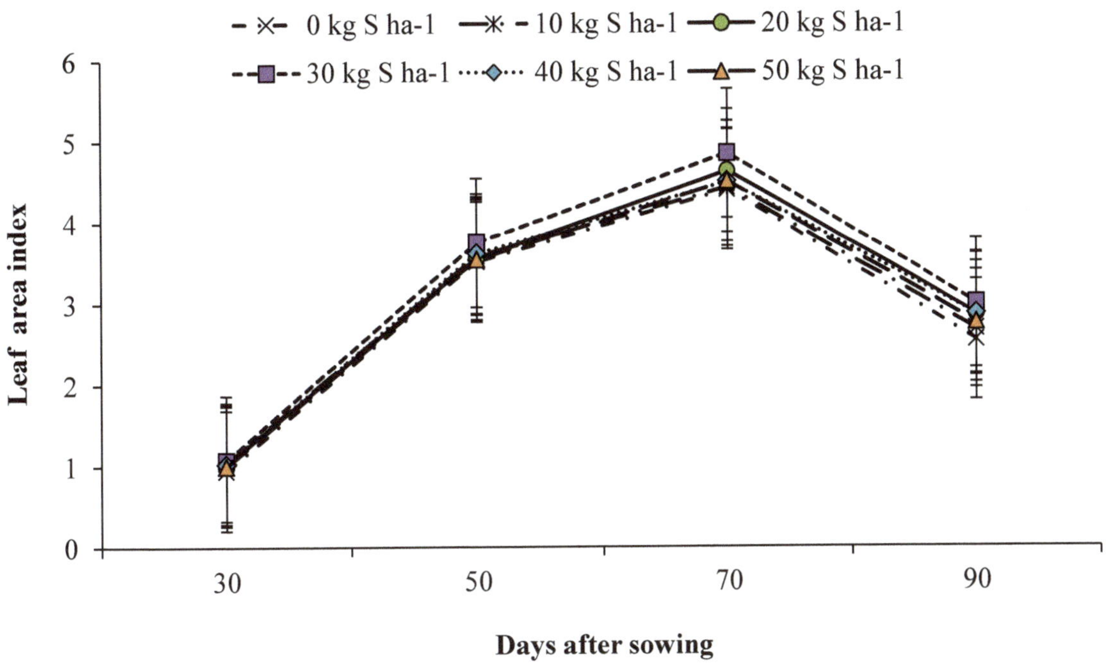

Fig. 2: Change in leaf area index in response to elemental sulfur

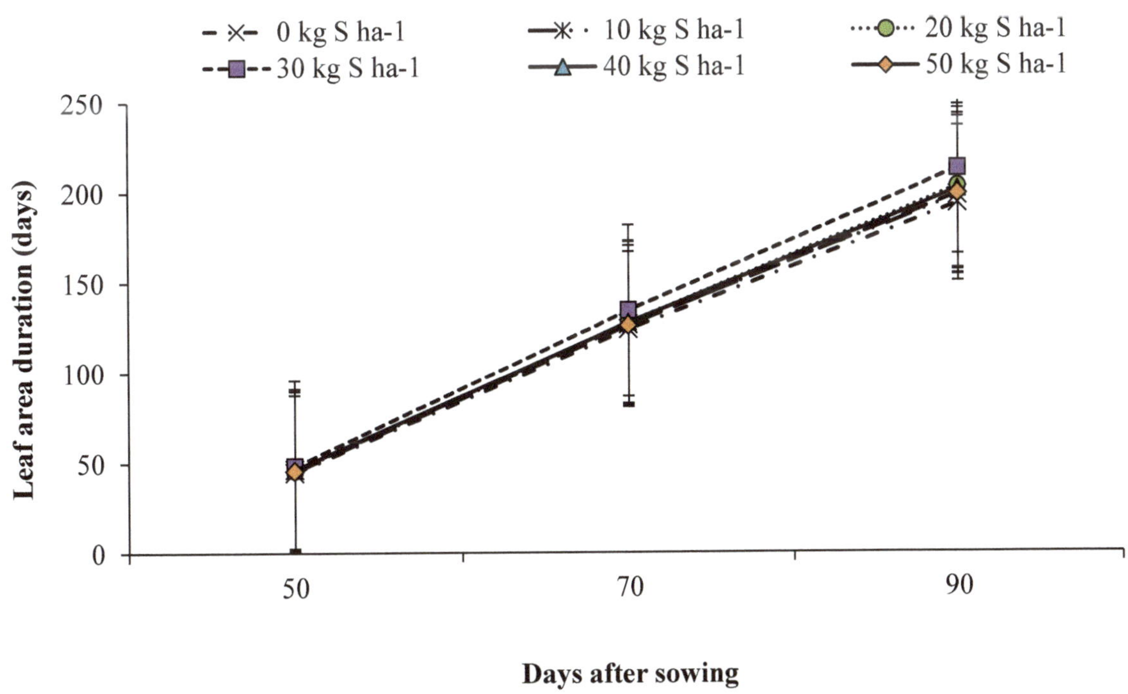

Fig. 3: Change in leaf area duration in response to elemental sulfur

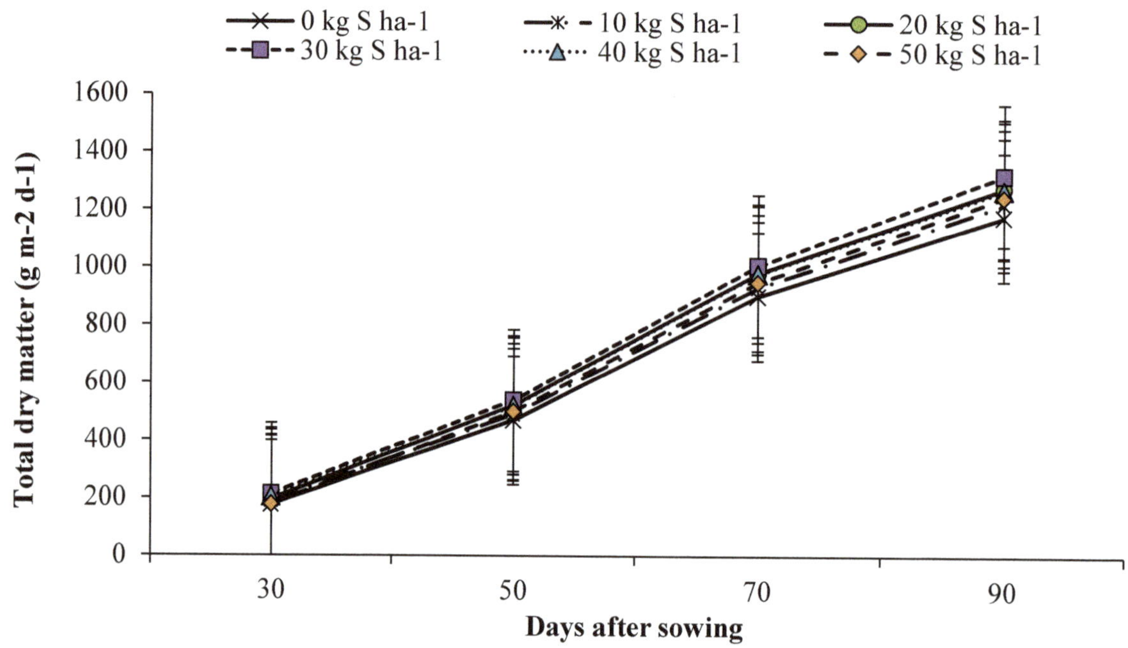

Fig. 4: Change in total dry matter in response to elemental sulfur

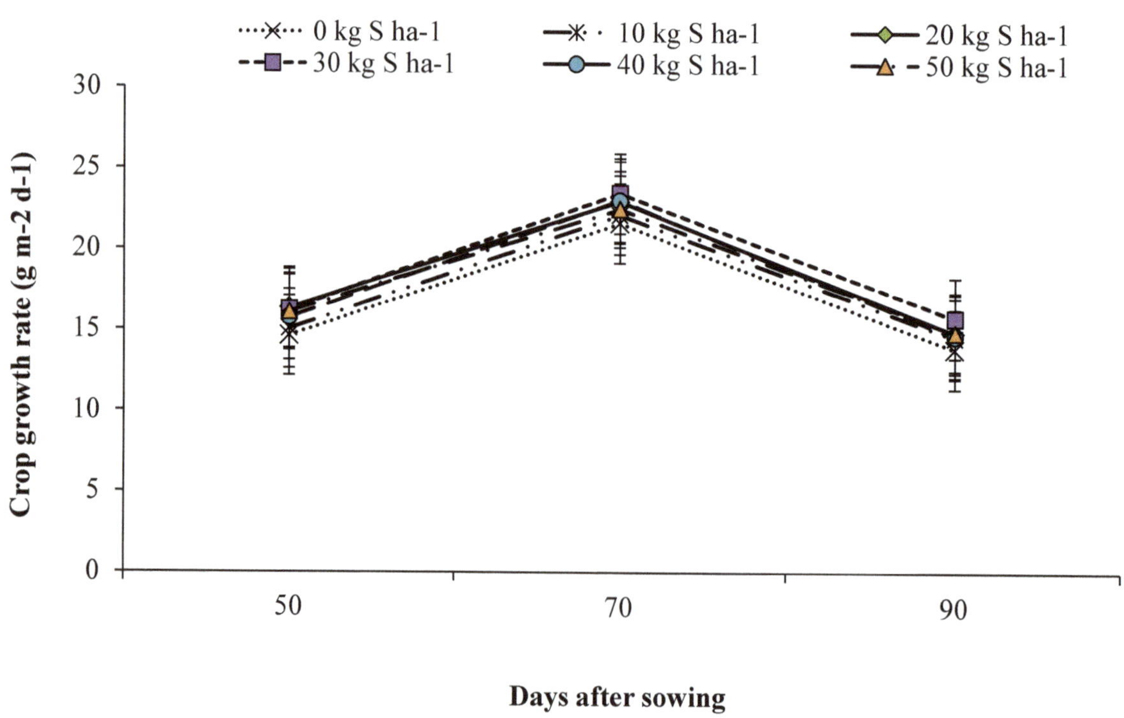

Fig. 5: Change in crop growth rate in response to elemental sulfur

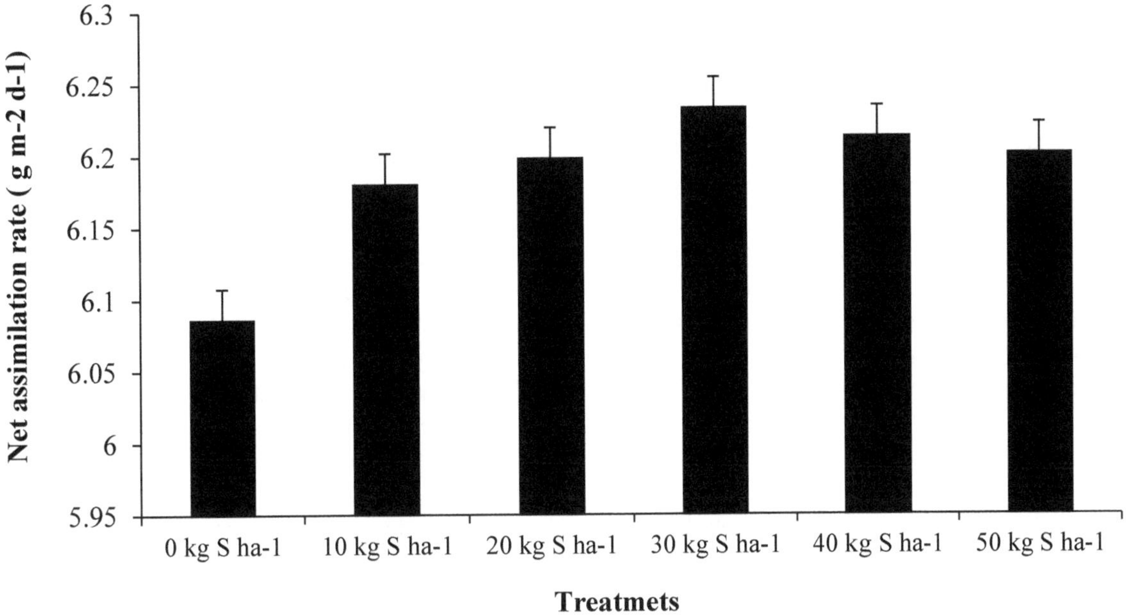

Fig. 6: Change in net assimilation rate in response to elemental sulfur

CONCLUSION

Good growth and development of any plant results in better productivity. However, numerous factors affect productivity also. It is concluded that elemental sulfur significantly boosted maize performance by increasing growth and development. However, application of elemental sulfur (S^0) @ 30 kg ha^{-1} has more influence on growth, rather beyond this rate, plant may get stressed. Further investigation particularly for maize should be related to depth of S application and method for improving S use efficiency.

REFERENCES

Andrade FH, 1995. Analysis of growth and yield of maize, sunflower and soybean grown at Balcarce, Argentina. Field Crops Res., 41: 1-12.

Aulakh MS, Pasriacha NS and Dev G, 1977. Response of different crops to sulfur fertilization in Punjab. Fertilizer News. Sep. 1977.

Bressani R, 1990. Chemistry, technology and nutritive value of maize tortillas. Food Rev. Int. 6: 225-264.

Craig JA, Batie J, Chen W, Freeland SB, Kinkima M and Lanahan MB, 2004. Expression of starch hydrolyzing enzymes in corn. In: Tumbleson, M. (Ed.) Corn: Feedstock of the future. (Corn utilization and technology conference proceedings) Corn Utiliz. Tech., Indianapoles.

Daniela O, Pietkiewicz S, Ciesinski M, Kucinska K and Gowsdozki G, 2008. Biomass accumulation and absorption of photosynthetic active radiation by rapeseed depending on sulfur fertilization. World J. Agric Sci., 4: 133-136.

El-Tarabily KA, Abdou AS, Maher ES and Satoshi M, 2006. Isolation and characterization of sulfur-oxidizing bacteria, including strains of Rhizobium from calcareous sandy soils and their effects on nutrient uptake and growth of maize. Aust. J. Agric. Res., 57: 101-111.

Fischer KS and Palmer AFE, 1984. Tropical maize. The physiology of tropical field crops. Bath, Avon: J. Wiley & Sons Ltd., 213-248.

GOP. 2012. Economic Survey of Pakistan 2011-2012, Finance Division, Economic Advisory Wing, Islamabad, Pakistan, pp.22

Havlin JL, Beaton JD, Tisdale SL and Nelson WL, 2005. Soil fertility and fertilizers: An introduction to nutrient management. 7thedn. New Jersey, Pearson Prentice Hall, p 528.

Hunt, R. 1978. Plant growth analysis. Edward Arnold, U.K. 26-38.

Jamal A, Fazli IS, Ahmad S, Abdin MZ and Yun SJ, 2005. Effect of sulphur and nitrogen application on growth characteristics, seed and oil yield of soybean cultivars. Korean. J. Crop Sci., 50: 340-345

Khan NA and Mobin M, 2005. The influence of gibberellic acid and sulfur fertilization rate on growth and S-use efficiency of mustard (*Brassica juncea*). Plant and soil, 270: 269-274.

Mengel, K., Kirkby, E.A., 2001: Principles of Plant Nutrition, 5th ed., Kluwer Academic Publishers, Dordrecht.

Monteith JL, 1981. Climate variation and growth of crop. Q.J.R. Meterel.soc. 107:749-774

Poonia KL, 2000. Effect of planting geometry, nitrogen and sulfur on growth and yield of sunflower (*Helianthus annuus*L.). J. Eco-Physiol., 3: 59-71.

Rahman MM, Soaud AA, AL-Darwish FH, Golam F and Sofian-Azirun MS, 2011. Growth and nutrient uptake of maize plants as affected by elemental sulfur and nitrogen fertilizer in sandy calcareous soil. Afr. J. Biotechnol., 10: 12882-12889.

Rehman H and Ghani A, 1989. Calibration of soil test value of potassium with the yield apple. Pak. J. Soil Sci. 4: 37-40.

Schonhof I, Blankenburg D, Müller S and Krumbein A, 2007. Sulfur and nitrogen supply influence growth, product appearance, and glucosinolate concentration of broccoli. Journal of Plant Nutrition and Soil Science, 170: 65-72.

Steel RGD, Torrie JH and Dickey DA, 1997. Principles and procedures of statistics: A biological approach. 3rd Ed. McGraw Hill, Inc. Book Co. N.Y. USA. pp. 352-358.

Watson DJ, 1952. The physiological basis of variation in yield. Adv. Agron. 4: 101-145.

PERMISSIONS

All chapters in this book were first published in AJAB, by Life Sciences Society; hereby published with permission under the Creative Commons Attribution License or equivalent. Every chapter published in this book has been scrutinized by our experts. Their significance has been extensively debated. The topics covered herein carry significant findings which will fuel the growth of the discipline. They may even be implemented as practical applications or may be referred to as a beginning point for another development.

The contributors of this book come from diverse backgrounds, making this book a truly international effort. This book will bring forth new frontiers with its revolutionizing research information and detailed analysis of the nascent developments around the world.

We would like to thank all the contributing authors for lending their expertise to make the book truly unique. They have played a crucial role in the development of this book. Without their invaluable contributions this book wouldn't have been possible. They have made vital efforts to compile up to date information on the varied aspects of this subject to make this book a valuable addition to the collection of many professionals and students.

This book was conceptualized with the vision of imparting up-to-date information and advanced data in this field. To ensure the same, a matchless editorial board was set up. Every individual on the board went through rigorous rounds of assessment to prove their worth. After which they invested a large part of their time researching and compiling the most relevant data for our readers.

The editorial board has been involved in producing this book since its inception. They have spent rigorous hours researching and exploring the diverse topics which have resulted in the successful publishing of this book. They have passed on their knowledge of decades through this book. To expedite this challenging task, the publisher supported the team at every step. A small team of assistant editors was also appointed to further simplify the editing procedure and attain best results for the readers.

Apart from the editorial board, the designing team has also invested a significant amount of their time in understanding the subject and creating the most relevant covers. They scrutinized every image to scout for the most suitable representation of the subject and create an appropriate cover for the book.

The publishing team has been an ardent support to the editorial, designing and production team. Their endless efforts to recruit the best for this project, has resulted in the accomplishment of this book. They are a veteran in the field of academics and their pool of knowledge is as vast as their experience in printing. Their expertise and guidance has proved useful at every step. Their uncompromising quality standards have made this book an exceptional effort. Their encouragement from time to time has been an inspiration for everyone.

The publisher and the editorial board hope that this book will prove to be a valuable piece of knowledge for researchers, students, practitioners and scholars across the globe.

LIST OF CONTRIBUTORS

Muhammad Younas
Senior Subject Matter Specialist (FM&FE), Adaptive Research Farm, Sheikhupura

Momna Abdul Rehman
Assisant Research Officer (FM&FE), Adaptive Research Farm, Sheikhupura

Umer Farooq
Department of Agronomy, Faculty of Agriculture, Gomal University, Dera Ismail Khan-Pakistan

Ejaz Ahmed Khan
Department of Agronomy, Faculty of Agriculture, Gomal University, Dera Ismail Khan-Pakistan

Muhammad Nawaz
Adaptive Research Farm, Sheikhupura, 2Directorate of Adaptive Research, Lahore, Punjab

Anwar Javaid Wahla
Adaptive Research Farm, Sheikhupura, 2Directorate of Adaptive Research, Lahore, Punjab

Muhammad Aslam
Senior Subject Matter Specialists (Agronomy)

Ashiq Hussain Sanghi
Senior Subject Matter Specialists (Plant Protection)

Md. Rayhan Shaheb
Senior Scientific Officer, On-Farm Research Division, Bangladesh Agricultural Research Institute (BARI), Sylhet-3100, Bangladesh

Mahmudul Islam Nazrul
Senior Scientific Officer, On-Farm Research Division, Bangladesh Agricultural Research Institute (BARI), Sylhet-3100, Bangladesh

M. Ataur Rahman
Principal Scientific Officer, Regional Wheat Research Station, Bangladesh Agricultural Research Institute (BARI), Gazipur, Bangladesh

A. M. Shrestha Malla
Botany Department, Tri-Chandra College, Tribhuvan University, Kathmandu, Nepal

Shaheen, N. Sabir and M. Zafar
Department of Soil and Environmental Sciences. The University of Poonch, Rawalakot, Azad Jammu & Kashmir-Pakistan

Mehran Khan
Department of Plant Pathology, The University of Agriculture, Peshawar Pakistan

Abdur Rafi
Department of Plant Pathology, The University of Agriculture, Peshawar Pakistan

Haidar Ali
Department of Horticulture, The University of Agriculture Peshawar-Pakistan

Gohar Ayub
Department of Horticulture, The University of Agriculture Peshawar-Pakistan

Sultan Mahmood
Department of Plant Breeding and Genetics, PMAS Arid Agriculture University Rawalpindi

Saad Imran Malik
Department of Plant Breeding and Genetics, PMAS Arid Agriculture University Rawalpindi

Shuaib Kaleem
Agriculture Adaptive Research Complex, Dera Ghazi Khan, Pakistan

M. Shabbir
Agriculture Adaptive Research Complex, Dera Ghazi Khan, Pakistan

Waqar Islam
College of Plant Protection, Fujian Agriculture & Forestry University, Fuzhou, Fujian 350002, China

Asif Riaz
University of Agriculture, Faisalabad, Pakistan

M. E. Morojele
Small Grain Institute, Private Bag X29. Bethlehem. 9700. South Africa

D. N. M. Mbewe
The University of Zambia, Lusaka, Zambia

Muhammad Younas
Department of Horticulture, The University of Agriculture, Peshawar-Pakistan

Rifat Ali
Department of Plant Pathology, The University of Agriculture, Peshawar-Pakistan

Ehsanullah
Department of Agronomy, University of Agriculture, Faisalabad

Umair Ashraf
Department of Crop Science and Technology, College of Agriculture, South China Agricultural University, Guangzhou, China

K. V. Pavani, V. Divya, I. Veena, M. Aditya and G. V. S Devakinandan
Department of Biotechnology, Gokaraju Rangaraju Institute of Engineering and Technology, Bachupally, Hyderabad 500090, Telangana, India

Adetunji Charles Oluwaseun
Nigerian Stored Product Research Institute, Km 3 Asa dam road, P.M.B. 1489, Ilorin, Nigeria

Williams Jospeh
Nigerian Stored Product Research Institute, Km 3 Asa dam road, P.M.B. 1489, Ilorin, Nigeria

N.S. Kottearachchi
Department of Biotechnology, Faculty of Agriculture and Plantation Management, Wayamba University of Sri Lanka, Makandura, Gonawila (NWP), Sri Lanka

R.K. Peiris
Department of Biotechnology, Faculty of Agriculture and Plantation Management, Wayamba University of Sri Lanka, Makandura, Gonawila (NWP), Sri Lanka

Shabana Ehsan
Soil Chemistry section, ISCES, Ayub Agricultural Research Institute, Faisalabad

Ifra Saleem
Soil Chemistry section, ISCES, Ayub Agricultural Research Institute, Faisalabad

Mashamba Philipo
The Nelson Mandela African Institution of Science and Technology (NM-AIST), Arusha, Tanzania

Clara Mollay
The Nelson Mandela African Institution of Science and Technology (NM-AIST), Arusha, Tanzania

S. Bandopadhyay
Department of Botany, Sonamukhi College, Bankura, W.B., India

Anser Ali
Faculty of Agriculture, Ghazi University, Dera Ghazi Khan, Pakistan

Nasir Abbas
Faculty of Agriculture, Ghazi University, Dera Ghazi Khan, Pakistan

Atique ur Rehman
Department of Agronomy, Bahauddin Zakariya University, Multan

Muhammad Farooq
Department of Agronomy, University of Agriculture, Faisalabad

Aqleem Abbas
Department of Plant Pathology, The University of Agriculture Peshawar Pakistan

Muhammad Arif
Department of Plant Pathology, The University of Agriculture Peshawar Pakistan

Afaq Ahmad, Ihsan Ul Khaliq and Maid Zaman
Department of Plant Pathology,The University of Agriculture,Peshawar-Pakistan

Sumera Javad
Department of Botany

Sana Khalid
Department of Botany

Aliya Noreen
Department of Environmental Sciences, Karakoram International University Gilgit-Pakistan

Ghulam Raza
Department of Environmental Sciences, Karakoram International University Gilgit-Pakistan

Banashree Sarma, Prema Devi, Nirmali Gogoi, Yendrembam Menaka Devi
Department of Environmental Science, Tezpur University, Napaam, Assam - 784028, India

Asghar, K.
Department of Plant Breeding and Genetics, Sindh Agriculture University Tandojam-70060, Pakistan

Z.A Soomro
Department of Plant Breeding and Genetics, Sindh Agriculture University Tandojam-70060, Pakistan

Hakim Ali Sahito, Abdul Waheed Solangi, Abdul Ghani Lanjar, Abdul Hafeez Solangi and Sajjad Ali Khuhro
Department of Entomology, Sindh Agriculture University, Tandojam- Sindh

Rafaqat Hussain Shah
Department of Agronomy, Faculty of Agriculture, Gomal University, Dera Ismail Khan, Pakistan

Mohammad Safdar Baloch
Department of Agronomy, Faculty of Agriculture, Gomal University, Dera Ismail Khan, Pakistan

Muhammad Asif Aziz
Department of Entomology

Muhammad Irfan
Department of Entomology

Mohsin Tanveer
Department of Agronomy, University of Agriculture, Faisalabad, Pakistan

Ehsanullah
Department of Agronomy, University of Agriculture, Faisalabad, Pakistan

Tasneem Khaliq
Agro-climatology Lab., University of Agriculture, Faisalabad, Pakistan

Amjed Ali
Agro-climatology Lab., University of Agriculture, Faisalabad, Pakistan
University College of Agriculture, University of Sargodha, Sargodha, Pakistan

Index

CPSIA information can be obtained
at www.ICGtesting.com
Printed in the USA
LVHW060852030122
707669LV00002B/110

9 781639 870332